CONCEPTS AND CHALLENGES IN
Life Science

Leonard Bernstein • Martin Schachter • Alan Winkler • Stanley Wolfe

STANLEY WOLFE
Project Coordinator

GLOBE BOOK COMPANY
Englewood Cliffs, New Jersey

AUTHORS
Alan Winkler
Leonard Bernstein
Martin Schachter
Stanley Wolfe
Project Coordinator

Cover Photos
Background: Grass (Ralph Clevenger/West Light)
Inset: Bear (John Hyde/Bruce Coleman)
Illustrations: Ebet Dudley
Technical Art: Gary Tong
Photo Researcher: Rhoda Sidney
Content Reviewers:
Dr. Jeane J. Dughi
Senior Coordinator, Science
Department of Instruction
Norfolk Public Schools
Norfolk, Virginia

Sarah Longino
Biology Teacher
Colonial High School
Orlando, Florida

J.J. Olenchalk
Chairman, Science Department
Antioch Senior High School
Antioch, California

Harold Robertson
Science Resource Teacher
Los Angeles Unified School District
Los Angeles, California

Globe Book Company
A division of Simon & Schuster
Englewood Cliffs, New Jersey 07632

10 9 8 7 6 5 4

ISBN: 1-55675-818-9

Contents

UNIT 4

CELLS, TISSUES, AND ORGANS *77-98*

UNIT 5

CLASSIFICATION *99-114*

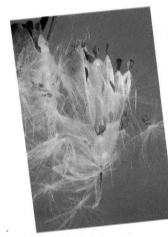

UNIT 16

HEALTH AND DISEASE   *311-334*

UNIT 17

REPRODUCTION AND DEVELOPMENT

335-352

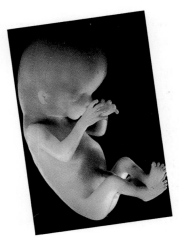

UNIT 18

HEREDITY AND GENETICS *353-380*

SCIENTIFIC METHODS AND SKILLS

CONTENTS

STUDY HINT As you read each lesson in Unit 1, write the lesson title and lesson objectives on a sheet of paper. After you complete each lesson, write the sentence or sentences that answer each objective.

What is life science?

Objective ▶ Identify and describe what is studied in some of the branches of life science.

TechTerms

▶ **specialization** (SPESH-uh-lih-zay-shun): studying or working in only one part of a subject

Studying Life Science Science is an organized collection of knowledge about the world. It also is a way of finding out why things happen as they do. It is a way of solving problems by testing possible answers to see if they work. The knowledge of science is based on observations.

The study of the areas of science that deal with living things is called life science. Life science is like a tree. It is made up of many different branches. One branch is biology (by-AHL-uh-jee). Biology is the study of living things. Table 1 lists some of the life sciences. Two—botany and zoology—are part of biology.

◼️ *Analyze:* What are four branches of life science?

Specialization As more and more is learned about the world, people must choose specific subjects to study. This is called **specialization**

(SPESH-uh-lih-zay-shun). A person who studies or works in one part of a subject is called a specialist (SPESH-ul-ist). There are many life science specialists. For example, some zoologists (zoh-AHL-uh-jists) study only one group of animals. Some scientists study diseases that affect only animals. Other scientists study diseases in plants.

▉▶ *Describe:* What is meant by specialization?

Importance of Life Science Life science is part of your everyday life. The study of living things affects your life in many ways. The medicine you take for a cold was developed based on scientific study. The causes and warning signs of cancer were learned from scientific research. Operations can be done because doctors know about the parts of the human body and how they work.

The kinds of foods you eat were grown by using information about plants. The making of some foods also uses knowledge of life science. Many cheeses could not be made without molds. Pickles could not be made without bacteria. People had to learn about bacteria and molds to use them to make these foods.

▉▶ *Explain:* How was life science part of your life today?

Table 1 Branches of Life Science		
BRANCH	**WHAT IS STUDIED**	**CAREERS**
Anatomy (uh-NAT-uh-mee)	parts that make up living things	Doctor Physical therapist
Physiology (fiz-ee-AHL-uh-jee)	how the parts of living things work	Physiologist Chiropractor
Botany (BAHT-un-ee)	plants	Horticulturist Florist
Zoology (zoh-AHL-uh-jee)	animals	Veterinarian Marine biologist
Microbiology (my-kroh-by-AHL-uh-jee)	microscopic living things	Microbiologist Pathologist
Ecology (ee-KAHL-uh-jee)	interaction of living things and their surroundings	Ecologist

LESSON SUMMARY

▶ Science is an organized collection of knowledge about the world.

▶ The study of the areas of science that deal with living things is called life science.

▶ Specialization is the study of only one part of a subject.

▶ People are affected everyday by the discoveries of life scientists.

▶ Growing and making food use information about living things.

CHECK *Complete the following.*

1. Botany and _____ are part of biology.

2. The knowledge of science is based on _____ .

3. Life science is made up of many different _____ .

4. Any life scientist who studies only one small branch of life science is a _____ .

APPLY *Use the table on page 14 to complete the following.*

5. Ecology is the study of living things and their _____ .

6. The study of how a living thing is put together is _____ .

7. The study of how the parts of a living thing work is _____ .

8. **Classify:** In which branch of life science would you study each of the following?
   a. whales and birds   **b.** corn and barley
   c. the heart and lungs   **d.** bacteria and microorganisms.

9. **Infer:** What area of life science would you need to know about to study the problem shown in the drawing?

Skill Builder

Building Vocabulary Look at the list of careers in Table 1 on page 14. Use reference materials to find out what the people in each of these careers study or do.

♦♦♦♦ CAREER IN LIFE SCIENCE ♦♦♦♦♦♦♦♦♦♦♦♦♦♦♦♦♦♦♦♦♦♦♦♦♦♦♦♦♦♦♦♦♦♦♦♦♦

WILDLIFE BIOLOGIST

Do you like working with animals? Do you enjoy working outdoors? If so, you may enjoy working as a wildlife biologist. Wildlife biologists work to protect the natural environments of animals. They also help to save animals from becoming extinct.

There are many jobs in wildlife biology. Some wildlife biologists work as game wardens. They help to enforce hunting and fishing laws. You need a high school diploma to become a game warden. Other wildlife biologists work in wildlife management and research. They may work on a game preserve or in a national park or forest. These jobs usually require a college degree.

For more information about wildlife biologists, you can write to the Department of the Interior, Fish and Wildlife Service, 18th and C Street, N.W., Washington, DC 20210 or to conservation organizations.

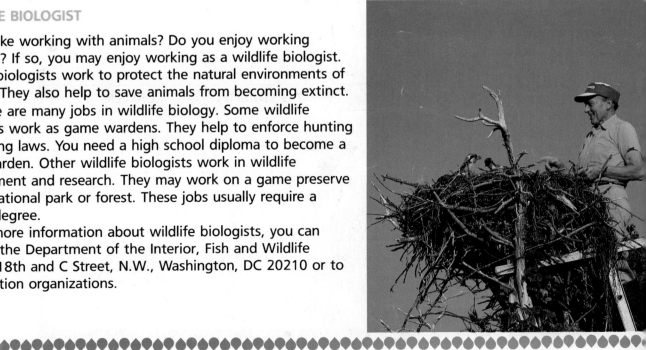

What are science skills?

Objective ▶ Identify and use science skills to solve problems and answer questions.

TechTerm

▶ **hypothesis** (hy-PAHTH-uh-sis): suggested solution to a problem

Science Skills Scientists use many skills to gather information. These skills are sometimes called science skills. You use science skills, too. You probably used some science skills today. When you use most science skills, you use your five senses. The five senses are seeing, hearing, touching, smelling, and tasting.

Eleven science skills are used in this book. You will even see skill symbols for nine skills. These symbols are shown below. They will let you know when you are using a skill. Researching and communicating also are important skills. You will use these skills, too. Soon, you will be thinking like a scientist.

Analyze: Which science skill makes observations more exact?

Researching Have you ever done research for a science project? When you do research, you look for something again. You study or investigate. You can do research by reading books, magazines, and newspapers. You can also perform experiments to do research. Experimenting is a kind of research.

Identify: What are two ways to do research?

Communicating When you talk to someone, you are communicating, or sharing ideas and information. If you write a letter, you are communicating. Scientists communicate all the time. They write books and magazine or newspaper articles about their work. If you read about a scientist and a new discovery, the scientist has communicated with you. Sharing information is very important to scientists.

Describe: What are you doing when you communicate with someone?

Think Scientifically

Observing When you observe, you use your senses. You must pay close attention to everything that happens.

Measuring When you measure, you compare an unknown value to a known value. Measuring makes observations more exact.

Inferring When you infer, you form a conclusion based upon what you think explains an observation.

Classifying When you classify, you group things based upon how they are alike.

Organizing When you organize, you work in an orderly way. You put your information in order.

Predicting When you predict, you state ahead of time what will happen based upon what you already know.

Hypothesizing When you hypothesize, you state or suggest a solution to a problem. A **hypothesis** (hy-PAHTH-uh-sis) is a suggested solution to a problem based upon what is already known or observed.

Modeling When you model, you use a copy of what you are studying to help explain it. A model can be a three-dimensional copy, a drawing, or a diagram.

Analyzing When you analyze, you study information carefully.

LESSON SUMMARY

▶ Scientists use many skills to gather information.

▶ Eleven science skills are used in this book.

▶ Researching includes talking, reading, and experimenting.

▶ Communicating means sharing information.

▶ Other science skills are observing, measuring, inferring, classifying, organizing, predicting, hypothesizing, modeling, and analyzing.

CHECK *Answer the following.*

1. What are your five senses?

2. Name two ways a scientist can communicate a new discovery to people.

Complete the following.

3. When you group things based upon how they are alike, you are _____ them.

4. A suggested solution to a problem is a _____ .

5. If you put information into a table, you are _____ the information.

APPLY *Complete the following.*

6. **Describe:** Describe two ways in which you used science skills today. What skills did you use? How did they help you to solve a problem?

Match each skill to its symbol.

7. ▲   **a.** organizing

8. 📁   **b.** modeling

9. ◬   **c.** classifying

10. ❯   **d.** predicting

11. Which skill do you think is the most important? Give reasons for your choice.

Ideas in Action...................................

IDEA: You make or use many different measurements every day.

ACTION: Describe five situations in which you use measurements during a day.

ACTIVITY

ORGANIZING DATA

You will need a sheet of graph paper, lined paper, and a pencil.

1. Study each set of data.

2. Decide the best way to organize each set of data. You may want to use a table, some kind of graph, a diagram, or another way you think will work.

3. Be sure to give each table, graph, or diagram a title. Tables should have headings for each column.

Questions

1. How did you organize each set of data?

2. Compare the way you organized the data with the ways two classmates organized the data.

Data 1: Animals
fishes; snakes; coral snake; tuna; trout; boa constrictor; birds; robin; rattlesnake; blue jay; bass; sparrow; duck; swordfish

Data 2: Blood Types in a Given Population	
O	45%
A	40%
B	10%
AB	05%

Data 3: Uses of the Peanut
livestock food; peanut butter; salad oil; machine oil; glue; textile fiber; soap; face powder; shaving cream; shampoo; margarine; packing oil for fish; explosives; medicines; insulation; candy; plastic filler.

1-3 What is scientific method?

Objective ▶ Describe how to use scientific method to solve a problem.

TechTerms

▶ **data** (DAY-tuh): information

▶ **hypothesis** (hy-PAHTH-uh-sis): suggested answer to a problem

▶ **scientific method:** model, or guide, used to solve problems and to get information

Scientific Method Scientists solve problems like you do. Suppose a house plant on a window sill looked like it was dying. What would you do? You might check the soil to see if it was dry or maybe had too much water. You might see if a cold draft was coming from the window. You could check to see if too much heat was coming from a radiator. If you did one of these things, you started to use **scientific method.** Scientific method is a model, or guide, used to solve problems.

Scientists do not have one scientific method. They combine some or all of the science skills to solve different problems. Scientists also follow certain steps. Because each problem is different, the steps may be used in any order. You can write a laboratory report using these steps.

▶ *Define:* What is scientific method?

Work Scientifically

▶ **Identify and State the Problem** Scientists often state a problem as a question.

▶ **Gather Information** Scientists read and communicate with one another. In this way, they can learn about work that has already been done.

▶ **State a Hypothesis** Scientists state clearly what they expect to find out. They state a **hypothesis** (hy-PAHTH-uh-sis), or a suggested answer to a problem.

▶ **Design an Experiment** To test a hypothesis, scientists design an experiment.

▶ **Make Observations and Record Data** During an experiment, scientists make careful observations. The information that they get is their **data** (DAY-tuh).

▶ **Organize and Analyze Data** Scientists organize their data. Graphs, tables, charts, and diagrams are ways to organize data. Then the data can be analyzed, or studied.

▶ **State a Conclusion** A conclusion is a summary that explains the data. It states whether or not the data support the hypothesis. The conclusion answers the question stated in the problem.

LESSON SUMMARY

▶ Scientists use a model, or guide, called scientific method to solve problems.

▶ Scientists use different science skills, and different steps of scientific method, to solve different problems.

▶ The steps of scientific method are: Identify and state the problem; Gather information; State a hypothesis; Design an experiment; Make observations and record data; Organize and analyze data; State a conclusion.

CHECK *Complete the following.*

1. A model used to solve problems is called _____ .

2. Scientists often state a problem as a _____ .

3. When scientists state what they expect to find out, they are stating a _____ .

4. Scientists design experiments to test _____ .

5. Data must be organized and _____, or studied.

6. A _____ is a summary that explains the data.

7. Another word for information is _____ .

8. Tables, _____, diagrams, and charts are ways to organize data.

APPLY *Complete the following.*

9. **Explain:** How are a hypothesis and a conclusion related?

10. List the steps in scientific method.

11. Why do you think it is important for scientists to gather information before they begin work on a problem?

Skill Builder

Writing a Laboratory Report Write a laboratory report describing the following experiment. Be sure to include the steps of scientific method in your report.

Place a few drops of iodine on a piece of bread, a rubber ball, a slice of orange, a piece of potato, a piece of lettuce, a few drops of milk, and piece of paper. Note the color changes that occur. If any substance contains starch, it will turn black. The iodine turns black on the bread and potato. There is no color change on the other things.

SCIENCE CONNECTION

PURE AND APPLIED SCIENCE

At times, science can be thought of as two sciences. In one, the scientist is interested mainly in research. This is pure science. In the other, the science is used to solve everyday problems. This is applied science. Applied science often is called technology. When the life science of biology is used to solve everyday problems, it is called biotechnology (by-oh-teck-NAHL-uh-jee).

Biotechnology is used in industry, medicine, and agriculture. Bacteria are used to tan leather. The leather is used to make clothing and shoes, belts, and jackets. Bacteria also are used to make cheeses, yogurt, pickles, and sauerkraut.

How would you like to eat tomatoes as big as grapefruits and grapefruits as big as watermelons? Biologists are working on it. Turkeys with only white meat and others with only dark meat also are being developed using biotechnology.

Biotechnology also is used in the medical field. Hospitals use machines to keep people alive. A respirator helps a person to breathe. A heart-lung machine takes over the jobs of the heart and lungs. Can you think of any other ways biotechnology is being used to make our lives better?

What is an experiment?

Objective ▶ Explain why a controlled experiment is important.

TechTerms

▶ **control:** part of the experiment in which no change is made

▶ **controlled experiment:** two experiments exactly alike except for one change in one of them

▶ **experiment** (ik-SPER-uh-munt): something that is done to test a hypothesis or prediction

▶ **variable:** anything that can be changed in an experiment

Experimenting Scientists often do **experiments** (ik-SPER-uh-munts). An experiment is something that is done to find out why things happen. Experiments are used to test hypotheses. Experiments are done in laboratories or outdoors. If an experiment is conducted outdoors, it is called a field study. Field studies are done in many areas of the world.

▶ *Explain:* What is the purpose of an experiment?

Controlled Experiment Most experiments in biology are **controlled experiments.** In a controlled experiment, two setups are used. Everything about the two setups is exactly the same. One setup is left alone. This experiment is called the **control.** In the other setup, one thing is changed. The thing that is changed is called the **variable.** This part of the controlled experiment is the test, or experimental setup. Both experiments are then observed. Any differences in the results of the two setups would be due to the one thing that was changed.

▶ *Define:* What is a control?

Analyzing an Experiment Look at the experiment in the figure. The problem for this experiment is: How does fertilizer affect plant growth?

CONTROL   **TEST PLANT**

Fertilizer

After 4 weeks

In the experiment, two plants of the same kind and size are placed in identical pots. Both plants are given the same amount of water. Both plants get the same amount of light. The control plant is left alone. Fertilizer is added to the soil of the test plant. At the end of the experiment, the plants are measured. The number of leaves on each plant are counted. The test plant is larger than the control plant. It also has many more leaves.

▶ *Analyze:* What is the variable in this experiment?

LESSON SUMMARY

► Scientists do experiments to test hypotheses.

► Experiments are done in laboratories and out-doors.

► In a controlled experiment, there are two setups—one called the control and one called the test, or experimental, setup.

► Analyzing the results of an experiment helps answer the problem.

CHECK *Complete the following.*

1. A _____ is an experiment that is done out-doors.

2. In a controlled experiment, nothing is done to the _____ .

3. An experiment is performed to test a _____ .

4. In a controlled experiment, the outcome of an experiment is affected by the _____ .

5. An experiment that involves the observation of mountain lions in the hills of the western United States is an example of a _____ .

APPLY *Complete the following.*

6. What is the purpose of a control?

7. **Explain:** Why is it important to change only one variable during an experiment?

8. **Compare:** When would a field study be better than a laboratory experiment?

Designing an Experiment...............

Design an experiment to solve the following problem.

PROBLEM: Can healthy plants be grown without sunlight?

Your experiment should:

1. List the materials you need.
2. Identify any safety precautions that should be followed.
3. List a step-by-step procedure.
4. Include a control.
5. Describe how you would record your data.

PEOPLE IN SCIENCE

JANE GOODALL (1934–present)

Jane Goodall is a British zoologist. Jane was interested in animals at an early age. She did field studies on chimpanzees in Africa. She worked with a famous fossil hunter, Dr. Louis Leakey. Leakey convinced Goodall to set up a camp in the Gombe National Park in northwest Tanzania. She observed the behavior of chimpanzees in the park. Through daily contacts with the chimpanzees, she gained their trust. She studied them at close range. She kept detailed reports on their behavior. Her field studies showed that some earlier information on these animals was incorrect.

Scientists thought that chimpanzees ate mostly fruit and vegetables. Sometimes they ate insects and small rodents. Dr. Goodall discovered that chimpanzees often hunted and ate large animals. She also observed the chimpanzees using crude tools. She watched them using tree stems to catch termites. Dr. Goodall pointed out that chimpanzees use tools more than any other animal except humans. She wrote many books on her field studies of chimpanzees. She also did field work on the behavior of African hyenas, jackals and wild dogs.

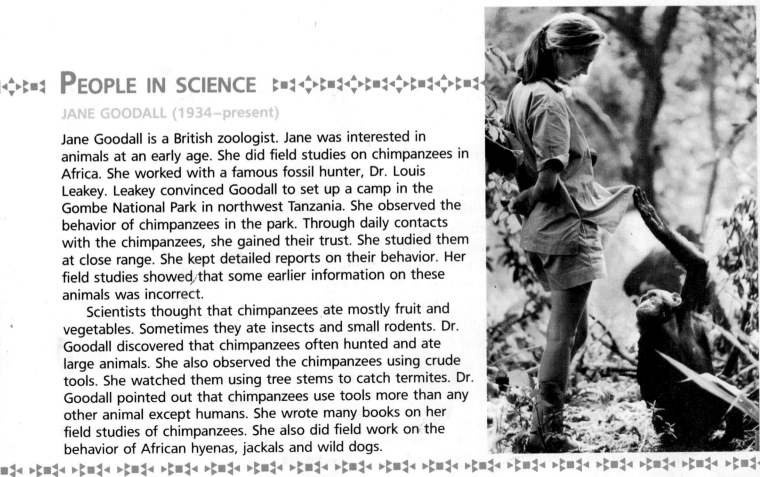

What is the metric system?

Objective ▶ Identify various metric units used to make measurements.

TechTerms

- ▶ **gram:** basic metric unit of mass
- ▶ **liter:** basic metric unit of volume
- ▶ **mass:** amount of matter in an object
- ▶ **meter** (MEE-tur): basic metric unit of length

Scientific Measurements The metric system is an international system of measurement. It is used in most countries. Everyday measurements are given in metric units (YOU-nits). A unit is an amount used to measure something. In the United States, the English system and the metric system are used.

Since 1960, scientists have used a more modern form of the metric system. This measurement system is called SI. The letters "SI" stand for Systems International. Many of the units in SI are the same as in the metric system.

▶ *Name:* What are two internationally used measurement systems?

Metric Units In the metric system, length and distance are measured by a unit called the **meter** (MEE-tur). The meter (m) is the basic unit of length. The basic unit of **mass** in the metric system is the **gram** (g). Mass is the amount of matter in an object. There is more matter in a nail than a tack. The nail has more mass than the tack. The **liter** (L) is the basic unit of volume in the metric system. The amount of space something takes up is its volume.

▶ *Name:* What is basic unit of mass in the metric system?

Changing Size The metric system is based on units of 10. This makes it easy to use. Each unit in the metric system is ten times greater or smaller than the next unit. To change the size of a unit,

you add a prefix to the unit. The prefix makes the unit larger or smaller. Table 1 shows prefixes and their meanings. Table 2 compares some metric units. It also shows the symbols of the units.

Table 1 Prefixes and Meanings

PREFIX	MEANING
kilo- (KILL-uh)	one thousand (1000)
hecto- (HEC-tuh)	one hundred (100)
deca- (DEC-uh)	ten (10)
deci- (DESS-ih)	one tenth (1/10)
centi- (SEN-tih)	one hundredth (1/100)
milli- (MILL-ih)	one thousandth (1/1000)

Table 2 Metric Units

1000 millimeters (mm) = 1 meter (m)
100 centimeters (cm) = 1 meter (m)
1000 grams (g) = 1 kilogram (kg)
1 liter (L) = 1000 milliliters (mL)

▶ *Describe:* How do you change the size of a metric unit?

Measuring Length A meter stick is used to measure length. A meter stick is divided into 100 equal parts by numbered lines. The distance between two numbered lines is equal to one centimeter (cm). Each centimeter is divided into 10 equal parts. Each of these parts is one millimeter (mm). One millimeter is 1/1000 of a meter.

◾ *Analyze:* How many millimeters are in 1 meter?

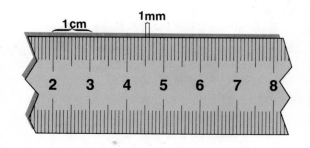

LESSON SUMMARY

▶ The metric system is an international system of measurement.

▶ Scientists use a more modern form of the metric system called SI.

▶ Prefixes are used to change the size of a metric or an SI unit.

▶ The meter, gram, and liter are basic metric units.

▶ A meter stick is used to measure length.

CHECK *Answer the following.*

1. Name three systems of measurement.

2. Which systems of measurement are used by scientists?

3. What is the prefix meaning 1000?

Complete the following.

4. The metric system is based on units of _____ .

5. A _____ added to a unit changes the size of the unit.

6. One centimeter is _____ of a meter.

7. The _____ is the basic metric unit of length.

APPLY *Complete the following.*

8. Two towns are 15 km apart. What is the distance in meters?

9. A rug is 4 m wide. How much is this in centimeters?

Calculate: Complete the following equations.

10. 1000 g = _____ kg

11. _____ m = 100 cm

12. 1 L = _____ mL

13. 1 cm = _____ mm

Skill Builder..

Sequencing Arrange the following lengths in order from shortest to longest: 40 millimeters, 10 kilometers, 6 hectometers, 50 centimeters, 6 millimeters, 3 centimeters, 3 kilometers, 2 hectometers.

ACTIVITY

MEASURING LENGTH

You will need a metric ruler.

1. **Observe:** Study your metric ruler carefully.

2. Use the metric ruler to find the width and length of your desk, your textbook, your fingernail, and a sheet of notebook paper. Measure to the nearest millimeter.

3. Use the metric ruler to find the lengths of the objects shown in this activity. Measure to the nearest millimeter.

4. **Organize:** Put your data in a table.

5. **Sequence:** Arrange all the objects in order from smallest to largest.

Questions

1. How long is your metric ruler?

2. How many millimeters are shown on the ruler?

3. Which object is the longest?

4. Which object is the shortest?

How do you measure using the metric system?

Objective ▶ Understand how to measure different things using the metric system.

TechTerms

▶ **degree Celsius:** metric unit of temperature

▶ **temperature:** measure of how hot or cold something is

Measuring Area Do you know how people find the area of the floor of a room? They measure the length and width of the room. Then, they multiply the two numbers. You can find the area of any rectangle by multiplying its length by its width. Area is expressed in square units, such as square meters (m^2) or square centimeters (cm^2).

▶ *Calculate:* What is the area of a rectangle 2 cm × 3 cm?

Measuring Volume A graduated cylinder can be used to measure the volume of a liquid in milliliters. Volume can also be measured in cubic centimeters. Imagine a small box shaped like a cube. Each side is 1 cm long. The volume of the cube is 1 cubic centimeter (cm^3). One cubic centimeter is the same as 1 milliliter. The volume of any box can be found by multiplying its length by its width by its height.

If you have a box that is 10 cm on each side, its volume would be 1000 cm^3. A liter is the same as 1000 cm^3. One liter of liquid will fill the box exactly.

▶ *Analyze:* How many milliliters of water would fill a 12-cm^3 box?

Measuring Mass Mass is measured with an instrument called a balance. A balance works like a seesaw. It compares an unknown mass with a known mass. Known masses are used on one side. The object is placed on the other side. The mass of the object is equal to the total of the known masses.

▶ *Name:* What instrument is used to measure mass?

Measuring Temperature The **temperature** of anything is a measure of how hot or cold it is. You also can say it is the amount of heat energy something contains. Temperature is measured with a thermometer. Temperature can be measured on one of three scales. They are the Fahrenheit (FAHR-uhn-hyt) scale, the Celsius (SEL-see-us) scale, and the Kelvin (K) scale. The Fahrenheit scale is part of the English system of measurement. The Celsius scale is usually used in science. Each unit on the Celsius scale is a **degree Celsius** (°C). The degree Celsius is a metric unit. The Kelvin scale is a part of SI. The Kelvin scale begins at absolute zero, or 0 K. There is no heat energy at absolute zero.

Table 1 Comparing Temperatures			
	°C	K	°F
Absolute zero	−273	0	−459
Freezing Point (Water)	0	273	32
Room Temperature	22	295	72
Human Body Temperature	37	310	98.6
Boiling Point (Water)	100	373	212

▶ *Observe:* What is the freezing point of water on the Celsius scale?

LESSON SUMMARY

▶ The area of a rectangle is found by multiplying its length by its width.

▶ Volume can be measured in cubic centimeters.

▶ A balance is used to measure mass.

▶ Temperature is a measure of how hot or cold something is.

▶ Temperature is measured with a thermometer.

▶ Three temperature scales are the Fahrenheit scale, the Celsius scale, and the Kelvin scale.

CHECK *Complete the following.*

1. The area of a rectangle is found by multiplying its length by its _____ .

2. One liter of liquid is equal to _____ cubic centimeters.

3. The symbol for cubic centimeters is _____ .

4. An instrument used to measure mass is a _____ .

5. An instrument used to measure temperature is a _____ .

6. Three temperature scales are the _____ scale, the Celsius scale, and the Kelvin scale.

7. The _____ scale begins at absolute zero.

APPLY *Use Table 1 on page 24 to complete the following.*

▬ 8. **Analyze:** What is the boiling point of water on the Celsius temperature scale?

9. **Compare:** What is absolute zero on the Fahrenheit scale? On the Celsius scale?

▬ 10. **Analyze:** What symbol is used for each temperature scale?

Health & Safety Tip.........................

Mercury is poisonous if you inhale it. It also can be absorbed through the skin. If you break a thermometer, tell your teacher or your parents immediately. Never clean up a broken thermometer with your hands. Contact your local poison control center. Find out how to clean up mercury properly.

ACTIVITY

CALCULATING AREA AND VOLUME

You will need 3 boxes of different sizes, paper, and a metric ruler.

1. Copy Table 1 on a sheet of paper.

2. Measure the length, width, and height of each box in centimeters. Record each measurement in your table.

3. Calculate the volume of each box. Record each volume in the table.

4. Find the area of every rectangle shown in the diagram. Organize the data in a table.

Questions

1. Which of your three boxes has the largest volume?

2. How many milliliters of liquid would fill each box?

3. How many rectangles are in the diagram?

4. What is the area of the largest rectangle?

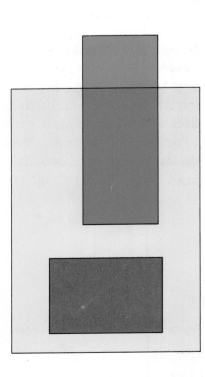

Table 1	Box Measurements			
BOX	LENGTH	WIDTH	HEIGHT	VOLUME
1				
2				
3				

1-7 What is a microscope?

Objective ► Identify and describe different kinds of microscopes.

TechTerms

- **lens** (LENZ): piece of curved glass that causes light rays to come together or spread apart as they pass through
- **magnify:** to make something look larger than it is
- **microscope** (MIKE-roh-scope): tool that makes things look larger than they really are

Microscopes One of the most important tools used to study living things is the **microscope** (MIKE-roh-scope). "Micro" means very small. "Scope" means to look at. A microscope is a tool used to make things look larger than they really are. There are different kinds of microscopes.

Define: What is a microscope?

Lenses A **lens** (LENZ) is a curved piece of glass. Some lenses have one curved surface and one flat surface. Others have two curved surfaces.

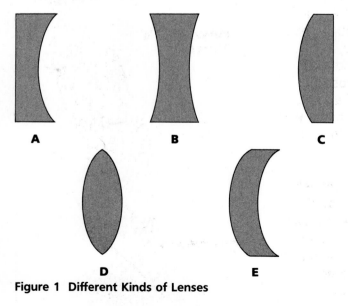

Figure 1 Different Kinds of Lenses

A lens brings light rays together or spreads them apart. Light that passes through a lens is bent. The bending of the light rays causes the object to look either larger or smaller. If the object looks larger, the lens has **magnified** (MAG-ni-fyd) the object.

Observe: Look at Figure 1. Which lenses have two curved surfaces?

Light Microscopes The microscopes you use in the classroom are light microscopes. Light microscopes have one or more lenses in them. These microscopes use light and lenses to magnify things.

Have you ever used a magnifying glass? If you have, you have used a simple microscope. A simple microscope has only one lens. Other microscopes have two or more lenses. A compound microscope has two or more lenses. The two lenses make things larger than does one lens.

Describe: How many lenses does a compound microscope have?

Electron Microscope An electron microscope uses electrons to magnify objects. It does not use light. Electrons are the particles that light up your television screen. A standard microscope can make an object appear 50 to 400 times larger. An electron microscope can magnify objects up to 300,000X their normal size. One type of electron microscope, scans the object and produces a three-dimensional image. Electron microscopes are very useful but they are very expensive.

Identify: What kind of microscope can magnify an object up to 300,000X?

26

LESSON SUMMARY

► A microscope makes objects appear larger.

► A lens is a curved piece of glass that bends light.

► A magnifying glass is a simple microscope.

► A compound microscope uses two or more lenses to magnify objects.

► An electron microscope uses electrons which magnify objects up to 300,000X.

CHECK *Answer the following.*

1. What is the difference between a compound microscope and a simple microscope?

2. What is a lens?

3. What kind of microscope can magnify things up to 300,000 X?

4. What household appliance uses electrons?

APPLY *Answer the following.*

► 5. **Infer:** Why do you think a compound microscope is better than a simple microscope?

► 6. **Infer:** Which of your senses is aided by using a microscope?

Skill Builder

Making a Simple Microscope Dip the part of a key with a hole into a glass of water. Make sure that a drop of water stays in the hole. Look through the drop of water to read the small print in a book. Move the key up and down very slowly. What happens? Which part of the microscope is the drop of water?

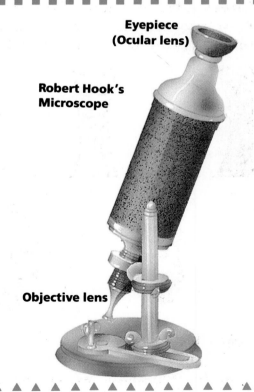

⏏⏏⏏ LOOKING BACK IN SCIENCE ▽▽▽▽▽▽▽▽▽▽▽▽▽▽▽▽▽

DEVELOPMENT OF THE MICROSCOPE

The ancient Greeks used magnifying glasses in the 2nd century BC. They used them as "burning glasses". The ancient Romans made simple microscopes out of rock crystals. Over 3000 years ago, glass balls filled with water were used as magnifying glasses. In 1590, two Dutch eyeglass makers, Hans and Zacharias Janssen developed the first compound microscope. Their microscope had two lenses, one at each end of a tube. The microscope made objects appear larger, but the image was fuzzy and distorted. Poor lenses caused this problem. Robert Hooke, an English scientist made and used a compound microscope in 1665. He looked at thin slices of cork.

About 1670, a Dutch merchant, Anton van Leeuwenhoek made a lens of fine quality. The lens produced a clear image. Leeuwenhoek made more than 500 lenses. They were used as simple microscopes. Some lenses could magnify objects 250X.

In 1931, two German scientists developed the first electron microscope. Today, electron microscopes are used in many science laboratories. Electron microscopes have a major drawback. They cannot be used to look at living things.

Eyepiece
(Ocular lens)

Robert Hook's
Microscope

Objective lens

What are the parts of a microscope?

Objective ▶ Identify and describe the functions of the parts of a compound microscope.

The Compound Microscope The first compound microscope was invented about 1590 by two Dutch lens makers, Hans and Zacharias Janssen. Many scientists made and used them. Much of what is known about living things would not be possible without the microscope.

All compound microscopes have the same basic parts. Using a microscope can be a lot of fun. It is easy to use if you know its parts and what they do.

▷ *Identify:* Who invented the first compound microscope?

Parts of the Compound Microscope As you read about each part of the compound microscope, find the part on the drawing.

▶ **Eyepiece** The eyepiece is located at the top of the microscope. It holds the ocular lens.

▶ **Body tube** The body tube is a hollow tube through which light passes. It holds the lenses apart.

▶ **Nosepiece** The nosepiece holds the objective lenses.

▶ **Objective lens** There are several objective lenses. Each lens has a different magnification power.

▶ **Arm** The arm supports the body tube, and is used to carry the microscope.

▶ **Coarse-adjustment knob** This knob turns and is used to raise or lower the body tube to focus the microscope.

▶ **Fine-adjustment knob** The knob also raises or lowers the body tube. It is used to bring objects into sharp focus.

▶ **Stage** The stage is the place where the object you are looking at is put.

▶ **Stage clips** The stage clips hold down the slide on the stage.

▶ **Diaphragm** The diaphragm (DY-uh-fram) changes the amount of light entering the body tube.

▶ **Light source** The light source is located beneath the diaphragm. It sends light toward the hole in the stage. A light source can be an electric light built into the microscope or a mirror that reflects light into the microscope.

▶ **Base** The base is the bottom part of the microscope. It often is shaped like a horseshoe. With the one hand holding the arm and one hand under the base, you carry a microscope properly.

▷ *Describe:* How do you carry a microscope properly?

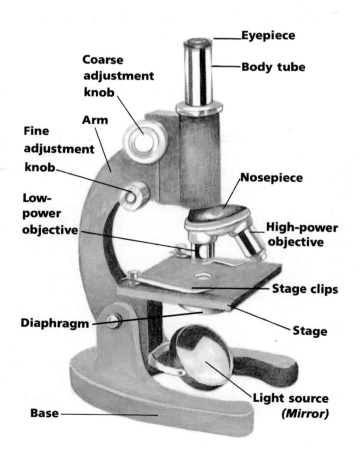

Labels: Eyepiece, Body tube, Coarse adjustment knob, Arm, Fine adjustment knob, Low-power objective, Nosepiece, High-power objective, Diaphragm, Stage clips, Stage, Base, Light source (Mirror)

LESSON SUMMARY

► The first compound microscope was invented by Hans and Zacharias Janssen.

► All compound microscopes have the same basic parts.

CHECK *Complete the following.*

1. The lens at the top of the microscope is found in the _____ .

2. When looking through a microscope the _____ lens is closest to your eye.

3. The nosepiece holds the _____ lenses.

4. The coarse- and fine- adjustment knobs help _____ the microscope.

5. The object to be viewed is placed on the _____ of the microscope.

6. The amount of light entering the microscope is controlled by the _____ .

APPLY *Answer the following.*

7. **Explain:** Why is a compound microscope called a light microscope?

8. **Infer:** Why do you think it is more important to get a clearer image than one that is fuzzy but greatly magnified?

9. **Infer:** How does changing the objective lenses affect what is seen through the microscope?

Use the microscope shown on page 28 to answer the following.

10. How many objective lenses are on the microscope?

11. What is the light source?

Use the microscope shown on page 28 to answer the following.

....
Skill Builder.......................................

Calculating To find the magnification of a microscope multiply the number found on the ocular lens by the number found on the objective lens. Find the magnification of the microscopes listed below.

	Ocular	Objective
a.	5X	10X
b.	5X	43X
c.	10X	10X
d.	10X	20X
e.	5X	20X

⋮⋮⋮ TECHNOLOGY AND SOCIETY ⠿ ⠿ ⠿ ⠿

MICROSURGERY

An ambulance races across town to a hospital. Inside there is an accident victim. The victim has had a finger severed in a machine. A team of microsurgeons will work to reattach the finger. The surgeons will use microsurgery.

In microsurgery, special microscopes are used by the surgeons to look at the very small parts of the body as they operate. The parts would be too small to operate on with their unaided eyes. With these microscopes, the surgeons can even attach tiny nerve fibers and small blood vessels. They use tiny instruments developed by medical engineers. Some of these instruments are so small the surgeons can stick a needle into a blood cell that is only 0.008 millimeters across.

The special microscope used in microsurgery was developed in the mid-1950s. It was used for ear and eye operations. Today, microsurgery is used in many delicate operations. Some brain surgery is done using microsurgery techniques.

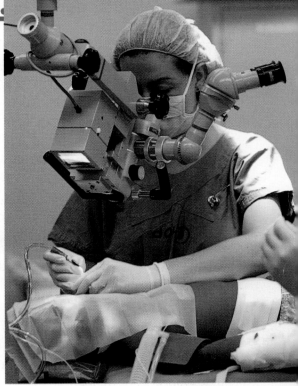

Challenges

STUDY HINT Before you begin the Unit Challenges, review the TechTerms and Lesson Summary for each lesson in this unit.

TechTerms

control (20)
controlled experiment (20)
data (18)
degree Celsius (24)
experiment (20)
gram (22)

hypothesis (16)
lens (26)
liter (22)
magnify (26)
mass (22)
meter (22)

microscope (26)
scientific method (18)
specialization (14)
temperature (24)
variable (20)

TechTerm Challenges

Matching *Write the TechTerm that best matches each description.*

1. studying or working in only one field
2. suggested solution to a problem
3. model or guide used to solve problems

4. information
5. test of a hypothesis
6. anything that can change the results of an experiment
7. to make something larger
8. tool that makes things appear larger

Identifying Word Relationships *Explain how the terms in each pair are related.* Write your answers in complete sentences.

1. degree Celsius, thermometer
2. matter, mass
3. lens, microscope
4. gram, balance
5. liter, volume
6. control, controlled experiment
7. meter, length
8. experiment, hypothesis
9. data, research

Content Challenges

Multiple Choice *Write the letter of the term or phrase that best completes each statement.*

1. A florist specializes in the branch of biology called
   a. botany. **b.** zoology. **c.** anatomy. **d.** physiology.

2. The interaction of living things and their surroundings is studied in the branch of life science called
   a. botany. **b.** zoology. **c.** anatomy. **d.** ecology.

3. The knowledge of science is based upon
   a. medicine. **b.** plants. **c.** observations. **d.** experimentation.

4. In the metric system, volume is measured in
   a. centimeters. **b.** square units. **c.** liters. **d.** grams.

5. When you use a copy of something to help explain it, you are using
   a. a model. **b.** an observation. **c.** a table. **d.** a prediction.

6. When you share information with someone, you are
   a. modeling. **b.** communicating. **c.** predicting. **d.** hypothesizing.

7. When you state ahead of time how or when something will happen, you are
   a. inferring. **b.** modeling. **c.** predicting. **d.** hypothesizing.

8. The first step in scientific method is to
   a. identify and state the problem.   **b.** record data.   **c.** gather information.   **d.** state a conclusion.

9. An experiment that is performed outdoors is
   a. an observation.   **b.** a field study.   **c.** a laboratory.   **d.** a controlled experiment.

10. The prefix kilo- means
    a. 10.   **b.** 100.   **c.** 1000.   **d.** 10,000.

11. A meter is equal to
    a. 1 kilometer.   **b.** 10 kilometers.   **c.** 1000 millimeters.   **d.** 10 millimeters.

12. A centiliter is equal to
    a. 10 liters.   **b.** 1/100 L.   **c.** 1/100 km.   **d.** 1 liter.

Completion *Write the term that best completes each statement.*

1. Plants are studied in the branch of biology called _____ .
2. When you classify, you group things based upon how they are _____ .
3. Graphs, tables, charts, and diagrams are ways to _____ information.
4. An experiment that is done outdoors is called a _____ .
5. The prefix deca- means _____ .
6. Length is measured with a _____ .
7. A graduated cylinder is used to measure the volume of a _____ .
8. Mass is measured with an instrument called a _____ .
9. The SI unit of temperature is the _____ .
10. The Kelvin temperature scale begins at _____ .
11. The formula for calculating area is _____ .
12. A magnifying lens is an example of a _____ .
13. A microscope with two lenses is a _____ microscope.
14. The amount of light that passes through a microscope is controlled by the _____ .

Understanding the Features..

Reading Critically *Use the feature reading selections to answer the following. Page numbers for the features are shown in parentheses.*

1. **Apply:** Do wildlife biologists do most of their work in a laboratory or as field studies? (15)
2. **Define:** What is technology? (19)
3. What kind of animals has Jane Goodall worked with most? (21)
4. Who was Dr. Louis Leakey? (21)
5. **List:** Identify three things that have been used to make microscopes. (27)
6. **Relate:** How have improved microscopes been helpful in the field of medicine? (29)

Concept Challenges...

Critical Thinking *Answer each of the following in complete sentences.*

1. **Contrast:** What are two differences between an electron microscope and a compound microscope?
2. **Infer:** How are microscopes useful to microbiologists?
3. **Apply:** What temperature scale would you use to measure very cold temperatures? Why?
4. **Relate:** How are the microscopes that are used today better than those that were used two hundred years ago?

Interpreting a Diagram *Answer the questions about the diagram.*

1. **Identify:** What is shown in the diagram?
2. What is the instrument shown used for?
3. What is the light source for the instrument shown?
4. **Observe:** How many objective lenses does this instrument have?
5. **Calculate:** If the magnifying power of the eyepiece lens is 10x and the magnifying power of the low-power objective is 10x, what is the total magnification of the microscope using the low-power lens?
6. **Describe:** How should a microscope be carried?
7. What part of a microscope holds the slide in place?
8. **Infer:** Is the magnification of this microscope greater than or less than the magnification of a hand lens?
9. **Contrast:** How does a compound microscope differ from a simple microscope?
10. **Define:** What is a lens?

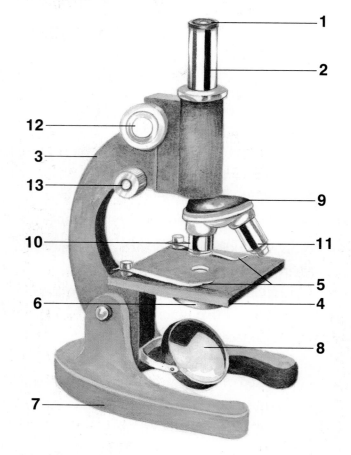

Finding Out More..

1. **Organize:** Make a poster that shows the tools used to measure length, mass, volume, and temperature. Label each tool on your poster and identify how it is used.
2. **Research:** Use library references to find out about the work done by Dian Fosse. How is the work done by Dian Fosse similar to the work done by Jane Goodall? Write your findings in a report.
3. **Communicate:** Research the work of Louis and Mary Leakey. What kind of fossils were these scientists most interested in? How have the discoveries of the Leakey's helped scientists learn more about humans who lived long ago? Present your findings in an oral report to the class.
4. Find ten products that use metric or SI units on their labels. Combine your labels with those of five of your classmates. Use your labels to make a bulletin board display for your classroom.
5. **Organize:** Create a table that shows and compares the basic units of mass, volume, temperature, length, and weight in the English, metric, and SI systems.

NEEDS OF LIVING THINGS

CONTENTS

STUDY HINT Before beginning each lesson in Unit 2, review the TechTerms for each lesson and read the Lesson Summary.

2-1. What are living things?

► List and describe the six characteristics of living things.

TechTerms

► **cell:** basic unit of structure and function in living things
► **organism** (AWR-guh-niz-um): any living thing
► **response** (ri-SPAHNS): reaction to a change

Organisms The world around you is made up of many different things. Some things, such as dogs and trees, are living. Living things are called **organisms** (AWR-guh-niz-ums). Other things, such as cars and radios, are nonliving.

▌▶ *Define:* What is an organism?

Characteristics of Organisms It is not always easy to decide if something is living or nonliving. Nonliving things may do some of the same things as organisms. For example, a robot may move and speak like a person. A robot, however, is not living. Plants and animals grow, or get larger. Icicles also may seem to grow, but icicles are not living.

Biologists use six characteristics to classify something as a living thing. All living things have these six characteristics.

► Organisms are made up of one or more **cells.** A cell is the basic unit of structure and function in living things. In fact, cells often are called the "building blocks of life."

► Energy is the ability to do work. Organisms use energy. Sunlight is the source of energy for most living things. Plants use the energy in sunlight to make food. Animals get energy from the sun by eating plants or animals that have eaten plants.

► Organisms are adapted (uh-DAP-tud), or suited, to their surroundings. All organisms have features that help them survive in their surroundings. For example, fishes have gills. Gills are organs that allow fishes to breathe in water.

► Organisms react to changes in their surroundings. Any reaction to a change is called a **response** (ri-SPAHNS). You might respond to the honking of a car's horn by jumping. A bright light may cause you to close your eyes.

► Organisms produce more organisms of their own kind. Dogs produce more dogs. Pine trees produce more pine trees. The production of new organisms allows each kind of organism to continue living on the earth.

► Organisms grow and develop. Living things change, or develop during their lifetimes. One way organisms change is by growing. Living things also may change in appearance.

▌▶ *Identify:* What is the source of energy for most living things?

LESSON SUMMARY

▶ Living things are called organisms.

▶ It is not always easy to tell living things from nonliving things.

▶ All living things have six characteristics. Living things are made up of cells, use energy, are adapted to their surroundings, produce more of their own kind, grow, and develop.

CHECK *Complete the following.*

1. Organisms are made up of one or more _____ .

2. Energy is the ability to do _____ .

3. Organisms _____ to changes in their surroundings.

4. Plants use the energy in _____ to make food.

5. Organisms are suited, or _____ , to their surroundings.

6. Organisms get larger, or _____ .

7. Growth is one of the ways that organisms _____ .

APPLY *Complete the following.*

8. **Apply:** How does a cow get energy from sunlight?

9. **Hypothesize:** Could an ant be the offspring of a fly? Explain your answer.

10. **Classify:** Flashlights and cars both use energy to work. Use the characteristics of living things to explain why cars and flashlights are not classified as living things.

Ideas in Action................................

IDEA: All living things respond to changes in their surroundings.

ACTION: Describe three examples of responses you have to changes in your surroundings. Identify the change that causes each response.

SCIENCE CONNECTION ◆○◆○◆○◆○◆○◆○◆○◆○◆○◆○◆○◆○◆○◆○◆

MATTER

Do you know what everything around you has in common? Everything around you is made up of matter. Matter is anything that has a mass and takes up space.

Matter is made up of tiny particles called atoms. Some substances are made up entirely of atoms of the same kind. These substances are called elements. Most living things are made up mainly of the elements carbon, hydrogen, oxygen, nitrogen, sulfur, and phosphorous.

When two or more elements combine chemically, they form a compound. Four kinds of compounds are important to living things. These compounds are carbohydrates, lipids, proteins, and nucleic acids. Carbohydrates are made up of carbon, hydrogen, and oxygen. They are important energy sources. Lipids are made up mostly of carbon and hydrogen. Lipids store energy. Proteins are the basic building blocks of all living things. Proteins are made up of carbon, hydrogen, oxygen, and nitrogen. Nucleic acids control cell activities. Nucleic acids also control the making of proteins.

Other Elements 4.5%
Nitrogen 3%

Hydrogen 10%

Carbon 18%

Oxygen 64.5%

2-2 What are adaptations?

Objective ▶ Identify and describe some adaptations of organisms.

TechTerms

▶ **adaptation** (ad-up-TAY-shun): trait of a living thing that helps it live in its environment

▶ **environment** (in-VY-run-munt): everything that surrounds a living thing

Environment All organisms live in an **environment** (in-VY-run-munt). The environment is everything that surrounds a living thing. All the living things that surround you are part of your environment. Your environment also is made up of nonliving things. Some of the nonliving things in the environment are air, water, sunlight, rocks, and soil.

▶ *Name:* What are some nonliving parts of the environment?

Adaptations Any trait of an organism that helps the organism live in its environment is called an **adaptation** (ad-up-TAY-shun). Adaptations make each kind of organism suited to life in its environment. Adaptations allow one kind of organism to live where other kinds of organisms cannot live. A cactus has adaptations that allow it to live in a dry, desert environment. A cactus has a thick, leathery stem that stores water. It also has spines. Spines are special leaves that keep a cactus from losing too much water.

Some living things are adapted to cold environments. Polar bears live in very cold environments. The thick fur and fat layers of a polar bear keep it warm. A polar bear would not be able to survive in a hot, humid jungle.

▶ *Infer:* How are fishes adapted to their environments?

People and Adaptations People have adaptations too. You have a thumb that can touch all of your other fingers. Your thumb allows you to use your hands to do many things. You can write, build things, draw, and so on. Your flexible thumb is an adaptation.

Unlike most organisms, people often adapt their environments to meet their needs. People build shelters to protect themselves from the weather. People control their indoor environments by using air conditioners and heating systems. People also wear different kinds of clothing depending on the weather.

▶ *Describe:* What are two ways that you adapt to a rainy day?

LESSON SUMMARY

► All organisms live in an environment.
► An adaptation is any trait of an organism that helps it live in its environment.
► People have adaptations.
► People often adapt their environments to meet their needs.

CHECK *Find the sentence in the lesson that answers each question. Then, write the sentence.*

1. What is the environment?
2. What nonliving things are part of the environment?
3. What is an adaptation?
4. To what kind of environment is a cactus adapted?
5. Why is your thumb an adaptation?
6. How do people adapt their indoor environments to meet their needs?

APPLY *Complete the following.*

7. **Identify:** Name two living and nonliving parts of your environment.
8. **Explain:** Give two examples of how you adapt to warm weather.

Match each organism with the environment to which it is most suited.

9. cold, snowy climate
10. hot, dry desert
11. the ocean
12. a pond
13. a heavily, wooded forest

a. frog
b. octopus
c. polar bear
d. cactus
e. raccoon

State the Problem.................

Study the illustration. Then, state the problem.

CAREER IN LIFE SCIENCE

ECOLOGIST

Do you enjoy studying plants and animals? Would you like to study how conditions in the environment affect living things? If so, you may enjoy a career as an ecologist (ee-KAHL-uh-jee). Ecologists are scientists who study the relationship between living things and their environments.

Ecologists study many different problems. Some ecologists study only one kind of plant or animal. Most ecologists, however, study entire communities. Some study land communities. Others specialize in water environments. An ecologist may manage forest land and wildlife refuges, evaluate the effects of pollution and mining on living things, study soil composition, or help control insect pests.

An ecologist may work in the field or in a research laboratory. Ecologists are employed by the government, industries, and universities. If you are interested in becoming an ecologist, you should have a good background in the sciences. A college degree is required for most jobs in ecology.

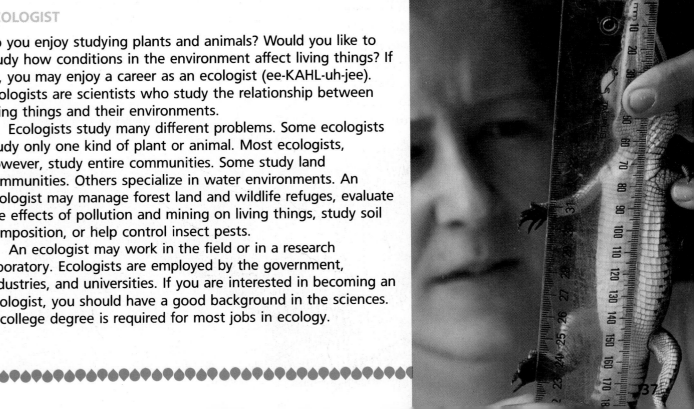

2-3 What are responses?

▶ Explain how stimuli and responses are related. ▶ Identify some kinds of behaviors.

TechTerms

- **behavior** (bi-HAYV-yur): ways in which living things respond to stimuli
- **hibernation** (HY-bur-nay-shun): inactive state of some animals during winter months
- **migration** (my-GRAY-shun): movement of animals from one living place to another
- **stimulus** (STIM-yuh-lus): change that causes a response

Stimulus and Response Living things respond to their environments. In the morning, your alarm clock rings. You respond by waking. The ringing alarm clock is a **stimulus** (STIM-yuh-lus). A stimulus is a change that causes a response. The plural of stimulus is stimuli (STIM-yuh-ly).

Living things respond to stimuli in different ways. Turn a plant so that its leaves face away from the sun. In a few days, the leaves of the plant will turn back toward the sun. The movement of the plant's leaves is a response. Plant responses usually are slower than animal responses.

▶ *Identify:* A flower slowly turns to face the sun. What is the stimulus and response for this action?

Behavior The way in which an organism responds to stimuli is called **behavior** (bi-HAYV-yur). A behavior that an organism is born with is called an instinct (IN-stinkt). Nest-building is an instinct in some kinds of birds. Birds do not have to be shown how to build nests. Other behaviors have to be learned. Tying your shoelaces is a learned behavior. You have many learned behaviors.

▶ *Analyze:* Foxes are taught how to hunt by other foxes. Is hunting a learned behavior or an instinct in these animals?

Animal Behaviors The movement of animals from one living space to another is called **migration** (my-GRAY-shun). Animals often migrate to warmer places during the cold months to find food. Animals also migrate to find a safe place to reproduce and raise their young.

Some animals spend the winter months in a sleeplike state called **hibernation** (HY-bur-nay-shun). During hibernation, an animal is not active. The body temperature of the animal lowers. The heartbeat of the animal slows. The animal does not need as much energy. As a result, the animal can live off the fat stored in its body. Chipmunks and squirrels are two kinds of animals that hibernate.

▶ *Define:* What is hibernation?

LESSON SUMMARY

▸ All living things respond to stimuli.

▸ Living things respond to stimuli in different ways.

▸ The ways in which living things respond to stimuli is called behavior.

▸ Migration and hibernation are two behaviors of animals.

CHECK *Write true if the statement is true. If the statement is false, change the underlined term to make the statement true.*

1. Behaviors that an animal is born with are <u>learned behaviors</u>.

2. The movement of animals from one living area to another is <u>hibernation</u>.

3. A change that causes a response is a <u>stimulus</u>.

4. The inactive state of some animals during the winter months is called <u>hibernation</u>.

5. The ways in which living things respond to stimuli is called <u>migration</u>.

6. During hibernation, an animal's activities <u>speed up</u>.

APPLY *Complete the following.*

➤ **7. Infer:** What happens to a squirrel's breathing rate during hibernation?

Classify each action described as a learned behavior or an instinct.

8. reading a book

9. blinking

10. a newborn crying

11. going to school

InfoSearch

Read the passage. Ask two questions about the topic that you cannot answer from the information in the passage.

Bird Migration Many kinds of birds migrate during the winter and summer months. Year after year, birds migrate over the same paths, or migration routes. Scientists are not completely sure how birds are able to follow the same route during their round-trip migrations. Some scientists think the birds use the sun as a landmark. Other scientists think birds use wind currents.

SEARCH: *Use library references to find answers to your questions.*

ACTIVITY

PUPIL RESPONSES TO LIGHT

You will need a penlight or flashlight.

1. Work with a partner. Look at the pupil of your partner's eye. The pupil is the dark circle in the middle of the colored part of the eye. Note the size of the pupil.

2. **Observe:** Quickly shine the penlight in your partner's eye. Observe what happens to the size of the pupil.

3. **Observe:** Take the light away. Observe what happens to the size of the pupil.

Questions

1. **Observe:** What happened to the size of the pupil when the light was shined on it? When did the pupil get smaller?

2. **a.** What is the stimulus? **b.** What is the response?

3. **a. Infer:** What happened to the size of the pupil when you took the light away? **b.** Why do you think this happens?

4. **Hypothesize:** What would happen if the pupil did not change size?

5. **Predict:** When would the size of the pupil be the largest?

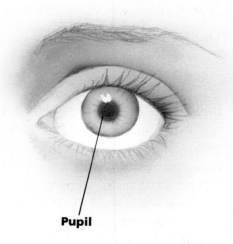

Pupil

Where do living things come from?

Objective ▶ Recognize that all life comes from existing life.

TechTerm

▶ **spontaneous generation** (spahn-TAY-nee-us jen-uh-RAY-shun): idea that living things come from nonliving things

Spontaneous Generation Do you believe that living things can grow from straw? Hundreds of years ago, people believed that mice came from straw. They also believed that worms and flies grew from rotting meat. The idea that living things came from nonliving things is called **spontaneous generation** (spahn-TAY-nee-us jen-uh-RAY-shun). Until the 1600s, many people believed in spontaneous generation.

▶ *Infer:* Why do you think people who lived hundreds of years ago believed that frogs came from rotting wood and water?

Francesco Redi Francesco Redi was an Italian doctor. He lived during the seventeenth century. Redi did not think that living things came from nonliving things. He thought that living things could come only from other living things. To test his hypothesis, Redi performed an experiment.

▶ *Identify:* What did Redi do to test his hypothesis about spontaneous generation?

Redi's Experiment Redi put some spoiled meat into several jars. Some jars he left uncovered. Some jars he covered with a thin net. Other jars were sealed tightly with lids. The jars with the lids were Redi's control. The setup for Redi's experiment is shown below.

After a few days, Redi observed wormlike animals on the meat in the uncovered jars. He also observed the wormlike animals on the net covering some of the jars. There were no wormlike animals in the jars with lids.

The wormlike animals that Redi observed were maggots (MAG-uts). Maggots are a stage in the life cycle of a fly. The maggots hatched from eggs that flies had laid on the meat and net. Redi showed that maggots did not come from the meat. Today, scientists know that flies often lay eggs on spoiled meat. The meat is food for the maggots. Scientists know that all living things come from other living things of the same kind.

▶ *Infer:* Why did maggots not appear in the covered jars?

EXPERIMENTAL SETUP

A — Open jar
Sealed jar
SEVERAL DAYS LATER
Flies in jar
Maggots on meat
No flies in jar
No maggots on meat
C — Jar covered with veil
Maggots on veil
No flies in jar

► The idea that living things come from nonliving things is called spontaneous generation.

► Francesco Redi was a seventeenth century doctor who believed that living things could not come from nonliving things.

► Redi performed an experiment using flies and spoiled meat to disprove the idea of spontaneous generation.

► Redi's experiment showed that flies do not come from spoiled meat.

CHECK *Complete the following.*

1. What is spontaneous generation?
2. Who was one of the first people to disprove the idea of spontaneous generation?
3. What are maggots?
4. Why do flies often lay eggs on spoiled meat?
5. What was the control for Redi's experiment?

APPLY *Complete the following.*

6. Why is it important to use a control in an experiment?
7. What did Redi's experiment show?

Look at the illustration. Why do you think people from Redi's time believed that mice came from straw?

Designing an Experiment.................

Design an experiment to solve the problem.

PROBLEM: Do flies develop from garbage?

Your experiment should:

1. List the materials you need.
2. Identify safety precautions that should be followed.
3. List a step-by-step procedure.
4. Describe how you would record your data.

PEOPLE IN SCIENCE

LOUIS PASTEUR (1822-1895)

Did you have a glass of milk this morning? If so, you may have noticed the term "pasteurized" on the container. Pasteurization is a process by which germs growing in certain food products are killed. The process is named after the scientist who developed it, Louis Pasteur.

Louis Pasteur is one of the great scientists in the history of science. He was a French chemist and microbiologist. Among his many contributions to science, Pasteur performed a simple experiment that disproved spontaneous generation.

In his experiment, Pasteur poured a nutrient broth into swan-necked flasks. The swan-necks prevented dust particles and microorganisms from entering the flasks. He boiled the flasks to kill any microorganisms in the broth. After several days, Pasteur saw no microorganisms growing in the flasks. He then tipped the flasks, allowing dust and microorganisms living in the air to enter the broth. In a few days, microorganisms were growing in the flasks. Pasteur demonstrated that living things did not arise spontaneously from the broth or from the air. They grew from microorganisms carried in the air.

2-5 How does life continue on the earth?

Objective ▶ Describe the two kinds of reproduction.

TechTerms

▶ **asexual reproduction** (ay-SEK-shoo-wul ree-pruh-DUK-shun): reproduction needing only one parent

▶ **offspring:** new organisms produced by a living thing

▶ **reproduction:** process by which living things produce new organisms like themselves

▶ **sexual** (SEK-shoo-wul) **reproduction:** reproduction needing two parents

Reproduction Every day millions of living things die. Organisms do not live forever. Why is there still life on the earth? Before most organisms die, they produce new organisms like themselves. The process by which organisms produce new organisms is called **reproduction** (ree-pruh-DUK-shun). Reproduction does not keep individual organisms alive. Reproduction only continues each kind of living thing by producing new individuals.

▶ *Define:* What is reproduction?

Offspring All living things reproduce their own kind. The new living things that organisms produce are called **offspring.** The offspring of horses are horses. They are called foals. The offspring of oak trees are seeds called acorns. The acorns grow into oak trees.

▶ *Define:* What are offspring?

Asexual Reproduction Simple organisms and some plants produce offspring by **asexual** (ay-SEK-shoo-wul) **reproduction.** Asexual reproduction is reproduction that requires only one parent. The simplest form of asexual reproduction is called fission (FISH-un). In fission, new organisms are produced when the parent organism splits in

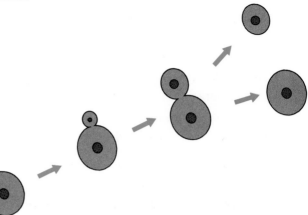

Figure 1 Budding

two. Another kind of asexual reproduction is budding. Budding is the growth of a new organism from the parent organism. In asexual reproduction, each new offspring is an exact copy of its parent.

▶ *Name:* What is the simplest form of asexual reproduction?

Sexual Reproduction Most living things reproduce by **sexual** (SEK-shoo-wul) **reproduction.** Sexual reproduction is reproduction needing two parents. During sexual reproduction, cells from two parents join. A new organism develops from the joined cells. This new organism is not exactly like either of its parents. Instead, the offspring has some features of each parent.

▶ *Contrast:* How are sexual and asexual reproduction different?

Figure 2 Cat and offspring

LESSON SUMMARY

▶ Reproduction is the process through which living things produce new organisms like themselves.

▶ Living things reproduce their own kind.

▶ Reproduction that needs only one parent is called asexual reproduction.

▶ Most living things reproduce by sexual reproduction.

CHECK *Complete the following.*

1. Asexual reproduction involves only- _____ parent.

2. The simplest form of asexual reproduction is _____ .

3. New living things produced by organisms are called _____ .

4. Sexual reproduction needs two _____ .

5. During fission, the parent organism _____ in two.

APPLY *Complete the following.*

▶ 6. **Predict:** What might happen to a kind of organism that produces few offspring?

▲ 7. **Model:** Draw a diagram showing the process of budding. Label your drawing.

Use the diagram to answer the following.

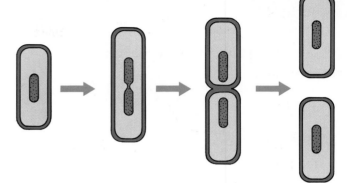

◉ **Observe:** What process is being shown?

▬ **Analyze:** Is this a type of asexual or sexual reproduction?

Skill Builder..............................

Researching The offspring of many animals have specific names. Using library references, find out what the offspring of each of the following animals are called: goats, pigs, giraffes, gorillas, tigers, deer, bears, cows, and ducks. Organize your findings in a table.

000193

TECHNOLOGY AND SOCIETY

CLONING ANIMALS

Imagine walking into a pet shop, selecting an adult dog, and ordering up an exact copy. Someday, this may be possible due to cloning. Cloning is the production of organisms with identical traits.

Scientists have been able to clone entire animals from a single cell. Scientists have cloned cattle that are good milk and meat producers. Cloning is done by taking apart the cells of an organism that is just beginning to develop. Each cell is then put into an egg from which the nucleus has been removed. The new egg begins developing. When the egg has finished developing, the offspring is identical to its parent.

Why would people want to clone animals? Cloning is one way to produce animals with specific traits. Cloning also may be used to protect animals that are nearing extinction. Endangered animals would be cloned in zoos in order to add to their populations. The organisms would then be sent back into their natural environments

What are life processes?

Objective ▶ Name and describe the life processes.

TechTerms

▶ **digestion** (di-JES-chun): process of breaking down food so that it can be used by living things

▶ **excretion** (iks-KREE-shun): process of getting rid of wastes

▶ **ingestion** (in-JEST-shun): process of taking in food

▶ **respiration** (res-puh-RAY-shun): process of getting energy from food

▶ **transport:** process of moving nutrients and wastes in a living thing

Life Processes All living things carry out life processes. Life processes are the things an organism must do to stay alive. The life processes also are features of living things. Being alive means carrying out all the life processes.

▶ *Define:* What are life processes?

Nutrition Living things need food because it provides them with nutrients. Nutrients are materials needed for growth and energy. Animals take food into their bodies. Taking in food is called **ingestion** (in-JEST-shun). Plants make their own food. Plants also take in some nutrients from the soil.

Food needs to be changed before an organism can use the nutrients in food. The process of changing food into a useable form is **digestion** (di-JES-chun). You have an organ system that digests your food. It is your digestive system.

ingestion
+ digestion
nutrition

▶ *Define:* What is digestion?

Respiration Living things get energy from food by a process called **respiration** (res-puh-RAY-shun). During respiration, oxygen is used to break food apart. This process produces energy. Carbon dioxide and water also are produced. These are waste products of respiration.

▶ *Name:* What are the waste products of respiration?

Excretion Your body makes many waste products. Some waste products are formed during digestion. Others are formed during respiration. All organisms must get rid of waste products formed by the life processes. Getting rid of waste products is called **excretion** (iks-KREE-shun).

▶ *Name:* Name two life processes that produce waste products.

Transport Once food is digested, nutrients must be carried to all parts of a living thing. Waste products must be carried away. The moving of nutrients and waste products is called **transport.**

▶ *Infer:* What materials are transported inside a living thing?

LESSON SUMMARY

▶ All living things carry out life processes.

▶ Taking in food is called ingestion.

▶ The process of changing food so it can be used is digestion.

▶ Living things get energy from food by respiration.

▶ Living things get rid of waste products by the process of excretion.

▶ The moving of nutrients and waste products is called transport.

CHECK *Write true if the statement is true. If the statement is false, change the underlined term to make the statement true.*

1. Taking in food is called <u>digestion</u>.

2. Two waste products of respiration are carbon dioxide and <u>water</u>.

3. The moving of waste products and nutrients is called <u>excretion</u>.

4. <u>Nutrients</u> are needed for growth and energy.

5. During respiration, <u>nitrogen</u> is used to break food apart.

6. <u>Animals</u> take in some nutrients from the soil.

APPLY *Complete the following.*

7. Why does being alive mean carrying out all the life processes?

8. **Sequence:** Describe what happens to food after it is ingested by an animal. Include a discussion of digestion, respiration, excretion, and transport in your description.

Health & Safety Tip

Food provides you with the nutrients you need for energy, growth, and repair of body tissues. Proper nutrition is important for staying healthy. Find out which nutrients you need. The Food and Drug Administration (FDA) has made a list of the U.S. Recommended Daily Allowance (U.S. RDA) for each nutrient. Find out how much of each nutrient you need. Organize your findings in a chart.

Skill Builder

Researching The process by which plants use the sun's energy is used to make food is called photosynthesis. Find out the chemical equation for photosynthesis. Write the equation using both symbols and words. Then explain the relationship between photosynthesis and respiration.

▼.▼.▼. LOOKING BACK IN SCIENCE ▼.▼.▼.▼.▼.▼.▼.▼.▼.▼.▼.▼.▼.▼.

TRANSPORT IN HUMANS

Today, scientists know that blood is transported through blood vessels. The blood vessels are connected and form a circular path. In the fourth century BC, people believed that the blood vessels carried both air and blood. In the second century, a Greek physician named Galen proved that arteries carried only blood. He still believed that air entered the body from the right side of the heart. Galen did not know the blood moved in a circular path. Galen and others thought that blood mixed with air in the lower parts of the heart. They thought there were holes in the dividing walls of the heart.

In the 16th century, scientists began to think that blood was transported through the heart and lungs in a circular path. In the 17th century, William Harvey, an English physician, explained how blood was transported in the human body. He proved that the heart did not have a wall with holes in it. Harvey could not show that the large blood vessels were connected by smaller blood vessels called capillaries. He did not have a microscope that was powerful enough to see capillaries.

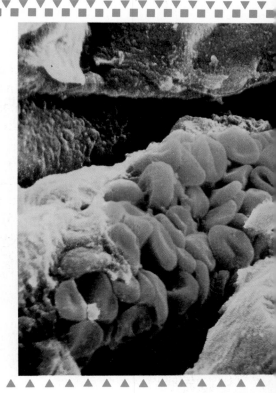

Objective ▶ Identify and describe the needs of living things.

TechTerm

▶ **homeostasis** (hoh-mee-oh-STAY-sis): ability of a living thing to keep conditions inside its body constant

Food and Water All organisms need food for growth and energy. Organisms also need water. Without water, all plants and animals would die. Plants use water to make food. About two-thirds of your body is water.

Most substances dissolve in water. These dissolved substances can then be transported throughout a living thing. Most chemical changes in living things cannot take place without water.

■ *Analyze:* What is the most common substance in your body?

Air Without air, most living things would die in minutes. Air is a mixture of gases. Oxygen is one of the gases in air. Oxygen is needed by most living things to change food into energy. Land organisms get oxygen from the air. Water organisms get oxygen from water. The oxygen is dissolved in the water.

■ *Analyze:* How does a fish get oxygen?

Temperature The temperature of the environment is important to living things. Living things need a proper temperature to carry out their life processes. Most living things could live only within a small temperature range if it were not for **homeostasis** (hoh-mee-oh-STAY-sis). Homeostasis is the ability of a living thing to keep conditions inside its body constant. When changes in temperature or other parts of the environment occur, homeostasis keeps things working properly inside a living thing.

▥▶ *Define:* What is homeostasis?

Living Space All organisms need living space. An organism's living space must provide enough air, water, sunlight, food, and shelter. In any environment, living space is limited. All the organisms in the environment compete for living space. They compete for food, water, sunlight, and so on.

▥▶ *Name:* List five needs of living things.

LESSON SUMMARY

▶ Organisms need food and water.

▶ Most chemical changes in living things cannot take place without water.

▶ Oxygen is needed by most living things.

▶ Living things need a proper temperature to carry out life processes.

▶ All organisms need living space.

CHECK *Write true if the statement is true. If the statement is false, change the underlined term to make the statement true.*

1. Most chemical changes in living things <u>cannot</u> take place without water.

2. In any environment, living space is <u>limited</u>.

3. <u>Animals</u> can make their own food.

4. Fishes get oxygen from the <u>air</u>.

5. <u>Homeostasis</u> keeps things working properly inside a living thing.

APPLY *Complete the following.*

▶ 6. **Infer:** How do you think shivering and perspiring are related to homeostasis?

7. What are the nonliving parts of the environment?

8. **Hypothesize:** What do you think happens to organisms when their living space is destroyed?

Ideas in Action..............................

IDEA: You use water every day.
ACTION: Identify ways in which you use water each day.

Health & Safety Tip..........................

Most living things need oxygen to live. These living things are aerobic (er-OH-bik) organisms. Some kinds of organisms grow only in places without oxygen. These living things are called anaerobic (an-er-OH-bik) organisms.

The organism that causes a deadly food poisoning called botulism is anaerobic. It can grow in improperly canned foods. Cans containing the botulism organism often bulge. Use library references to find out what kinds of foods are most likely to contain the botulism organism. Write your findings in a brief report.

LEISURE ACTIVITY

COOKING

Do you enjoy working in the kitchen? Would you like to prepare a meal that appeals to both the eyes and to the taste? If so, you may enjoy learning how to cook. For many people, cooking is an enjoyable, relaxing activity.

You can learn how to cook by observing an experienced cook at work. You can also try recipes from a cookbook. A good cookbook is an important tool for new cooks. Cookbooks usually have similar foods grouped together. Many have pictures showing the way to prepare foods and the finished recipe.

When you first start cooking, try making simple foods. Make sure you have all the ingredients you will need beforehand. As you become more experienced, try more difficult dishes. Ask your friends and relatives for their special recipes. Soon you may be developing "secret recipes" of your own.

STUDY HINT Before you begin the Unit Challenges, review the TechTerms and Lesson Summary for each lesson in this unit.

TechTerms...

adaptation (36)
asexual reproduction (42)
behavior (38)
cell (34)
digestion (44)
environment (36)
excretion (44)

hibernation (38)
homeostasis (46)
ingestion (44)
migration (38)
offspring (42)
organism (34)
reproduction (42)

respiration (44)
response (34)
sexual reproduction (42)
spontaneous generation (40)
stimulus (38)
transport (44)

TechTerm Challenges...

Matching *Write the TechTerm that best matches each description.*

1. ability of a living thing to keep conditions inside its body constant
2. inactive state of some animals during winter months
3. idea that living things come from nonliving things
4. process of getting energy from food
5. reproduction needing two parents

6. new organisms produced by living things

Identifying Word Relationships *Explain how the words in each pair are related. Write your answers in complete sentences.*

1. ingestion, digestion
2. environment, adaptation
3. reproduction, asexual reproduction
4. behavior, migration
5. cell, organism
6. excretion, transport
7. stimulus, response

Content Challenges...

Multiple Choice *Write the letter of the term that best completes each statement.*

1. Oxygen is used to break food apart during
   a digestion. **b.** ingestion. **c.** nutrition. **d.** respiration.

2. Two kinds of animals that hibernate are
   a. chipmunks and squirrels. **b.** birds and chipmunks. **c.** bears and birds. **d.** birds and squirrels.

3. Offspring that are not exactly like either parent are produced by
   a. budding. **b.** sexual reproduction. **c.** asexual reproduction. **d.** fission.

4. All of the following are nonliving parts of the environment *except*
   a. soil. **b.** sunlight. **c.** trees. **d.** water.

5. The source of energy for most living things is
   a. oxygen. **b.** sunlight. **c.** water. **d.** soil.

6. All living things
   a. make their own food. **b.** hibernate. **c.** migrate. **d.** are made up of cells.

7. The simplest form of asexual reproduction is
   a. spontaneous generation. **b.** budding. **c.** fission. **d.** respiration.

8. Getting rid of waste products is called
   a. transport. **b.** excretion. **c.** response. **d.** ingestion.

9. A change that causes a response is
   a. a stimulus. **b.** behavior. **c.** an adaptation. **d.** a reaction.

10. Behaviors that organisms are born with are called
    a. learned behaviors. **b.** instincts. **c.** stimuli. **d.** adaptations.

11. The ability of an organism to keep conditions inside its body constant is called
    a. homeostasis. **b.** hibernation. **c.** migration. **d.** reproduction.

12. The idea that living things could be produced from nonliving things is called
    a. asexual reproduction. **b.** Redi's theory. **c.** spontaneous generation. **d.** Pasteur's theory.

True/False *Write true if the statement is true. If the statement is false, change the underlined term to make the statement true.*

1. Energy is the ability to do work.
2. Animals often hibernate to find a safe place to reproduce.
3. In budding, new organisms are produced when the parent organism splits in two.
4. The waste products of respiration are carbon dioxide and oxygen.
5. When changes in temperature occur, homeostasis keeps things working properly inside a living thing.
6. Biologists use six characteristics to classify something as a living thing.
7. In any environment, living space is unlimited.
8. Plant responses usually are faster than animal responses.
9. During hibernation, animals live off fat stored in their bodies.
10. The environment is everything that surrounds a living thing.
11. Polar bears are adapted to a cold environment.
12. During hibernation, the body temperature of an animal raises.
13. Budding and fission are two types of sexual reproduction.
14. New living things produced by organisms are called cells.
15. Animals take in some nutrients from the soil.

Understanding the Features..

Reading Critically *Use the feature reading selections to answer the following. Page numbers for the features are shown in parentheses.*

1. **Define:** What is an ecologist? (37)
2. **Infer:** Why might people want to clone animals? (43)
3. **List:** What four kinds of compounds are important to living things? (35)
4. What did William Harvey explain? (45)
5. How did Pasteur disprove spontaneous generation? (41)
6. How are cookbooks usually organized? (47)

Interpreting a Diagram *Use the diagram to complete the following.*

1. **Observe:** Which jars attracted flies?
2. **Analyze:** Which jar is the control?
3. In which jar did maggots grow?
4. Why did maggots not grow in all the jars?
5. Where did the maggots come from?
6. How did the results of this experiment disprove spontaneous generation?

Critical Thinking *Answer each of the following in complete sentences.*

1. **Explain:** Why is a robot not an organism?
2. How are growing and developing related?
3. What are three products of respiration?
4. Why must living things compete with each other?

5. **Infer:** Why does sexual reproduction produce offspring that are not exactly like either of their parents?

Finding Out More...

1. **Organize:** Using library references, find out how animals are adapted to living in the desert and how plants are adapted to living in very cold environments. Write your findings in a report. Include pictures in your report.
2. **Experiment:** Design an experiment to show how plants respond to light. Place a plant in sunlight with its leaves facing away from the sun. Observe the plant for the next week. Write a laboratory report describing your observations.

3. **Observe:** Choose five kinds of organisms that live in your neighborhood. Construct a chart explaining two ways that each organism is adapted to its environment.
4. **Communicate:** Using library references, research the work of the Italian scientist Lazzaro Spallanzani. Find out about his experiment to test spontaneous generation. Explain Spallanzani's experiment in an oral report.

ECOLOGY

CONTENTS

STUDY HINT Before beginning Unit 3, write the title of each lesson on a sheet of paper. Below each title, write a short paragraph explaining what you think each lesson is about.

3-1 What is ecology?

TechTerms

- **ecology** (ee-KAHL-uh-jee): study of living things and their environments
- **environment** (in-VY-run-munt): everything that surrounds an organism
- **interact:** to act upon each other

Ecology Everything that surrounds a living thing makes up its **environment** (in-VY-run-munt). Living things are affected by their environments. Living things also have an effect on their environments.

The study of living things and their environments is **ecology** (ee-KAHL-uh-jee). Scientists who study ecology are ecologists. Ecologists study the relationships between living things and their environments. They also study how living things are adapted, or suited, to their environments.

▶ *Define:* What is ecology?

Importance of the Environment All living things need materials to carry out their life processes. Organisms get all the materials they need from their environments. Some materials, called nutrients (NOO-tree-unts), are used by living things for growth and energy. Green plants get nutrients and water from soil. They take carbon dioxide from the air. They use sunlight and carbon dioxide to grow and make food for energy. Plants also need oxygen. Some animals get nutrients and energy by eating plants. Some animals eat other animals. Most animals get oxygen from the air. Fish get oxygen that is dissolved in water.

▶ *List:* What are some materials that organisms get from their environment?

Interactions Look at Figure 1. It shows a plain in Africa. There are many different kinds of organisms that live on the plain. The plain is the environment for these organisms. The organisms on the plain **interact,** or act upon each other. The organisms also interact with the nonliving parts of the environment. The gnus (NOOS) eat grasses growing in soil. The wastes produced by the gnus enrich the soil. The enriched soil makes the grasses grow better. The grasses, soil, and gnus each have an effect on each other. An interaction also takes place between the lions and the gnus. Lions eat gnus. If there are many lions hunting gnus, the number of gnus will go down. With fewer gnus, some lions will die from a lack of food. As a result, the number of lions will go down.

▶ *Infer:* What will happen to the number of gnus if the number of lions goes down?

Figure 1 African plain

LESSON SUMMARY

▶ Everything that surrounds a living thing makes up its environment.

▶ Ecology is the study of the relationships between living things and their environments.

▶ Living things get all the materials that they need from their environments.

▶ Organisms interact with each other and with the nonliving parts of their environments.

CHECK *Answer the following.*

1. What is the study of living things and their environments?

2. What materials do living things use for growth and energy?

3. What specialists study the relationships between living things and their environments?

Complete the following.

4. Plants use sunlight and _____ to grow and make food.

5. All the organisms in an environment _____ with each other.

APPLY *Complete the following.*

6. What materials do you get from your environment?

7. **Infer:** What need would a green plant not be able to get growing in a dark cellar?

Skill Builder

📋 *Classifying* One interaction in an environment is predator and prey. Use a dictionary to find out what the terms "predator" and "prey" mean. Use this information to classify the organisms in each pair as either the predator or the prey.

1. lion, zebra
2. fly, frog
3. snake, rabbit
4. robin, worm
5. mouse, cat
6. cheetah, antelope

❖❖❖ CAREER IN LIFE SCIENCE ❖❖❖❖❖❖❖❖❖❖❖❖❖❖❖❖❖❖❖❖❖❖❖❖❖

AIR POLLUTION TECHNICIAN

Do you enjoy being outdoors? Are you good at putting things together? Can you think like a detective to solve problems? If you answered yes to these questions, you might enjoy a career as an air pollution technician. Air pollution technicians study the chemical make up of air samples. They perform tests on the air samples to identify the kinds of pollutants (puh-LOOT-unts) in the air. Pollutants are harmful substances that enter the environment. Air pollution technicians keep careful records of the test results. They also analyze the test data. They write reports that summarize their findings. Sometimes, they suggest ways to remove a pollutant from the air.

Air pollution technicians work outdoors when collecting air samples. They work in a laboratory to test the samples. The technicians must have good mathematics and science skills. They also must be able to communicate their findings in written reports. Most air pollution technicians have a two-year degree from a technical school.

For more information: write to the Air Pollution Control Association, P.O. Box 2861, Pittsburgh, PA 15230.

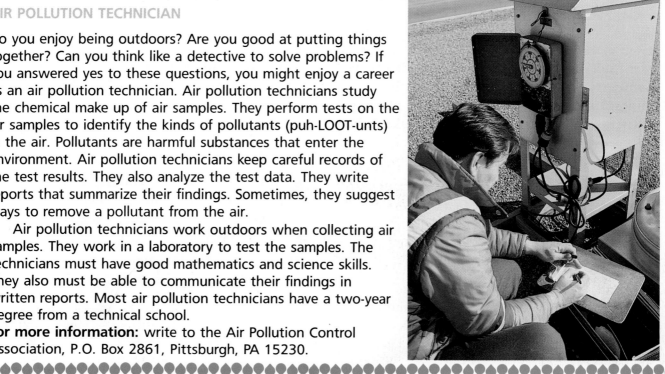

3-2 What is an ecosystem?

Objective ► Describe the parts of an ecosystem.

TechTerms

► **community:** all the populations that live in a certain place

► **ecosystem** (EE-koh-sis-tum): living and non-living things in an environment, together with their interactions

► **population:** group of the same kind of organism living in a certain place

Populations A **population** is all of the same kind of organism living in a certain place. Different populations may live in the same environment. Look at the rotting log. There are five mushrooms growing on the log. These five mushrooms make up one population of the log. Count the number of mice living in the log. There are four. The mouse population is four.

👁 *Observe:* What is the fern population on the rotting log?

Communities All the populations living on the rotting log make up a **community.** A community is all the populations that live in a certain place. It includes all the different kinds of living things that live together.

▌▶ *Define:* What is a community?

Ecosystem The organisms in a community interact with the nonliving parts of the environment. The organisms in the community also inter-

act with each other. The living and nonliving things in an environment, together with their interactions, make up an **ecosystem** (EE-koh-sis-tum).

There are many kinds of ecosystems. They are different sizes and have different makeups. An ecosystem can be as large as a desert or as small as a rotting log. Ecosystems can be rivers, lakes, or ponds. Even a puddle of water can be an ecosystem.

▌▶ *Identify:* What is an ecosystem?

A Self-Supporting Unit An ecosystem is a self-supporting unit. Four processes occur in an ecosystem to make it self-supporting.

► **Production of Energy** The sun is the source of energy in most ecosystems.

► **Transfer of Energy** Energy is transferred from the sun to plants that make their own food. The stored energy in plants is transferred to animals that eat the plants. Energy is transferred to other animals when they eat the plant-eating animals.

► **Breaking Down of Materials** When organisms die, their bodies decompose, or break down. The chemicals are reused by other living things.

► **Recycling** The materials needed by organisms in an ecosystem are recycled, or used over and over.

▌▶ *List:* What are the four processes that make an ecosystem self-supporting?

LESSON SUMMARY

▶ A population is made up of all organisms of the same kind that live in a certain place.

▶ All the populations that live in a certain place make up a community.

▶ An ecosystem is made up of the living and nonliving parts of an environment together with their interactions.

▶ Ecosystems have different sizes and makeups.

▶ An ecosystem is a self-supporting unit.

CHECK *Complete the following.*

1. All the different kinds of organisms that live in a pond make up a _____ .

2. All the living and nonliving things in an environment together with their interactions make up _____ .

3. In most ecosystems, the major source of energy is the _____ .

4. When organisms die, their bodies _____ .

5. When animals eat green plants, the stored _____ in the plants is transferred.

APPLY *Answer the questions about the pond shown.*

6. **Measure:** How many different populations live in the pond?

7. **Observe:** What is the turtle population?

8. Explain why the pond is an ecosystem.

InfoSearch

Read the passage. Ask two questions that you cannot answer from the information in the passage.

Commensalism Different kinds of relationships exist between organisms in an ecosystem. Any relationship between two different organisms in which one organism is helped and the other organism is unaffected is called commensalism (kuh-MEN-sul-izm). Orchids growing on trees is an example of commensalism. The orchids have a place to grow with lots of sunlight. The trees are neither harmed nor helped.

SEARCH: Use library resources to find answers to your questions.

SCIENCE CONNECTION

THE GAIA HYPOTHESIS

About 3.5 percent of Earth's oceans is made up of salt. Oceans get salt from rainwater. The rainwater picks up salt from soil and carries it into the oceans. This process has been going on for millions of years. Yet, the oceans do not get saltier. Keeping the salt content at 3.5 percent is good for organisms that live in the oceans. Even a small increase in salt content would be deadly to most kinds of ocean life.

The British scientist James Lovelock has a theory to explain why the salt content of the oceans does not change. He thinks the earth acts like a living organism. He named this organism *Gaia.* The name comes from the Greek goddess *Gaia* or Mother Earth. Lovelock's ideas are called the Gaia Hypothesis.

According to the Gaia Hypothesis, living things on Earth interact with nonliving things. These interactions have kept Earth's environment almost the same for millions of years. According to Lovelock, this interaction has made Earth suitable for life for almost 4 billion years.

What are habitats and niches?

Objective ► Explain how organisms may have the same habitat but not the same niche.

TechTerms

- **habitat** (HAB-i-tat): place where an organism lives
- **niche** (NICH): organism's role, or job, in its habitat

Habitat The place where an organism lives is its **habitat** (HAB-i-tat). The habitat of an organism has the food and water the organism needs to live. An organism's habitat also provides shelter and a place to reproduce.

There are many different kinds of habitats. Habitats can be very large or very small. There are land habitats and water habitats. An entire ocean is the habitat of a whale. The habitat of a woodpecker is the trees in a forest. An anthill also is a habitat.

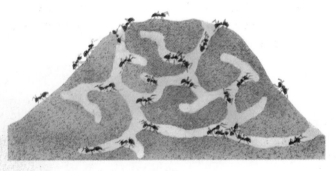

► *Infer:* What is your habitat?

Niche What is your role or job in life? Did you answer that you are a student? Student is the job or role that you do where you live. Organisms also have jobs or roles in their communities. The job or role of each organism is its **niche** (NICH). It includes everything an organism does and everything it needs.

▕▕▕► *Identify:* What is a niche?

Different Niches Many kinds of organisms share the same habitat. Tigers and deer both live in the same habitats in Asia. Each kind of organism has different needs and roles in the habitat. Tigers hunt deer and other animals. Deer eat grasses. Tigers have one kind of shelter. Deer have another. The activities of these two kinds of organisms are different. They have the same habitat, but they do not have the same niche.

Two populations cannot have the same niche for very long. For example, one animal may be able to hunt better and also find shelter better than another animal. One animal may be able to run away from its enemies faster than another animal. The population of animals that hunt or run better will survive and reproduce. After a while, members of the other populations will be crowded out.

► *Infer:* Do a lobster and a starfish share the same habitat? The same niche?

LESSON SUMMARY

▶ The place where an organism lives is its habitat.

▶ There are many different kinds of habitats.

▶ An organism's niche is its role or job in its habitat.

▶ Organisms can share the same habitat.

▶ If two populations share the same niche, the population best-suited to the role will survive and reproduce.

CHECK *Complete the following.*

1. The trees in a forest are the _____ of woodpeckers.

2. An organism's habitat has the food and _____ the organism needs to live.

3. The role of an organism in its community is its _____ .

Write true if the statement is true. If the statement is false, change the underlined term to make the statement true.

4. An anthill is an example of a niche.

5. If two populations have the same niche, the population best-suited to the niche will survive and reproduce.

APPLY *Complete the following.*

6. **Predict:** Two populations of moths live in the same habitat. They will share the same niche. What will happen to these populations?

7. **Infer:** The habitat of a rattlesnake is the hot, dry desert. Could this organism live in a cold, Arctic habitat? Why?

Classify: *In which kind of habitat do each of the organisms listed below live—salt water, fresh water, air, or land?*

8. pigeon	11. water lily	14. bat
9. shark	12. seaweed	15. squirrel
10. garter snake	13. guppy	

Skill Builder

Communicating Construction of a 116 million dollar dam was stopped in the mid-1980s because of a population of tiny fishes called snail darters. The dam was being built on a river in Tennessee. The river was the only habitat for the fish. After several years, it was decided to finish building the dam after all. The dam is expected to change the environmental conditions of the river. What do you think is the fate of the snail darter? Would you have let the dam be built? Write an essay supporting your answers.

LEISURE ACTIVITY

NATURE PHOTOGRAPHY

Do you enjoy taking pictures? Are you interested in plants and animals? Do you like to take hikes and camp outdoors? If you answered yes to these questions, you may enjoy nature photography as a hobby. All you need to begin taking nature photographs is a 35-mm camera. Depending upon the kinds of pictures you want to take, you also may need different kinds of lenses.

You should learn about animals you wish to photograph. You should know where the animal lives, where it sleeps, and what the animal eats. You also should learn how to move quietly so that you do not scare animals away.

State and national parks, recreation areas, seashores, and wildlife preserves all are great places for photographing plants and animals. Nature photographs also can be taken in your own yard. Just put out some food for small animals, such as squirrels and birds. Keep a safe distance from the food. When the animals come to eat, snap away!

What are limiting factors?

Objective ▶ Recognize why organisms live where they do.

TechTerms

▶ **carrying capacity:** largest amount of a population that can be supported by an area

▶ **limiting factors:** conditions in the environment that put limits on where an organism can live

▶ **range** (RAYNJ): area where a type of animal or plant population is found

Limiting Factors Certain conditions in the environment limit where an organism can live. These conditions are called **limiting factors.** Suppose a plant needs warm temperatures and a lot of water. It cannot live where it is cold and dry. Temperature and amount of water are limiting factors for the plant. Other limiting factors for plants are the amount of sunlight and the type of soil. Animals also are limited by conditions in the environment. Limiting factors for animals include temperature, water, food supply, and shelter.

▶ **List:** What are four limiting factors for plants?

Plants as Limiting Factors The number of green plants in a community is a limiting factor for the animals in the community. Suppose many green plants in a meadow died. Some of the mice that eat the plants would starve. With fewer mice

to eat, some of the owls also would starve. Even though owls do not eat plants, the number of owls is affected by the number of plants. The sizes of animal populations are limited by the sizes of plant populations.

▶ **Predict:** How would the number of mice and owls change if more plants grew than usual?

Carrying Capacity The largest amount of a population that can be supported by an area is the area's **carrying capacity.** An area has different carrying capacities for different populations. When a population becomes too large, some of its members must move. They must find a similar habitat with fewer organisms. The spread of organisms from one area to another is called dispersal (dis-PUR-sul). Many animals have no trouble dispersing. They move by walking, running, crawling, swimming, or flying.

▶ **Define:** What is carrying capacity?

Range The area where a kind of population lives is called its **range** (RAYNJ). The size of an organism's range is determined by its limiting factors. For example, black bears can eat a large variety of foods. As a result, the range of black bears is large. Giant pandas eat only bamboo. The range of the giant panda is very small.

▶ **Relate:** What is the relationship between an organism's diet and its range?

LESSON SUMMARY

▶ All organisms are limited by conditions in the environment.

▶ Animal populations are limited by the number of green plants in an area.

▶ The largest amount of a population that can be supported by an area is the area's carrying capacity.

▶ The range of a population is determined by its limiting factors.

CHECK *Choose the word that completes each sentence.*

1. Carrying capacity is the (largest/smallest) amount of a population that can be supported by an area.

2. The (type/size) of a range is determined by an organism's limiting factors.

3. The size of all other populations in the area is limited by the (plant/animal) populations.

4. A limiting factor for plants is the amount of (sunlight/shelter).

APPLY *Complete the following.*

▶ 5. **Infer:** Why are humans able to live almost anywhere on Earth?

6. **Relate:** Do changes in the size of plant populations have an effect on humans? Explain.

InfoSearch

Read the passage. Ask two questions that you cannot answer from the information in the passage.

Territories A territory (TER-uh-towr-ee) is an area that an animal claims as its own against other animals of the same kind. Having a specific territory gives an animal the space it needs to get food and reproduce. Some kinds of animals claim a territory as a group. Wolves hunt in groups called packs. Each pack claims and protects its own territory.

SEARCH: Use library references to find answers to your questions.

◆○◆ SCIENCE CONNECTION ◆○◆○◆○◆○◆○◆○◆○◆○◆○◆○◆○◆○◆○◆○◆○◆○

WORLD POPULATION GROWTH

About 20,000 years ago, almost 3 million people lived on Earth. Today the population of the United States is about 250 million. The total world population is over 5 billion. How can the earth carry a population that is so much larger than it was only 20,000 years ago?

People began to grow crops about 10,000 years ago. The world population increased rapidly after people began to grow crops. The development of agriculture increased the carrying capacity of the land. Land used for growing crops can support more people than land used for hunting.

In modern times, methods of food production have improved. Medical advances have allowed more people to live longer. Increased population has resulted in limited food supplies for some areas on Earth.

No one knows the total carrying capacity of the earth. Some areas may be close to their carrying capacities now. As a result, population growth is a serious problem facing the world today.

3-5 What cycles take place in nature?

Objective ► Explain the oxygen-carbon dioxide cycle and the nitrogen cycle.

TechTerms

► **cycle** (SY-kul): something that happens over and over in the same way

► **nitrogen-fixing bacteria**: bacteria that can use nitrogen in soil to make nitrogen compounds

Cycles in Nature There are many cycles (SY-kuls) in nature. A **cycle** is something that happens over and over in the same way. Rocks move through a cycle changing from one kind of rock to another. Water, oxygen, carbon dioxide, and nitrogen also cycle through the environment.

||||► *List:* What are five substances that cycle through the environment?

The Water Cycle Water falls to the earth as rain, snow, sleet, and hail. Some of this water flows to oceans, lakes, and rivers. Some water evaporates, or changes to a gas. Some water soaks into the ground and is used by plants. All animals drink water. Water is given off by animals and plants during respiration. This water evaporates into the air.

||||► *Identify:* How does water get into the air?

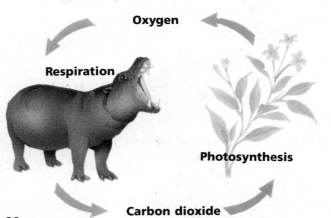

Oxygen-Carbon Dioxide Cycle Oxygen and carbon dioxide cycle through an ecosystem. During photosynthesis, plants take in carbon dioxide and give off oxygen. Animals use oxygen during respiration. They give off carbon dioxide. Respiration and photosynthesis work together to recycle oxygen and carbon dioxide in an ecosystem.

||||► *Name:* What organisms add oxygen to the air?

Nitrogen Cycle The using and reusing of nitrogen in an ecosystem is called the nitrogen cycle. Animals and plants need nitrogen to make proteins. However, plants and animals cannot use the nitrogen in the air. Special bacteria called **nitrogen-fixing bacteria** change nitrogen gas into

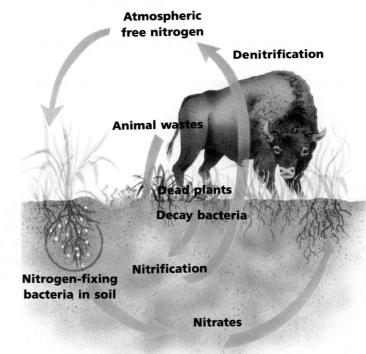

nitrogen compounds. These nitrogen compounds are called nitrites (NY-tryts) and nitrates (NY-trayts). Nitrates and nitrites can be used by living things. When living things die, decay bacteria break down the compounds. Nitrogen gas is given off.

||||► *Define:* What are nitrogen-fixing bacteria?

LESSON SUMMARY

▶ Rocks, oxygen, carbon dioxide, water, and nitrogen cycle through the environment.

▶ Water is cycled through the environment, falling as a liquid and then evaporating as a gas.

▶ Oxygen and carbon dioxide are cycled through the environment by the processes of photosynthesis and respiration.

▶ Nitrogen-fixing bacteria change nitrogen gas into compounds that can be used by living things.

CHECK *Write true if the statement is true. If the statement is false, change the underlined term to make the statement true.*

1. Something that happens over and over in the same way is called a cycle.

2. Water is given off by plants and animals during photosynthesis.

3. During photosynthesis, plants take in carbon dioxide and give off oxygen.

4. Animals take in carbon dioxide during respiration.

5. Animals and plants need nitrogen to make food.

6. Nitrates and nitrites are two kinds of nitrogen compounds.

APPLY *Complete the following.*

7. **Hypothesize:** If all the green plants on Earth died, what would happen to the oxygen content of air?

8. **Relate:** How are plants and animals dependent upon each other?

9. Animals give off nitrogen as liquid and solid wastes. What do you think happens to this nitrogen?

InfoSearch

Read the passage. Ask two questions that you cannot answer from the information in the passage.

Composition of Air Air is a mixture of gases. Gases in the air include nitrogen, oxygen, and carbon dioxide. Nitrogen makes up about 78% of air. Oxygen makes up about 21% of air. Argon, a gas used to make electric light bulbs, makes up a little less than 1% of air. Other gases, including carbon dioxide, make up the remaining 0.03% of air.

SEARCH: Use library references to find answers to your questions.

◆○◆ SCIENCE CONNECTION ◆○◆○◆○◆○◆○◆○◆○◆○◆○◆○◆○◆○◆○◆○◆○◆○◆

THE ROCK CYCLE

Rocks are classified into three groups. Rocks formed when liquid volcanic rock cools are classified as igneous (IG-nee-us) rocks. Rocks formed when rock particles are cemented together are classified as sedimentary rocks. Metamorphic rocks are formed when heat and pressure change igneous or sedimentary rocks.

Rocks can change from one kind to another. The process by which rocks change form is called the rock cycle. For example, a rock can be broken apart by wind and water. The pieces of rock may then become cemented together. A sedimentary rock is formed. Great heat and pressure can change the sedimentary rock into metamorphic rock. If the metamorphic rock is forced deep beneath the earth's surface, it may melt into a liquid. If this liquid rock cools, an igneous rock is formed.

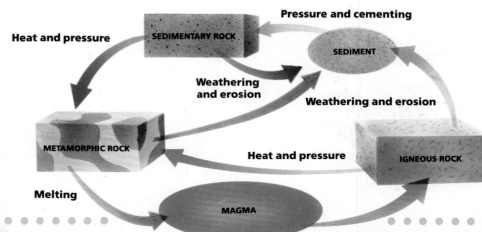

3-6 What are producers and consumers?

Objective ▶ Identify producers and different feeding levels of consumers in an ecosystem.

TechTerms

- **consumer** (kun-SOO-mur): organism that obtains food by eating other organisms
- **decomposer** (dee-kum-POHZ-er): organism that breaks down the wastes or remains of other organisms
- **producer** (pruh-DOOS-ur): organism that makes its own food
- **scavenger** (SKAV-in-jur): animal that eats only dead organisms

Producers Organisms that make their own food are called **producers** (pruh-DOOS-urz). Producers use energy from the sun to make food. On land, the main producers are plants. The main producers in lakes and oceans are algae (AL-jee).

◉ *Observe:* Name all the producers shown on this page.

Consumers Most organisms get food by eating other organisms. An organism that eats other organisms is a **consumer** (kun-SOO-mur). Rabbits eat grass and other plants. Rabbits are primary consumers. A primary consumer is an organism that eats producers. Consumers that eat primary consumers are secondary consumers. Weasels eat small plant-eating animals, such as rabbits. Weasels are secondary consumers. Consumers that eat secondary consumers are tertiary consumers. Hawks eat small meat-eating animals, such as weasels. Hawks are tertiary consumers. Some animals, such as hawks, are both secondary and tertiary consumers. Most people are primary, secondary, and tertiary consumers.

📂 *Classify:* Ducks eat plants and small insects or animals. Are ducks primary, secondary, or tertiary consumers?

Scavengers and Decomposers Some animals feed upon dead animals. These animals are **scavengers** (SKAV-in-jurz). Scavengers eat animals that have died or been killed by other animals. Vultures, hyenas, and certain ants, beetles, and worms are scavengers. Scavengers are both secondary and tertiary consumers. Organisms that break down the wastes or remains of organisms are **decomposers** (dee-kum-POHZ-ers). Decay bacteria are decomposers. Decomposers return materials from dead organisms to the soil.

▥▶ *List:* List three organisms that are scavengers.

LESSON SUMMARY

▶ Producers are organisms that make their own food.

▶ Consumers get food by eating other organisms.

▶ The three types of consumers are primary consumers, secondary consumers, and tertiary consumers.

▶ Scavengers and decomposers eat or break down the remains and wastes of other organisms.

CHECK *Complete the following.*

1. Secondary consumers are eaten by _____ .

2. Producers use energy from the _____ to make their own food.

3. Organisms that eat the remains of dead animals are _____ .

4. The producers in lakes and oceans are _____ .

5. Organisms that eat only producers are _____ consumers.

APPLY *Complete the following.*

6. **Analyze:** What is one way in which organisms in an ecosystem depend upon each other?

7. **Explain:** Give an example of when humans are primary, secondary, or tertiary consumers.

8. **Classify:** The organisms shown in the photograph are tube worms. Tube worms live deep in the ocean. They feed upon bacteria that are able to make their own food using chemicals contained in ocean water. Are tube worms producers or consumers?

....
Skill Builder.......................................

Building Vocabulary Three terms used to describe consumers are "herbivore", "carnivore", and "omnivore". Use a dictionary to identify the meaning of each term. Then, classify the organisms listed as herbivores, carnivores, or omnivores.

1. black bear	4. hawk	7. cat
2. wolf	5. human	8. cow
3. horse	6. giraffe	

∶∶ TECHNOLOGY AND SOCIETY

HYDROPONICS

Plants use energy from the sun to make food. They also use minerals such as nitrogen and calcium. Most plants get these minerals from soil.

A method of growing plants without soil has been developed. This method is called hydroponics (hy-druh-PAHN-iks). The roots of plants are placed in a special liquid. Minerals the plants need are contained in the liquid. Plants grown this way develop as well as plants grown in soil.

Plants grown by hydroponics are protected from harsh weather and from insect and animal pests. The plants can be checked and cared for more easily than if they were outdoors. In addition, the amounts of minerals provided for the plants can be carefully controlled. Hydroponics is a way of growing crops in areas with poor soil. As people look for new and better ways of growing crops, hydroponics is sure to attract more interest.

What are food chains, webs, and pyramids?

Objective ▶ Explain and construct models to show how organisms are related by how they get their food.

TechTerms

▶ **energy pyramid** (PIR-uh-mid): way of showing how energy moves through a food chain

▶ **food chain:** way of showing the food relationships among a group of organisms

▶ **food web:** way of showing how food chains are related

Food Chains The organisms in an environment are related by how they get their food. A **food chain** is a way of showing these food relationships. Every organism is part of a food chain.

Figure 1

Figure 1 shows a food chain. The arrows in the food chain show the direction that food and energy move along the chain. This food chain starts with grass seeds. A producer, or its seeds and fruit, always is at the start of a food chain.

◉ *Observe:* In the food chain shown, which organism is eaten by the snake?

Food Webs A **food web** is a more complete way of showing food relationships. A food web shows how a number of food chains are related. Figure 2

shows a food web. Notice that the owl can eat either a field mouse or a frog. Many organisms in a food chain eat more than one type of food. Many organisms also are a food source to more than one organism.

◉ *Observe:* In the food web shown, which organisms eat field mice?

Energy Pyramids An **energy pyramid** (PIR-uh-mid) shows how energy moves through a food chain. Figure 3 shows an energy pyramid. Notice

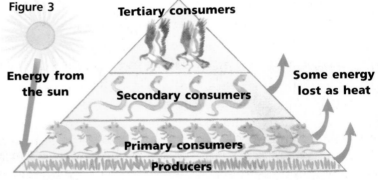

Figure 3
Tertiary consumers
Energy from the sun
Some energy lost as heat
Secondary consumers
Primary consumers
Producers

that the producers are at the bottom of the pyramid. The producer layer has most of the food and energy in the pyramid. Animals gain only a small amount of energy from the food they eat. So, as you move up the pyramid, the amount of energy decreases. Less energy means fewer organisms can be supported at each level. The tertiary consumers are at the top of the pyramid. They use the least amount of food and energy of all the organisms in the system.

◼ *Analyze:* Which consumers use more food and energy, primary or secondary consumers?

Figure 2

LESSON SUMMARY

▶ The organisms in an environment are related by how they get their food.

▶ Every organism is linked to other organisms in a food chain.

▶ Food webs show how a number of food chains are related.

▶ An energy pyramid shows that the amount of energy decreases at each level of a food chain.

CHECK *Complete the following.*

1. A food chain always begins with a _____ .

2. The top of an energy pyramid contains _____ .

3. Arrows show the flow of _____ in a food chain.

4. The relationships between food chains are shown in a _____ .

5. The layer of an energy pyramid that has the most food and energy contains _____ .

APPLY *Complete the following.*

6. **Infer:** Minnows are small fish that eat plant materials. Bass are large fish that feed upon minnows. Would you expect to find more minnows or bass in a pond? Why?

7. **Analyze:** Which organism in the food chain shown is a producer?

8. **Classify:** Is the butterfly a primary, secondary, or tertiary consumer?

9. **Model:** Draw a food chain that shows a butterfly, a bird, and a plant.

10. **Model:** Draw food chains that include you as a primary, secondary, and tertiary consumer.

Skill Builder ....

Modeling Draw an energy pyramid on a sheet of paper. Decide which of these organisms belong at each level: wolf, grass, snake, rabbit. Label each level of the pyramid with the correct organism.

ACTIVITY

MODELING FOOD CHAINS

You will need 12 index cards.

1. Write the names of the organisms listed in the box on 12 separate index cards.

2. Mix up the cards. Choose six cards.

3. Using as many of the six cards as possible, arrange the cards into a food chain. Record the food chain.

4. Return the cards to the pile. Shuffle the cards.

5. Repeat Steps 3 and 4 as many times as you can for 10 minutes.

Questions

1. How many food chains did you complete?

2. How many organisms were in your longest food chain?

3. **a. Classify:** Which organisms on your cards are producers? **b.** Which organisms on your cards are consumers?

4. **Organize:** Chose several of your food chains and develop a food web.

ORGANISMS
Tree
Grass
Rabbit
Field Mouse
Snake
Owl
Shrub
Elk
Cricket
Mountain Lion
Hawk
Frog

What is succession?

Objective ▶ Describe how communities of organisms develop.

TechTerms

▶ **climax** (KLY-maks) **community:** last community in a succession

▶ **succession** (suk-SESH-un): gradual change in organisms that occurs when the environment changes

Succession Environments do not always stay the same. When the environment changes, its populations are slowly replaced by new populations. This slow change in organism populations is called **succession** (suk-SESH-un).

A change in one group of organisms causes a change in another group of organisms. The first populations to change are the plant populations. As the plant populations change, different animal populations move in.

▶ *Identify:* What is succession?

Open Field to Forest Suppose a fire burns a forest to the ground. This disaster causes succession to occur in several stages.

▶ **Open Field** The first organisms to grow in the burnt area are grasses and weeds. They grow from roots and seeds left in the soil. As the grasses grow into a thick field, or meadow, small animals move into the area.

▶ **Shrub Land** Shrubs and trees begin to grow. Since these shade the grasses, some of the original grasses die. More shrubs grow and different kinds of animals move in. A new community replaces the community that lived in the grassy field.

▶ **Pine Forest** Pine trees begin to grow in the area. Since the pine trees shade parts of the area, some of the shrubs die. Eventually, a pine forest develops. The community that lives in the pine forest is different from that of the shrub land.

▶ **Hardwood Forest** Soon hardwood trees, such as oak and maple, begin to grow in the area. They grow taller than the pine trees, so the pine trees do not get as much sunlight. Many pine trees die, and the forest becomes a hardwood forest. A different animal community develops.

Once the hardwood forest has formed, succession stops. The hardwood forest is a **climax** (KLY-maks) **community** or last community in a succession. It will remain in the area until the environment changes again.

▶ *Observe:* How many years did the succession process shown take, from weeds to the climax community?

OPEN FIELD   SHRUB LAND   PINE FOREST   HARDWOOD FOREST

0 Years   100 Years

LESSON SUMMARY

▶ As the environment in an area changes, the plant and animal populations also change.

▶ Populations change in a series of stages, ending with a climax community.

CHECK *Complete the following.*

1. The first stage of forest succession is the _____ .

2. The last stage in a succession is the _____ .

3. The first populations to change in succession are the _____ populations.

4. Succession begins after the environment has been _____ .

APPLY *Complete the following.*

▶ 5. **Predict:** Is the climax community the same everywhere?

6. **Infer:** Suppose a maple-beech forest burns. Why is it that maples and beeches are not the first community in the succession?

Use the diagram of pond succession to answer the following.

▶ 7. **Infer:** What is a bog?

8. **Sequence:** A pond could develop into a hardwood forest. List the stages in the development of a hardwood forest in the correct order beginning with a pond.

Health & Safety Tip..........................

Lightning causes many forest fires. However, some of the most damaging forest fires are caused by careless people. Find out what safety precautions people should take when camping or cooking outdoors.

 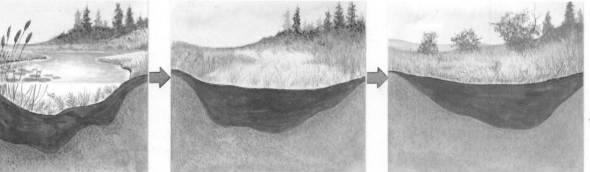

◦◆◦◆ SCIENCE CONNECTION ◆◦◆◦◆◦◆◦◆◦◆◦◆◦◆◦◆◦◆◦◆◦◆◦◆◦◆◦◆◦

SEED DISPERSAL

After a forest fire, plants may begin to grow from seeds in the ground. What if there were no seeds in the ground? Plants would still grow because seeds can travel.

Seeds often are carried far from their parent plant. The movement of seeds away from the parent plant is called seed dispersal.

The structure of some fruits helps in the dispersal of their seeds. For example, the coconut fruit floats. It can be carried great distances by water currents. The fruit of the maple tree is shaped like a pair of wings. Maple fruits are easily carried by the wind.

Some fruits are sweet and juicy. Birds and other animals eat these fruits as food. However, the seeds are not digested. They pass through the animal's digestive tract. The seeds are carried in the animal's body and then deposited on the ground.

What are biomes?

Objective ▶ Describe different biomes.

TechTerms

▶ **biome** (BY-ohm): large region of the earth with particular plant and animal communities

▶ **climate:** overall weather in an area over a long period of time

Biomes The overall weather in an area over a long period of time is the area's **climate.** Climate helps determine the kinds of plants and animals that live in an area. A **biome** (BY-ohm) is a large region of the earth that has characteristic plant and animal communities. Each biome has a different climate.

▰▶ *Define:* What is a biome?

Six Major Biomes Particular kinds of animals live in different biomes. Biomes often are named for their most common plants. Earth's land areas are divided into six major biomes.

▶ The *tundra* is found in the far north. The ground is permanently frozen. Mosses, grasses, small flowers and shrubs are the only plants that can survive in the tundra. Animals that live in the tundra include birds, wolves, foxes, and reindeer.

▶ South of the tundra lies an area with a cold climate. Enough rainfall falls there for trees to grow. *Coniferous* (kuh-NIF-uhr-us) *forests* made up of evergreens cover the area. Many animals such as moose, squirrels, rabbits, and beavers live in the coniferous forest biome.

▶ South of the coniferous forests is a region with a moderate climate. This area has a long growing season followed by a cold winter. *Deciduous* (di-SIJ-oo-wus) *forests* grow here. Unlike the needles of evergreens, the leaves of deciduous trees fall off in the winter. Many kinds of birds and animals live in the deciduous forest biome.

	Tundra
	Taiga
	Grassland
	Deciduous forest
	Desert
	Tropical rain forest

▶ *Tropical rain forests* occur in areas near the equator. These regions receive large amounts of rainfall and are hot all year. Tall trees and many kinds of plants grow well in tropical rain forests. Many different animals also live in this biome.

▶ Some areas have a moderate climate but not enough rain for tall trees to grow. These areas are *grasslands.* Large herds of animals feed on the many kinds of grasses that grow in this biome.

▶ *Deserts* form where temperatures are moderate or hot, but where there is very little rain. Very few plants can grow in desert biomes. Still, some kinds of animals such as lizards and coyotes find food and shelter there.

◢ *Analyze:* Look at the biome map. In what biome do you live?

LESSON SUMMARY

► The earth can be divided into biomes, large areas with similar plant and animal communities.

► There are six major land biomes: tundra, coniferous forest, deciduous forest, tropical rain forest, grassland, and desert.

CHECK *Write the letter of the term that best completes each statement.*

1. The coldest biome is the
   a. tropical rain forest.   **b.** grasslands.
   c. tundra.   **d.** coniferous forest.

2. The driest biome is the
   a. deciduous forest.   **b.** tundra.
   c. grasslands.   **d.** desert.

3. The biome that is hot and wet is the
   a. tropical rain forest. **b.** coniferous forest. **c.** deciduous forest. **d.** grasslands.

4. The biome where large herds of animals live is the
   a. deciduous forest. **b.** grasslands. **c.** tundra. **d.** coniferous forest.

APPLY *Complete the following.*

► 5. **Infer:** Why do more kinds of animals live in the deciduous forest than in the tundra?

6. **Contrast:** Deserts and tropical rain forests may both occur in hot areas. How are these biomes different?

InfoSearch

Read the passage. Ask two questions about the topic that you cannot answer from the information in the passage.

Taiga Another name for the coniferous forest biome is the taiga. The taiga has long, snowy winters. Animals that live in the taiga include moose, lynx, porcupines, and red squirrels. Some of these animals hibernate in winter. Many birds spend the summer in the taiga, feeding on the insects that live there.

SEARCH: Use library references to find answers to your questions.

LEISURE ACTIVITY

LEAF PEEPING

Fall is a beautiful time of year in the deciduous forest biome. As the weather gets cool and crisp, the leaves begin to lose their green color. Autumn colors appear. In some areas, the leaves are so bright that people travel there just to see the fall foliage. These tourists sometimes are called "leaf peepers".

Not all parts of the deciduous forest are good for leaf peeping. Some trees, such as oak trees, do not have colorful autumn leaves. Oak leaves simply turn brown in the fall. Some kinds of maple and birch trees turn bright red and gold. Areas where these trees grow are good spots for leaf peeping. It takes the right kind of tree and the right weather conditions to make colorful leaves.

The New England states are a great place for leaf peeping. Many tourists visit these states in the fall. Many parts of Pennsylvania, Delaware and Virginia also are good for leaf peeping.

3-10 What are natural resources?

Objective ▶ Distinguish between renewable and nonrenewable natural resources.

TechTerms

- **conservation** (kon-sur-VAY-shun): wise use of natural resources
- **natural resources** (REE-sowrs-ez): materials found in nature that are used by living things
- **nonrenewable resources:** natural resources that cannot be renewed or replaced
- **renewable resources:** natural resources that can be renewed or replaced

Natural Resources Living things use materials found in nature to survive. These materials are called **natural resources** (REE-sowrs-ez). Air and water are natural resources needed by almost all organisms. People use other natural resources such as oil, coal, and gas as fuels.

📁 *Classify:* What other natural resources can you name?

Renewable Resources Some natural resources can be reused or replaced. These resources are called **renewable resources.** Air, water, soil, and living things are renewable resources.

▷ *Define:* What is a renewable resource?

Nonrenewable Resources Some natural resources cannot be reused or replaced. These resources are called **nonrenewable resources.** Oil, coal, natural gas, and minerals are examples of nonrenewable resources. Nonrenewable resources need millions of years to form. Once existing supplies of these resources are used up, they cannot be replaced.

▷ *Contrast:* What is the difference between renewable and nonrenewable resources?

Conservation The wise use of natural resources is called **conservation** (kon-sur-VAY-shun). Conservation of all natural resources, including renewable resources, is important. Even though renewable resources are replaced by the environment, the supply of these resources is limited. People must be careful not to use renewable resources faster than they can be replaced.

Since nonrenewable resources cannot be replaced, it is especially important to use them wisely. One way of conserving nonrenewable resources is by using energy-efficient products. Driving cars that get more miles per gallon of gasoline is a way of conserving fuel.

▷ *Identify:* Why is it necessary to conserve renewable resources?

LESSON SUMMARY

▶ Organisms use materials found in nature to survive.

▶ Renewable resources can be reused or replaced.

▶ Nonrenewable resources cannot be reused or replaced.

▶ All natural resources must be used carefully and wisely.

CHECK *Complete the following.*

1. Once they are used up, nonrenewable resources _____ be replaced.

2. All organisms use _____ in order to survive.

3. Oil and gas are examples of _____ natural resources.

4. Sunlight is considered a _____ natural resource.

5. Using natural resources wisely is called _____ .

APPLY *Complete the following.*

6. **Relate:** What are two ways in which you can conserve natural resources?

▶ 7. **Infer:** Fuels such as oil and coal are still forming. Why are these fuels considered nonrenewable resources?

📁 8. **Classify:** Classify each thing shown in the drawings on page 70 as a renewable natural resource or a nonrenewable natural resource.

Ideas in Action.....................

IDEA: You use natural resources every day.
ACTION: List five natural resources that you use daily. Describe how you use each resource. Then, describe one way that you could conserve each resource.

Skill Builder.....................

📁 *Classifying* Write the headings "Renewable" and "Nonrenewable" on a sheet of paper. Classify each of the following resources as renewable or nonrenewable: copper, nutrients, food, oxygen, wood, fish, marble, gasoline, charcoal, methane gas, kerosene, propane gas. If necessary, use library books for reference.

TECHNOLOGY AND SOCIETY

NUCLEAR ENERGY

Nuclear (new-KLEE-ar) energy is the energy released when atoms are split. Most nuclear energy uses uranium as a fuel. When a uranium atom is split apart, energy is released as heat. This heat is used to produce steam. The steam turns large turbines that produce electricity.

Some people think that nuclear energy should be used more. They point out that nuclear energy does not pollute the air. There are large supplies of uranium and other nuclear fuels. Other people point out the problems with nuclear energy. Uranium is radioactive (ray-dee-oh-AK-tiv). An accident at a nuclear power plant could seriously harm the environment. Nuclear power plants produce dangerous wastes that must be stored for thousands of years. Nuclear power plants also release large amounts of heat into the environment. Much of this heat is released into streams and rivers. Organisms living in these environments are unable to survive in the heated water.

Citizens should be well informed about the helpful and harmful effects of nuclear energy. Such information will aid in making decisions about the future use of this energy source.

What is balance in an ecosystem?

Objective ▶ Recognize that every organism is part of an ever-changing environment.

TechTerms

▶ **endangered species** (in-DAYN-jurd SPEE-sheez): kinds of living things that are in danger of dying out

▶ **pollution:** release of harmful materials into the environment

Balance An environment is constantly changing. Sometimes, the changes work together to keep the environment the same. Then the environment is balanced. In a balanced environment, the size of a population may go up and then go down. However, the average size of the population remains the same over time. Sometimes, the balance in an environment is upset. A change in the balance of just one species can be harmful to the other organisms in the environment.

▶ **Define:** What is the balance of nature?

Natural Disturbances Natural causes such as a volcano or forest fire can upset the balance of an environment. These disturbances destroy organisms and their habitats. There is great loss of

wildlife. It may take many years for the environment to return to its original condition.

▶ **Infer:** What other natural disturbances can upset the balance of nature?

The Role of People People also can upset the balance of an environment. Often the actions of

people destroy the habitats of organisms. They cut down forests to create farms and towns. They build dams and dig mines. Unfortunately, the disturbances caused by people are often permanent. The environment cannot return to normal.

People also damage the environment by causing pollution. **Pollution** is the release of harmful substances into the environment. Harmful gases from cars, power plants, and factories pollute the air. Wastes and chemicals pollute the water. Improper disposal of waste materials pollutes the land. All of these acts upset the balance of nature.

▶ **Explain:** How do people upset the balance of nature?

Endangered Species Upsets in the balance of an environment have made it hard for many species to survive. Living things in danger of dying out are called **endangered species** (in-DAYN-jurd SPEE-sheez). The whooping crane, giant panda, elephant, bald eagle, and humpback whale all are endangered species.

▶ **Predict:** What will happen to the number of endangered species if people continue to pollute the environment?

LESSON SUMMARY

▶ Changes in an ecosystem work to keep it balanced.

▶ Natural disturbances can upset the balance of an ecosystem.

▶ People upset the balance of an ecosystem by damaging it or polluting it.

▶ Endangered species are living things in danger of dying out.

CHECK *Complete the following.*

1. Changes in the balance of an environment can be _____ to organisms that live there.

2. A volcano is an example of a _____ disturbance.

3. A species that is in danger of dying out is said to be _____ .

4. Disturbances caused by _____ may be permanent.

5. The addition of harmful substances to the environment results in _____ .

APPLY *Complete the following.*

6. **Sequence:** Put the following events in the order in which they most likely would occur: species becomes endangered, volcano erupts, species becomes extinct, environment is in balance

7. **Relate:** Some states have laws that require deposits on beverage containers. How do these laws help reduce land pollution?

....
Ideas in Action.................................

IDEA: Since people cause pollution, everyone must work to reduce or eliminate it.
ACTION: What are some ways that you can help solve the problem of land pollution?

....
State the Problem............................

Look closely at the photograph. What problem does the photograph show? How can this problem be corrected?

◀◆▶■◀ **PEOPLE IN SCIENCE** ▶■◀◆▶■◀◆▶■◀◆▶■◀◆▶■◀◆▶■◀◆▶■◀◆▶■◀◆▶■◀◆▶■◀◆▶■◀

RACHEL CARSON (1907–1964)

Rachel Carson was a biologist who worked for many years for the United States Fish and Wildlife Service. Her real love, however, was writing. She used her spare time to write about science. In 1952, Rachel Carson received the National Book Award for the best nonfiction book. The book was *The Sea Around Us* which became an international best seller.

Rachel Carson wrote many other books about the sea. She is best remembered for her book *Silent Spring.* This book was published in 1962. In the book, she made people realize that chemical poisons in insect sprays could harm other living things. She made people aware that these poisons could find their way into the plants and animals that people eat.

Rachel Carson was the first person to point out the danger of using pesticides. Pesticides are used to kill unwanted organisms. As a result of her book, chemical insect killers are used more carefully.

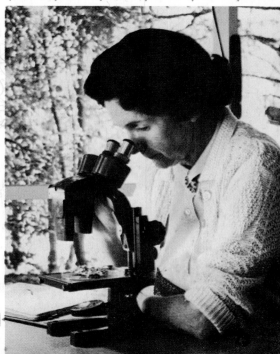

◀◀ ▶■◀ ▶■◀ ▶■◀ ▶■◀ ▶■◀ ▶■◀ ▶■◀ ▶■◀ ▶■◀ ▶■◀ ▶■◀

UNIT 3 Challenges

STUDY HINT Before you begin the Unit Challenges, review the TechTerms and Lesson Summary for each lesson in this unit

TechTerms..

biome (68)
carrying capacity (58)
climate (68)
climax community (66)
community (54)
conservation (70)
consumer (62)
cycle (60)
decomposer (62)
ecology (52)

ecosystem (54)
endangered species (72)
energy pyramid (64)
environment (52)
food chain (64)
food web (64)
habitat (56)
interact (52)
limiting factors (58)
natural resources (70)

niche (56)
nitrogen-fixing bacteria (60)
nonrenewable resources (70)
pollution (72)
population (54)
producer (62)
range (58)
renewable resources (70)
scavenger (62)
succession (66)

TechTerm Challenges..

Matching *Write the TechTerm that best matches each description.*

1. living and nonliving parts of an environment together with their interactions
2. overall weather in an area
3. living things in danger of dying out
4. to act upon each other
5. something that happens over and over in the same way
6. bacteria that change nitrogen into nitrogen compounds
7. materials found in nature that are used by living things
8. last community in succession

Applying Definitions *Explain the difference between the words in each pair.*

1. community, population
2. habitat, niche
3. limiting factors, carrying capacity
4. consumer, producer
5. scavenger, decomposer
6. food chain, food web
7. environment, ecology
8. renewable resources, nonrenewable resources
9. range, biome
10. pollution, conservation
11. food chain, energy pyramid
12. succession, balance

Content Challenges..

Multiple Choice *Write the letter of the term or phrase that best completes each statement.*

1. In a food chain, plants are classified as
   a. scavengers. **b.** producers. **c.** consumers. **d.** decomposers.

2. Organisms at the top of an energy pyramid are
   a. producers. **b.** primary consumers. **c.** scavengers. **d.** tertiary consumers.

3. Of the following, the only organism that can be a primary, secondary, and tertiary consumer is a
   a. tree. **b.** flower. **c.** human. **d.** grasshopper.

4. Oil and coal are examples of
   a. renewable resources. **b.** endangered species. **c.** nonrenewable resources. **d.** biomes.

5. Materials used by living things for growth and energy are
   a. nutrients. **b.** pollutants. **c.** scavengers. **d.** bacteria.

6. The major source of energy for most ecosystems is
   a. plants.   **b.** algae.   **c.** uranium.   **d.** the sun.

7. An anthill is an example of a
   a. biome.   **b.** habitat.   **c.** niche.   **d.** succession.

8. The spread of organisms from one area to another is called
   a. succession.   **b.** climate.   **c.** carrying capacity.   **d.** dispersal.

9. When water evaporates, it changes to
   a. a liquid.   **b.** a solid.   **c.** a gas.   **d.** rain.

10. Two processes involved in the oxygen-carbon dioxide cycle are respiration and
    a. digestion.   **b.** photosynthesis.   **c.** evaporation.   **d.** excretion.

11. Algae are classified as
    a. decomposers.   **b.** producers.   **c.** consumers.   **d.** scavengers.

12. The tundra and deciduous forest are
    a. biomes.   **b.** climax communities.   **c.** climates.   **d.** niches.

13. The first population to change in succession are the
    a. consumers.   **b.** scavengers.   **c.** producers.   **d.** bacteria.

14. Tropical rain forests are located
    a. near the equator.   **b.** north of the coniferous forest.   **c.** north of the grasslands.   **d.** in deserts.

True/False *Write true if the statement is true. If the statement is false, change the underlined term to make the statement true.*

1. An organism that eats grass is a <u>tertiary</u> consumer.
2. Vultures, hyenas, and ants are <u>decomposers</u>.
3. Living things are <u>renewable</u> resources.
4. The climate of the tundra is <u>cold</u> and dry.
5. The climate of the desert is warm and <u>wet</u>.
6. The number of plants in a community is a <u>limiting factor</u> for the community.
7. Nitrogen-fixing bacteria are part of the <u>water</u> cycle.
8. The organisms at the bottom of an energy pyramid are <u>tertiary consumers</u>.
9. The greatest amount of energy is at the <u>top</u> of an energy pyramid.
10. Both renewable and <u>nonrenewable</u> resources should be conserved.
11. The main producers in the oceans are the <u>fishes</u>.
12. During succession an ecosystem <u>is</u> balanced.
13. The movement of energy in an ecosystem can be shown by both a <u>food chain</u> and an energy pyramid.
14. The largest population that can be supported by an area is the area's <u>range</u>.

Understanding the Features...

Reading Critically *Use the feature reading selections to answer the following. Page numbers for the features are shown in parentheses.*

1. **Define:** What are pollutants? (53)
2. What percentage of the earth's oceans is made up of salt? (55)
3. **Infer:** Why is it important for animal photographers to learn about the animals they wish to photograph? (57)

4. **Generalize:** How is the population of the earth changing? (59)
5. **Identify:** What are the three classes of rocks? (61)
6. **Define:** What is hydroponics? (63)
7. **Hypothesize:** How is seed dispersal helpful in succession? (67)
8. **Identify:** In which time of year is leaf peeping done? (69)
9. **Identify:** What fuel is used for nuclear energy? (71)
10. **Infer:** In what branch of science do you think Rachel Carson worked? (73)

Concept Challenges..

Interpreting a Diagram *Use the diagram to complete the following.*

1. What is shown in the diagram?
2. **Classify:** Which organism in the diagram is a producer?
3. **Classify:** Which organisms in the diagram are primary consumers?
4. **Classify:** Which organisms in the diagram are secondary consumers?
5. **Classify:** Which organisms in the diagram are tertiary consumers?
6. What does a food web show?
7. **Model:** Draw one of the food chains shown in the diagram.
8. **List:** Name three organisms that could replace the producer shown in the food chain.

Critical Thinking *Answer each of the following in complete sentences.*

1. **Model:** Make an energy pyramid using one of the food chains shown in the diagram above.
2. **Compare:** How is a food chain similar to an energy pyramid?
3. **Classify:** Some bacteria are able to make their own food using chemicals from their environ- ments. Would you classify these bacteria as producers or consumers? Explain.
4. **Contrast:** How do scavengers differ from decomposers?
5. **Predict:** What would happen if decomposers did not break down the remains of dead organisms?

Finding Out More..

1. **Organize:** Make a list of twenty organisms that live in your community. In a table, classify each organism as a producer or a consumer.
2. **Model:** Cut out pictures of twenty kinds of plants and animals from magazines. Use the pictures to make a food web.
3. **Research:** Find out the names of two animals that are considered endangered. Use library resources to find out what is being done to save these animals.
4. **Organize:** Make a poster that shows how five kinds of natural resources can be conserved.

76

CELLS, TISSUES, AND ORGANS

CONTENTS

STUDY HINT Before beginning Unit 4, scan through the lessons in the unit looking for words that you do not know. On a sheet of paper, list these words. Work with a classmate to try to define each word on your list.

4-1 What are cells?

Objective ▶ State the cell theory in your own words.

TechTerms

▶ **cell:** basic unit of structure and function in living things

▶ **theory** (THEE-uh-ree): idea that explains something and is supported by data

Cells A brick house is made up of many bricks. A brick is the basic unit of structure of a brick house. The basic unit of structure in living things is a **cell.** All living things are made up of one or more cells. It is also the basic unit of function. Cells carry out all life processes. For example, a cell takes in food and breaks the food down. It breaks down a simple sugar called glucose (GLOO-kohs) to produce energy. This life process is called cellular respiration (res-puh-RAY-shun).

▷ *Define:* What is a cell?

Discovery of Cells The first person to observe cells was Robert Hooke. Hooke was an English scientist. He used a compound microscope to look at thin slices of cork. Cork is found in some plants. The cork seemed to be made up of many small boxes. Each box looked like a small room with walls around it. The boxes reminded Hooke of the rooms in which monks slept. These rooms were called cells. Hooke named the structures that made up the cork "cells."

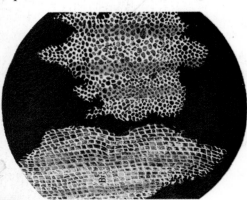

Figure 1 Cork cells

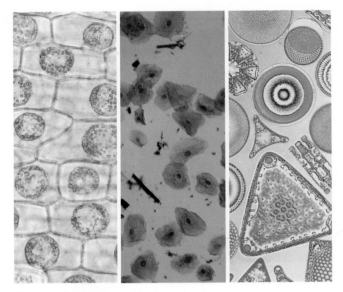

Figure 2 Three kinds of cells

Hooke saw only dead plant cells in cork. Anton van Leeuwenhoek (AN-tun van LAY-vun-hook) was the first person to observe living cells. Van Leeuwenhoek was a Dutch lensmaker. In 1675, he saw single-celled organisms in a drop of pond water. These living things were microscopic. They could not be seen without a microscope.

▷ *Name:* Who was the first person to see cells?

Cell Theory By 1800, better microscopes were being made. Many plants and animals were studied. Scientists had many ideas about cells. In the mid-1800s, these ideas were put together as a **theory** (THEE-uh-REE). A theory is an idea that explains something. The ideas in a theory are supported by data over and over. The theory that was developed is the cell theory. The cell theory states:

▶ All living things are made up of one or more cells.

▶ Cells are the basic units of structure in living things, and cells carry on all life processes.

▶ Cells come only from other living cells.

▷ *Restate:* What does the cell theory state?

LESSON SUMMARY

▶ Cells are the basic units of structure and function in living things.

▶ The first person to see cells was Robert Hooke.

▶ Anton van Leeuwenhoek was the first person to see living cells.

▶ The cell theory states that all living things are made up of one or more cells, cells are the basic units of structure and function in living things, and cells come only from other living cells.

CHECK *Find the sentence in the lesson that answers each question. Then, write the sentence.*

1. What is a theory?
2. What is the basic unit of structure in living things?
3. Who was the first person to see cells?
4. What did Anton van Leeuwenhoek see in a drop of pond water?
5. Where do cells come from?

APPLY *Complete the following.*

6. **Relate:** What is the relationship between improved microscopes and discoveries about cells?
7. **Apply:** Do cells carry on respiration? Explain.
8. **Apply:** Could new cells be produced by the cork that Hooke observed? Explain.

Ideas in Action

IDEA: Many things have basic units of structure. For example, a brick is the basic unit of structure of a brick house.
ACTION: Name three objects and give the basic units of structure of each object.

Skill Builder

Researching Anton van Leeuwenhoek named the single-celled organisms he observed "animalcules." Using library references, find out what microscopic organisms Leeuwenhoek saw. Why did Leeuwenhoek call these organisms "animalcules?" Write your findings in a report.

◄◈►■◄ PEOPLE IN SCIENCE ►■◄◈►■◄◈►■◄◈►■◄◈►■◄◈►■◄◈►■◄◈►■◄◈►■◄◈►■◄◈►■◄◈►■◄◈►■◄

SCHLEIDEN (1804–1881) AND SCHWANN (1810–1882)

The cell theory was formed by two scientists—Matthias Schleiden and Theodor Schwann. Schleiden was a German botanist. Schwann was a German zoologist. Although these scientists did not work together, they both contributed to the cell theory.

Matthias Schleiden was a professor of botany from 1839 until 1862. Schleiden studied many plants in order to learn more about living things. In 1838, Schleiden stated that all plants are made up of cells. He also recognized the role of the nucleus in cell division.

Theodor Schwann studied in Berlin. Schwann added to the work of Schleiden. In his work, Schleiden showed that the cell also is the basic unit of structure in animals. Schwann published his ideas in 1839.

Schwann also studied muscle cells and nerve cells in animals. He was the first person to describe the cells that cover and protect nerve cells. These protective cells are called Schwann cells.

◄◈►■◄ ►■◄◈◄ ►■◄◈◄ ►■◄◈◄ ►■◄◈◄ ►■◄◈◄ ►■◄◈◄ ►■◄◈◄ ►■◄◈◄ ►■◄◈◄ ►■◄◈◄ ►■◄◈◄ ►■◄◈◄ ►■◄◈◄ ►■◄◈◄ ►■◄◈◄ ►■◄◈◄

What are the main cell parts?

Objective ▶ Identify the main parts of a cell and describe their functions.

TechTerms

▶ **cell membrane** (MEM-brayn): thin structure that surrounds a cell

▶ **cytoplasm** (SYT-uh-plaz-um): all the living material inside a cell except the nucleus

▶ **nucleus** (NEW-klee-us): control center of a cell

Three Main Parts Most cells have three main parts. The three main parts of the cell are shown in Figure 1. They are the cell membrane (MEM-brayn), the nucleus (NEW-klee-us), and the cytoplasm (SYT-uh-plaz-um).

▶ *Name:* What are the three main parts of a cell?

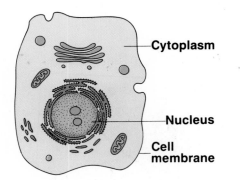

Figure 1 Three main parts of a cell

Nucleus The **nucleus** (NEW-klee-us) of a cell is round or egg-shaped. It usually is near the middle of a cell. The nucleus is darker than the rest of the cell. It is the control center of a cell. It controls all the life processes of a cell. The nucleus also controls cell reproduction.

▶ *Infer:* What would happen to a cell if the nucleus were taken out?

Cytoplasm The **cytoplasm** is all the living material in a cell except the nucleus. The nucleus floats in the cytoplasm. Most of a cell is made up

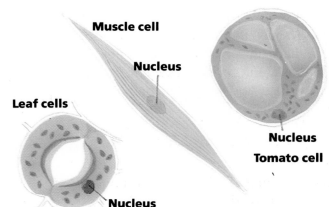

Figure 2 Nuclei in three cells

of cytoplasm. The cytoplasm looks like the white part of a raw egg. Most of the cell's activities take place in the cytoplasm.

▶ *Describe:* What takes place in the cytoplasm?

Cell Membrane The **cell membrane** is a thin structure that surrounds a cell. Sometimes, the cell membrane is called the plasma (PLAZ-muh) membrane. The cell membrane has three important jobs. It protects the inside of a cell. The cell membrane also supports and gives a cell its shape. The cell membrane controls the movement of materials into and out of a cell. Food, water, and oxygen move through the membrane into the cell. Wastes move out of the cell through the cell membrane.

▶ *List:* What are the three jobs of the cell membrane?

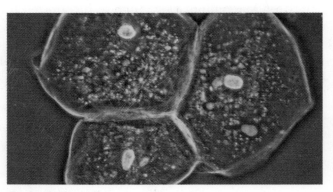

Figure 3 Skin cell

LESSON SUMMARY

▶ The three main parts of the cell are the cell membrane, the nucleus, and the cytoplasm.

▶ The nucleus is the control center of the cell.

▶ The cytoplasm is all the living material in a cell except the nucleus.

▶ The cell membrane surrounds a cell.

CHECK *Complete the following.*

1. All the life processes of a cell are controlled by the _____ .

2. The cell membrane also is called the _____ membrane.

3. The three main parts of the cell are the nucleus, the cell membrane, and the _____ .

4. The cell membrane controls the _____ of materials into and out of a cell.

5. Most of a cell is made up of _____ . .

APPLY *Complete the following.*

6. **Apply:** What part of a cell controls respiration?

▶ 7. **Infer:** The nucleus of a cell also is surrounded by a membrane. What do you think the jobs of the nuclear membrane are?

▲ 8. **Model:** Label the three main parts of the skin cell in the diagram below.

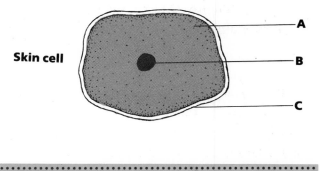

Skin cell

A
B
C

░░░░░░░░░░░░░░░░░░░░░░░░░░░░░░░░

Skill Builder.............................

Building Vocabulary A prefix is a word part placed at the beginning of a word. The meaning of a prefix usually remains the same. Many words in life science begin with the prefix "cyto-." Look up the meaning of "cyto-" in a dictionary. Write the meaning. List and define five words that begin with the prefix "cyto-" in your own words. Circle the part of each definition that relates to the meaning of the prefix "cyto-."

ACTIVITY

MODELING A CELL

You will need a package of gelatin, a grape, a small plastic bag, and a small plastic container.

1. Mix the gelatin by following the directions on the package.

2. Place the bag into the plastic container, so that the open end of the bag extends over the edge of the container.

3. Pour some gelatin into the plastic bag. Let it cool.

4. Before the gelatin hardens, put a grape in the center of the gelatin. Let the gelatin harden.

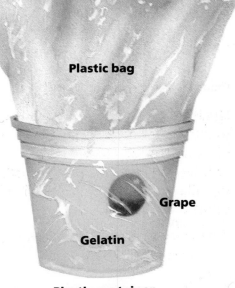

Plastic bag

Grape

Gelatin

Plastic container

Questions

▲ 1. **Model:** Draw a picture of your cell model. Label the parts.

2. **Relate:** What cell part is represented by the: **a.** plastic bag? **b.** gelatin? **c.** grape?

3. **a. Compare:** What cell part makes up most of your model? **b.** How does this compare to a real cell?

What are other cell parts?

Objective ▶ Name and describe the functions of the parts of a cell.

TechTerms

▶ **mitochondria** (myt-uh-KAHN-dree-uh): rice-shaped structures that produce energy for a cell

▶ **organelles** (or-guh-NELS): small structures in the cytoplasm that do special jobs

▶ **ribosomes** (RY-buh-sohms): small round structures that make proteins

▶ **vacuoles** (VAK-yoo-wohls): liquid-filled spaces in the cytoplasm

Organelles You can compare a cell to a factory. There are many machines in a factory. Each machine has a special job. The machines work together to keep the factory working. The "machines" of a cell are its **organelles** (or-guh-NELS). Organelles are small structures that float in the cytoplasm. Each organelle has a special job to do. They keep the cell working properly.

▷ **Define:** What are organelles?

Mitochondria One kind of organelle is the **mitochondria** (myt-uh-KAHN-dree-uh). Mitochondria are small, rice-shaped structures. In fact, they are so small they can be seen only with an electron microscope. Mitochondria are the "powerhouses" of the cell. They break down food to make energy for the cell. The energy is used by the cell to carry out its life processes.

▷ **Explain:** Why does a cell need energy?

Vacuoles The **vacuoles** (VAK-yoo-wohls) of a cell are liquid-filled spaces. The spaces are surrounded by a membrane. Vacuoles are like storage bins. They store food and wastes. Some vacuoles store extra water. They pump extra water out of a cell.

▷ **List:** What are three things that are stored in vacuoles?

Ribosomes Every cell has many small round structures in its cytoplasm. These structures are **ribosomes** (RY-buh-sohms). Ribosomes make proteins. The proteins are needed for growth.

▶ **Infer:** Why is "protein factory" a good name for a ribosome?

Transport Tubes Cells have very small networks of tubes in them. The tubes are like a tiny highway system for the cell. Substances in the cell are moved from one organelle to another in these tubes.

▷ **Describe:** What is the job of the transport tubes?

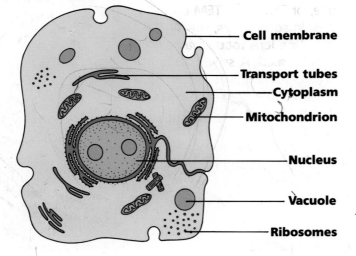

Cell membrane

Transport tubes

Cytoplasm

Mitochondrion

Nucleus

Vacuole

Ribosomes

LESSON SUMMARY

▶ Organelles are small structures in the cytoplasm.

▶ Mitochondria break down food to produce energy.

▶ Vacuoles are liquid-filled spaces that store food and wastes.

▶ Ribosomes make proteins.

▶ Cells have small networks of transport tubes.

CHECK *Write true if the statement is true. If the statement is false, change the underlined term to make the statement true.*

1. The "powerhouses" of the cell are <u>ribosomes</u>.

2. Small structures in the cytoplasm are <u>organelles</u>.

3. Ribosomes make <u>proteins</u>.

4. Liquid-filled spaces are <u>mitochondria</u>.

5. Ribosomes are small <u>rice-shaped</u> structures.

APPLY *Complete the following.*

6. **Compare:** Name the organelle that has a job similar to each of these common objects: railroad, cabinets, battery, electric company.

7. **Infer:** Think about the jobs performed by your muscles and your skin. Do you think there are more mitochondria in muscle cells or in skin cells? Explain.

▲ 8. **Model:** Draw a diagram of a cell. Label the parts of your diagram.

InfoSearch

Read the passage. Ask two questions about the topic that you cannot answer from the information in the passage.

Golgi Bodies Cells contain many flattened sacs called Golgi bodies. Golgi bodies can be seen only with an electron microscope. The Golgi bodies store and release chemicals from a cell. There are a lot of Golgi bodies in cells that produce large amounts of chemicals.

SEARCH: Use library references to find answers to your questions.

TECHNOLOGY AND SOCIETY

ELECTRON MICROSCOPES

Most cells can be seen only with a light microscope. However, an electron (ih-LEK-tron) microscope is used to observe small cell structures. Electrons are negatively charged parts of an atom. Electron microscopes use electrons to form images of very tiny objects.

There are two kinds of electron microscopes. One kind is the transmission electron microscope, or TEM. The TEM can magnify objects up to 200,000 times. In some ways, the TEM is like an upside-down light microscope. One lens focuses a beam of electrons through the object being studied. A second lens focuses the beam onto a film or screen. The beam focused on the screen forms an image. Images are viewed through a lens at the bottom of the microscope. Unfortunately, the TEM cannot be used to study living specimens.

The second kind of electron microscope is the scanning electron microscope, or SEM. The SEM can be used to study some living things. The SEM reflects electrons from the surface of the object being studied. A three-dimensional image is produced.

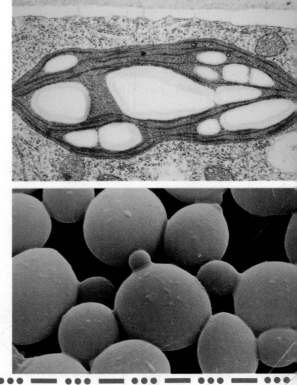

How do plant and animal cells differ?

Objective ▶ Compare plant cells and animal cells.

TechTerms

- **cellulose** (SEL-yoo-lohs): hard, nonliving material that makes up the cell wall of a plant cell
- **cell wall:** outer, nonliving part of a plant cell
- **chlorophyll** (KLOR-uh-fil): green material in chloroplasts that is needed for plants to make food
- **chloroplast** (KLOR-uh-plast): round, green structure in a plant cell that contains chlorophyll

Cell Wall All plant cells have a **cell wall.** Animal cells do not have a cell wall. The cell wall surrounds the cell membrane. The cell wall is nonliving. It is made up of a hard material called **cellulose** (SEL-yoo-lohs). Wood is made up mostly of cellulose.

The cell wall has three jobs. It protects a plant cell and gives the cell its shape. It also gives a plant cell support. Large plants, such as trees and bushes, do not need a skeleton because each cell has support from the cell wall.

▌▶ **Name:** What substance makes up most of the cell wall?

Vacuoles The number and size of vacuoles is different in plant and animal cells. Plant cells have only one or two vacuoles. The vacuoles are usually very large. Animal cells have many small vacuoles.

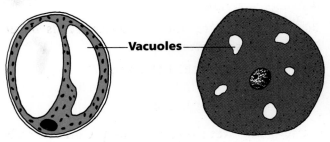

Vacuoles

👁 **Observe:** How many vacuoles does the plant cell have?

Chloroplasts Most plant cells have **chloroplasts** (KLOR-uh-plasts). Chloroplasts are round, green structures. They contain a green material called **chlorophyll** (KLOR-uh-fil). Chlorophyll gives a plant its green color. Chlorophyll is very important to plant cells. Plants need chlorophyll to make food. Animal cells do not have chloroplasts or chlorophyll.

▌▶ **Explain:** Why are most plants green?

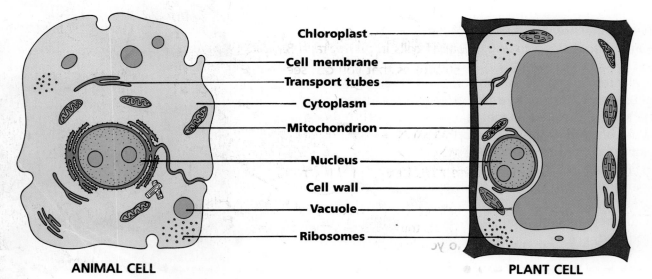

Chloroplast
Cell membrane
Transport tubes
Cytoplasm
Mitochondrion
Nucleus
Cell wall
Vacuole
Ribosomes

ANIMAL CELL

PLANT CELL

LESSON SUMMARY

▶ All plants have a cell wall.

▶ The cell wall protects a plant cell, and gives the cell its shape and support.

▶ Plant cells have only one or two large vacuoles.

▶ Most plant cells have chloroplasts.

CHECK *Complete the following.*

1. Why do large plants not need a skeleton?

2. What are the three jobs of the cell wall?

3. What are chloroplasts?

4. What is cellulose?

5. How do the vacuoles in animal cells differ from those in plant cells?

6. Why do plants need chlorophyll?

APPLY *Complete the following.*

7. **a. Compare:** How are plant and animal cells alike? **b.** How do plant and animal cells differ?

8. **Analyze:** Which of the following organisms contain cellulose? **a.** pine tree **b.** cow **c.** rabbit **d.** fern **e.** grass **f.** goldfish

9. **Infer:** Is wood mainly a living or a nonliving material? Explain.

InfoSearch

Read the passage. Ask two questions about the topic that you cannot answer from the information in the passage.

Pigments When sunlight strikes the leaf of a plant, the sunlight is absorbed by special pigments. Pigments are substances that absorb light. Different pigments absorb different wavelengths of light and reflect others. Chlorophyll is the main pigment needed for photosynthesis. Chlorophyll absorbs large amounts of red, orange, and blue light. It reflects yellow and green light. Because chlorophyll reflects yellow and green light, the leaf looks green.

SEARCH: *Use library references to find answers to your questions.*

ACTIVITY

ANALYZING CELLS

You will need a pencil and a sheet of paper.

1. Study photograph A.

2. **Measure:** Count the number of cells in photograph A.

3. Identify the cell parts and structures that you can see in the cells.

4. Study photograph B.

5. **Measure:** Count the number of cells in photograph B.

6. Identify the cell parts and structures that you can see in the cells.

Questions

1. **Compare:** What cell parts and structures are the same in the cells in both photographs?

2. **a. Analyze:** Which cells are plant cells? **b.** Which are animal cells? **c.** How do you know?

3. **a. Measure:** In one plant cell, count the number of chloroplasts. **b. Infer:** Does this cell make food for the plant? **c. Explain:** How do you know?

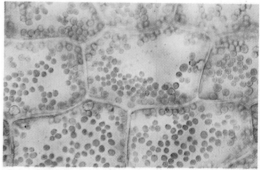

Photograph A

Photograph B

What is diffusion?

Objective ▶ Describe diffusion in cells.

TechTerms

▶ **diffusion** (dih-FYOO-shun): movement of material from an area where molecules are crowded to an area where they are less crowded

▶ **osmosis** (ahs-MOH-sis): movement of water through a membrane

Molecules and Diffusion A molecule (mahl-uh-KYOOL) is the smallest part of a substance that is still that substance. Molecules are made up of tiny parts called atoms. Molecules are always moving. Most molecules move from places where they are crowded to places where they are less crowded. The movement of molecules from crowded areas to less crowded areas is called **diffusion** (dih-FYOO-shun). For example, if you open a bottle of perfume in a room, the molecules of perfume will diffuse throughout the room. In the beaker, ink molecules will spread through the water until the molecules of ink are evenly distributed.

▶ *Identify:* What are molecules made up of?

Diffusion in Cells In order for a cell to carry on its life processes, oxygen and other substances must pass through the cell membrane. Wastes must be removed from the cell. The cell membrane has tiny holes in it. Substances can go in and out of a cell by moving through the holes. Some substances move into and out of a cell by diffusion.

Molecules also move in and out to keep the same amount of a substance on both sides of the cell membrane. However, the cell membrane only lets some substances in and out. If a molecule of a substance is too big, it cannot pass through the cell membrane.

▶ *Name:* What cell structure controls the materials that go in and out of a cell?

Osmosis The movement of water through a membrane is called **osmosis** (ahs-MOH-sis). Osmosis is a special kind of diffusion. Many substances dissolve in water before they move into a cell. Water moves through a cell membrane by osmosis.

▶ *Predict:* What would happen if water kept entering a cell and no water left the cell?

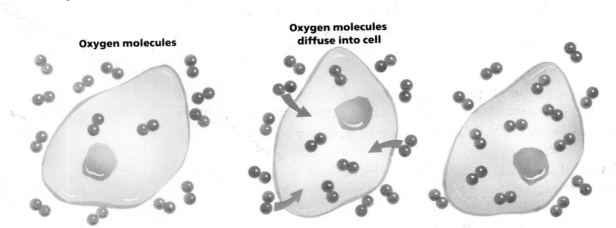

Oxygen molecules

Oxygen molecules diffuse into cell

LESSON SUMMARY

▶ The movement of molecules from crowded areas to less crowded areas is called diffusion.

▶ Diffusion in a cell occurs when substances move in and out of a cell.

▶ The cell membrane lets only some substances in and out of a cell.

▶ Osmosis is the movement of water through a membrane.

CHECK Complete the following.

1. Movement of molecules from crowded areas to less crowded areas is called _____ .

2. The movement of water through a membrane is called _____ .

3. The _____ controls what substances pass in and out of a cell.

4. Many substances dissolve in _____ before entering a cell.

APPLY Complete the following.

5. **Hypothesize:** Will a teaspoon of salt dissolve more quickly in a glass of fresh water or a glass of salt water? Explain.

6. **Predict:** In which direction will carbon dioxide move if the amount of carbon dioxide in the fluid outside a cell became greater than the amount of carbon dioxide in the fluid inside the cell?

7. **Explain:** Why would the smell of freshly baked bread eventually fill an entire house?

State the Problem

Study the model of diffusion carefully. What is wrong with the model? How could you correct it?

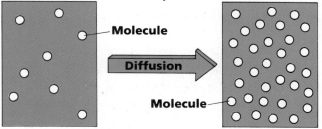

ACTIVITY

MEASURING DIFFUSION IN EGGS

You will need 3 eggs, 3 large plastic cups, vinegar, oil, water, a metric ruler, plastic wrap, a marker, and string.

1. **Measure:** Wrap string around the middle of each egg. Measure the lengths of string. Record the measurements.

2. Repeat step 1 for the lengths of the eggs.

3. Fill a cup half-full with vinegar. Add oil and water to each of the other cups.

4. Mark the liquid level on the outside of each cup.

5. Place an egg in each cup. Cover each cup with plastic wrap.

6. **Measure:** After three days, remove your eggs. Measure and record the size of each egg as you did in steps 1 and 2.

7. Mark the level of the liquid on the side of each cup. Measure and record the height of each of the marks.

Questions

1. a. **Observe:** In which cup did the volume of liquid change?   b. **Calculate:** How much did the volume change?

2. **Compare:** How did the size of the egg change as the volume of liquid changed?

3. **Relate:** Which substance's molecules were able to pass through the egg's membrane?

How do cells produce new cells?

Objective ▶ Describe cell division.

TechTerms

- **cell division:** process by which cells reproduce
- **chromosomes** (KROH-muh-sohms): cell parts that determine what traits a living thing will have
- **daughter cells:** new cells produced by cell division
- **mitosis** (my-TOH-sis): division of the nucleus

Cell Division Cells can reproduce. They produce new cells. Living things grow because their cells can reproduce and make new cells. The process by which cells reproduce is called **cell division.** The cells of all living things are produced by cell division.

▶ **Define:** What is cell division?

Mitosis The nucleus controls cell division. Inside the nucleus are **chromosomes** (KROH-muh-sohms). Chromosomes are cell parts that determine what traits a living thing will have. During cell division, each chromosome makes an exact copy of itself.

After the chromosomes in the nucleus form twin pairs, the nucleus divides. The division of the nucleus is called **mitosis** (my-TOH-sis). During mitosis, the nuclear membrane disappears. All the pairs of chromosomes line up across the center of the cell. The chromosomes of each pair separate and move to opposite ends of the cell. The cell begins to pull apart. It splits into two new cells.

▶ **Identify:** What controls cell division?

Daughter Cells The two new cells formed by cell division are called **daughter cells.** The nuclei of the two daughter cells are exactly alike. Each daughter cell is about half the size of the parent cell. In time, each of the daughter cells will divide to form two more daughter cells.

▶ **Name:** What are the two new cells formed by cell division called?

Cell Division in Plants Plant cells also reproduce by cell division. Like animal cells, plant cells make copies of their chromosomes and carry out mitosis. In animal cells, the cell pulls apart and forms two daughter cells. This does not happen in plant cells. In plant cells, a new cell wall and new cell membrane form down the middle of the cell. They form a wall between the two new nuclei. Two daughter cells are formed, one on each side of the new cell wall.

▶ **Identify:** What forms down the middle of a dividing plant cell?

MITOSIS AND CELL DIVISION IN ANIMAL CELLS

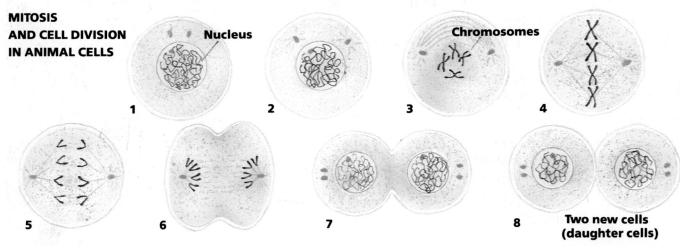

Nucleus

Chromosomes

1   2   3   4

5   6   7   8   Two new cells (daughter cells)

Mitosis and cell division in animals

LESSON SUMMARY

► The process by which cells reproduce is called cell division.

► During cell division, each chromosome makes an exact copy of itself.

► Division of the nucleus is called mitosis.

► The two new cells that are formed by cell division are called daughter cells.

► Plant cells also reproduce by cell division.

CHECK *Answer the following.*

1. What is cell division?

2. What is mitosis?

3. What controls cell division?

4. How can living things grow?

5. What are daughter cells?

6. What are chromosomes?

APPLY *Complete the following.*

▶ 7. **Infer:** Why do you think chromosomes are copied during cell division?

◢ 8. **Analyze:** Is cell division a form of reproduction? Explain.

9. **Sequence:** Place the descriptions of mitosis in the correct order: **a.** Pairs of chromosomes line up across the center of the cell **b.** chromosomes move to opposite ends of the cell **c.** the nuclear membrane disappears **d.** the chromosomes of each pair separate.

10. **Analyze:** Is plant or animal cell division shown in the diagram? Explain what is happening in each part of the diagram.

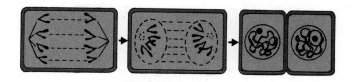

Health & Safety Tip................................

Cells usually divide and grow in an orderly manner. Cancer is a rapid, disorderly growth of cells. A cancerous growth crowds out and destroys healthy tissues. Use library references to find out the seven warning signs for cancer.

ACTIVITY

MODELING CELL DIVISION IN AN ANIMAL CELL

You will need eight pipe cleaners, a sheet of white paper, and a drawing compass.

1. Use the compass to make a large circle on a piece of white paper. Draw a small circle in the middle of the first circle.

2. Place four pipe cleaners in the inner circle.

3. Twist a pipe cleaner around each of the original pipe cleaners. Erase the inner circle.

4. Line up each pair of pipe cleaners in the middle of the large circle.

5. Separate the pipe cleaners. Move each pipe cleaner in a pair to opposite ends of the circle.

Questions

1. **a. Identify:** In your model, what does each circle represent? **b.** What do the pipe cleaners represent?

2. How many chromosomes does the parent cell have?

3. **Describe:** How could you show the formation of daughter cells on your model?

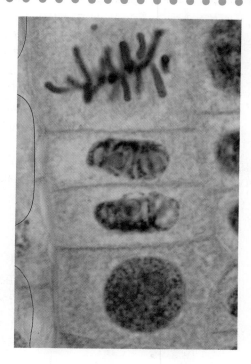

Why do cells have different shapes?

Objective ▶ Describe the structures and functions of different kinds of cells.

TechTerms

▶ **guard cell:** cell in a plant that helps control the passage of materials into and out of the stomates

▶ **red blood cell:** blood cell that carries oxygen

▶ **white blood cell:** blood cell that helps destroy germs

Cell Size and Shape In one-celled organisms, the one cell carries on all the life processes. Large animals and plants are made up of many cells. The cells are not all the same. They have different sizes and shapes. Look at the different shapes of the cells shown. Different kinds of cells have different jobs. The shapes of most cells help them to do their jobs.

▶ *Explain:* How are cells different?

Nerve Cells Nerve cells are the "telephone wires" of the body. They carry messages from one part of the body to another. The message carried by a nerve cell is called an impulse. Nerve cells are long and thin. Some nerve cells are the longest cells in your body. The longest cells are the nerve cells in a giraffe's leg. They are 2 m long.

▶ *Define:* What is an impulse?

Muscle Cells Muscle cells are long and thin. The shape of muscle cells can change. Muscle cells can become shorter. Some muscles are attached to bone. When cells in these muscles shorten, they make the bones move.

▶ *Explain:* What is the job of some muscles?

Blood Cells There are two main kinds of blood cells. **Red blood cells** are round. They do not have a nucleus. The job of a red blood cell is to carry oxygen. **White blood cells** help destroy germs in the body. White blood cells are shapeless, or do not have a definite shape.

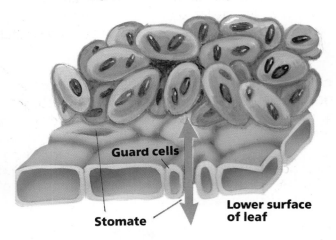

▶ *Infer:* Do you think red blood cells reproduce themselves? Why?

Guard Cells A stomate is a tiny opening on the lower surface of a plant leaf. A stomate takes in carbon dioxide from the air and gives off oxygen and water. Two bean-shaped cells called **guard cells** surround each stomate. Guard cells control the size of the stomate. When the guard cells swell, the stomate opens. When the guard cells shrink, the stomate closes.

▶ *Predict:* Does carbon dioxide enter a plant when the guard cells swell or shrink?

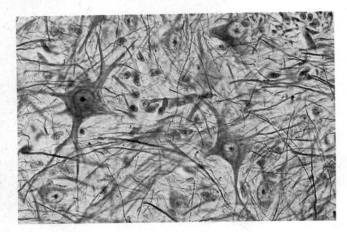

Guard cells

Stomate

Lower surface of leaf

LESSON SUMMARY

► Cells have different sizes and shapes.

► Nerve cells carry messages from one part of the body to another.

► When muscle cells become shorter, they make bones move.

► There are two main kinds of blood cells: red blood cells and white blood cells.

► Guard cells surround each stomate.

CHECK *Write true if the statement is true. If the statement is false, change the underlined term to make the statement true.*

1. Germs are destroyed by <u>red</u> blood cells.

2. Muscle cells can become <u>shorter</u>.

3. The job of a <u>white blood cell</u> is to carry oxygen.

4. The message carried by a nerve cell is called an <u>impulse</u>.

5. The largest cells in the body are <u>muscle</u> cells.

6. A <u>stomate</u> is a tiny opening on the lower surface of a plant leaf.

APPLY *Complete the following.*

7. **Relate:** How is the size and shape of a nerve cell related to its function?

8. **Contrast:** How do red blood cells differ from most other kinds of cells?

InfoSearch

Read the passage. Ask two questions about the topic that you cannot answer from the information in the passage.

Body Defenses Some white blood cells protect the body from disease by surrounding and digesting germs. These white blood cells are called phagocytes (FAG-uh-syts). Other white blood cells help destroy germs by making antibodies (AN-ti-bahd-ees). Antibodies are special proteins that destroy bacteria and viruses that invade the body.

SEARCH: Use library references to find answers to your questions.

CAREER IN EARTH SCIENCE

CYTOLOGY LAB TECHNICIAN

Scientists study cells to find answers to questions, such as "Why do some cells become diseased?" Research in cell biology may provide answers to many questions about human health. Cytology (sy-TAHL-uh-jee) lab technicians play an important role in this research.

Cytology lab technicians work with tissue samples. From these samples, the lab technicians prepare slides of cells. The slides are then analyzed under a microscope. Many cytology lab technicians also do tests on cells. They keep careful records of the test results.

Cytology lab technicians are employed by universities or hospitals. Some work in private industry. Most entry-level positions in this field require two years of college with courses in biology, chemistry, or physics. Many cytology lab technicians are hired after completing technical training schools. You can get more information about this career by speaking with your guidance counselor or writing to the personnel director of your local hospital.

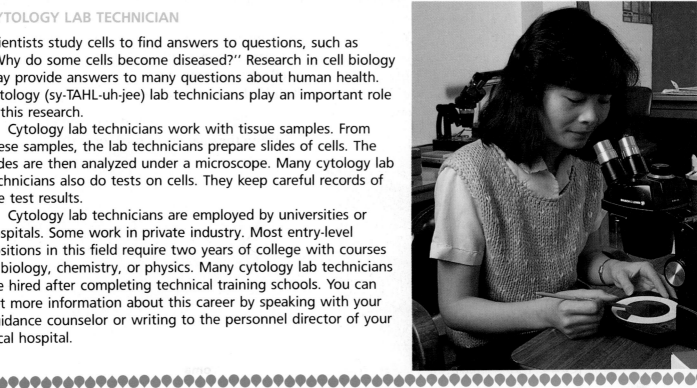

What are tissues?

Objective ▶ Describe the four main kinds of tissues.

TechTerms

- ▶ **connective tissue:** tissue that holds parts of the body together
- ▶ **epithelial** (ep-uh-THEEL-ee-uhl) **tissue:** tissue that covers and protects parts of the body
- ▶ **plasma** (PLAZ-muh): liquid part of blood
- ▶ **tissue:** group of cells that look alike and work together

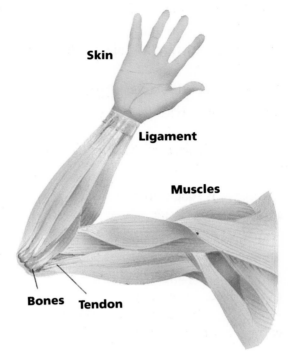

Skin

Ligament

Muscles

Bones   Tendon

Tissues On a baseball team, the players work together. They wear uniforms that make them look alike. In many-celled organisms, cells work as teams. A group of cells that look alike and work together make up a **tissue.** Tissues are named for the jobs they do. There are four main kinds of tissue.

▷ *Define:* What is a tissue?

Muscle Tissue Muscle tissue makes up muscles. Muscle tissue is made up of cells that can become shorter. There are different kinds of muscle tissue. One kind is attached to bones. When these muscles shorten, they pull on the bones and make the bones move.

▷ *Relate:* How do some muscles and bones work together?

Covering Tissue The skin that covers your body is made up of **epithelial** (ep-uh-THEEL-ee-uhl) **tissue.** Epithelial tissue is made up of cells that join tightly together. Epithelial tissue also covers many parts inside you body. It protects your body. It helps to keep germs out of your body.

▷ *Name:* What type of tissue is skin?

Connective Tissue Tissue that holds parts of the body together is **connective tissue.** It also sup-

ports and protects the body. Bone is a connective tissue. Other kinds of connective tissue are ligaments (LIG-uh-muhnts) and tendons. Ligaments connect bones to one another. Tendons connect muscles to bones.

Blood is a liquid connective tissue. It has blood cells that float in **plasma** (PLAZ-muh). Plasma is the yellow, liquid part of blood. Plasma carries food, gases, and other important substances to and from all the cells in the body.

▷ *List:* What are four kinds of connective tissue?

Nerve Tissue Nerve tissue is made up of nerve cells, or neurons. Nerve tissue carries messages. It causes muscles to act. It controls breathing, digestion, and heartbeat. Your brain and spinal cord are made up mostly of nerve tissue.

☛ *Infer:* What is the function of the brain?

LESSON SUMMARY

▶ A group of cells that look alike and work together make up a tissue.

▶ Muscle tissue is made up of cells that can become shorter.

▶ Epithelial tissue protects the body.

▶ Connective tissue holds parts of the body together.

▶ Ligaments and tendons are made up of connective tissue.

▶ Blood is a liquid connective tissue.

▶ Nerve tissue carries messages.

CHECK *Write the letter of the term that best completes each statement.*

1. The brain is made up mostly of
   a. connective tissue.   **b.** muscle tissue.
   c. nerve tissue.   **d.** epithelial tissue.

2. Epithelial cells make up
   a. muscles. **b.** skin. **c.** ligaments. **d.** bone.

3. Bones _____ the body.
   a. support   **b.** carry messages in   **c.** keep germs out of   **d.** cover

4. Muscles are connected to bones by
   a. plasma.   **b.** ligaments.   **c.** neurons.
   d. tendons.

5. Plasma is _____ in color.
   a. red   **b.** yellow   **c.** colorless   **d.** blue

6. Blood is _____ tissue
   a. a   connective   **b.** an   epithelial   **c.** a nervous   **d.** a solid

APPLY *Complete the following.*

7. **Compare:** What are the functions of muscle, connective, and nerve tissue?

▶ 8. **Infer:** Bones have a lot of calcium in them. Why do you think the proper amount of calcium in your diet is important for healthy bones?

Health & Safety Tip

Injuries to the brain and spinal cord often result from sports and automobile accidents. Make a poster showing ways that people can protect their brain and spinal cords from injury in these ways.

TECHNOLOGY AND SOCIETY

CORNEA TRANSPLANTS

Blindness often is caused by scarring of the cornea (KAWR-nee-uh). The cornea is the clear, outer layer of the eye. Light first enters the eye through the cornea. It is now possible to restore lost eyesight in some people with scarred corneas. This can be done through cornea transplant surgery. During a cornea transplant, the scarred cornea is replaced with clear corneal tissue. The healthy tissue comes from an eye donor who has died.

An eye bank is an agency that helps get donated eyes to patients waiting for transplants. An eye bank collects, prepares, and distributes the eyes to specially trained surgeons. Eyes must be removed shortly after a person has died. After removal, they are packed in special containers and put in cold storage. The eyes are examined at an eye bank to see if they can be used for a transplant. If they are useful, the eyes are made available for immediate use in any part of the United States.

The first eye bank was formed in New York City in 1944. Today, many cities and states have eye banks. For further information about eye banks and eye donations, contact the Eye-Bank for Sight Restoration, Inc., New York, New York.

What are organs and organ systems?

Objective ▶ Describe organs and organ systems.

TechTerms

▶ **organ** (OWR-gun): group of tissues that work together to do a special job

▶ **organ system:** group of organs that work together

Organs Groups of cells that work together form tissues. Different tissues work together too. A group of tissues that work together to do a special job is called an **organ** (OWR-gun).

Your body has many different organs. Each organ has a special shape and job. Your arm is an organ. It is made up of many different tissues. Muscle and bone tissues work together to move your arm. Nerve tissue carries impulses to the muscles tissue. These impulses make the muscles move your arm. Epithelial tissue forms the skin on your arm. Blood flows through your arm. It supplies all the cells in these different tissues with oxygen and food. Blood also carries wastes away.

▶ *Define:* What is an organ?

Organ Systems Organs do not work alone. Groups of organs work together. These groups of organs form **organ systems.** All the organs in an organ system work together to carry on a life process. All the organ systems of a living thing work together to keep the organism alive. The human body has ten organ systems. Each organ system carries out a life process. Some organ systems are listed in Table 1.

▶ *Analyze:* What is the job of the excretory system?

Glands Some organs and groups of cells make and give off substances used by the body. These special organs and groups of cells are called glands. Some glands produce chemicals that act as messengers. The blood carries these "messengers" to organs. The organs start to do their work when

Table 1 Organ Systems and Their Jobs		
SYSTEM	**MAJOR ORGANS**	**JOBS**
Skeletal	Bone	support and movement
Muscular	Muscles	movement
Digestive	Mouth, Stomach	digestion
Circulatory	Heart, Blood vessels	transport
Respiratory	Lungs	respiration
Excretory	Kidneys, Skin	removal of wastes
Nervous	Brain, Spinal cord	control

the chemical "messengers" arrive. The glands that produce chemical "messengers" make up an organ system. The organ system is the endocrine (EN-duh-krin) system. Many of the chemical "messengers" made by glands control the activities of other tissues and organs.

▶ *Define:* What is a gland?

Plant Organs Plant tissues also form organs. These organs have special jobs. The roots of a plant hold a plant in the soil. Roots also take in water and minerals needed by the plant. The stem of a plant is an organ, too. It carries food and water to other parts of a plant. The leaves of a plant are its food-making organs. A flower is a plant's organ of reproduction.

▶ *Name:* What are four plant organs?

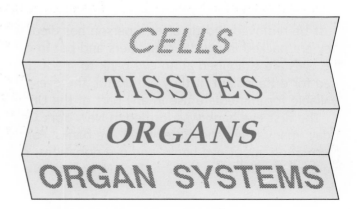

CELLS
TISSUES
ORGANS
ORGAN SYSTEMS

LESSON SUMMARY

▶ A group of tissues that work together to do a job is called an organ.

▶ Organs are made up of many different tissues.

▶ Groups of organs that work together are called organ systems.

▶ The human body has ten organ systems.

▶ The endocrine system is made up of glands that make and give off substances used by the body.

▶ Plant tissues form organs.

CHECK *Complete the following.*

1. The _____ of a plant carries food and water to other parts of the plant.

2. Groups of organs form _____ .

3. Each organ system carries out a life _____ .

4. A group of tissues that work together is called an _____ .

5. The human body has _____ organ systems.

6. Glands make and give off _____ "messengers."

APPLY *Use Table 1 on page 94 to complete the following.*

7. The brain and spinal cord make up the _____ system.

8. **Analyze:** The job of the heart and blood vessels is _____ .

9. **Analyze:** Bone is part of the _____ system.

Complete the following.

10. **Sequence:** Place the levels of organization in an organism from smallest to largest. **a.** organism **b.** cells **c.** organ **d.** tissues **e.** organ systems

11. **Apply:** What different kinds of tissues do you think work together in your leg?

Skill Builder......................................

Researching There are four main kinds of plant tissue. They are the epidermis, ground tissue, vascular tissue, and meristem tissue. Use library references to find out the main functions of each of these plant tissues. Organize your findings in a table.

◆○◆ SCIENCE CONNECTION ◆○◆

MACHINES

Humans are multicellular organisms. Multicellular organisms are very complex. They are made up of billions of cells. All these cells can work together because complex organisms have several levels of organization. Cells combine to form tissues. Tissues combine to form organs, which form organ systems.

Machines also have levels of organization. There are six basic, or simple, machines. They are the inclined plane, the wedge, the screw, the lever, the pulley, and the wheel and axle. All machines are made up of different combinations of these simple machines. For example, fixed and movable pulleys can be combined to form a pulley system. A block and tackle is a pulley system.

Complex machines are all around you. Cars, bicycles, typewriters, vacuum cleaners, and washing machines are complex machines. Cars and bicycles use levers, gears, wheels and axles, and other simple machines. How many other complex machines can you find in your home? Try to identify the simple machines they contain.

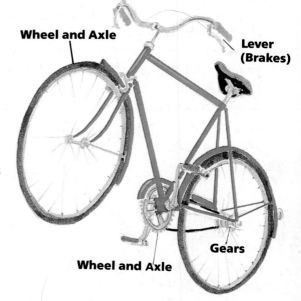

Wheel and Axle

Lever (Brakes)

Gears

Wheel and Axle

Challenges

STUDY HINT Before you begin the Unit Challenges, review the TechTerms and Lesson Summary for each lesson in this unit.

TechTerms..

cell (78)
cell division (88)
cell membrane (80)
cell wall (84)
cellulose (84)
chlorophyll (84)
chloroplast (84)
chromosomes (88)
connective tissue (92)
cytoplasm (80)

daughter cells (88)
diffusion (86)
epithelial tissue (92)
guard cell (90)
mitochondria (82)
mitosis (88)
nucleus (80)
organ (94)
organ system (94)
organelles (82)

osmosis (86)
plasma (92)
red blood cell (90)
ribosomes (82)
theory (78)
tissue (92)
vacuoles (82)
white blood cell (90)

TechTerm Challenges..

Matching *Write the TechTerm that best matches each description.*

1. group of organs that work together
2. control center of a cell
3. basic unit of structure and function in living things
4. liquid part of blood
5. outer part of a plant cell
6. liquid-filled spaces in the cytoplasm
7. material that makes up the cell wall
8. blood cell that helps destroy germs
9. controls passage of materials into and out of stomates
10. group of cells that work together

Identifying Word Relationships *Explain how the words in each pair are related. Write your answers in complete sentences.*

1. chlorophyll, chloroplast
2. diffusion, osmosis
3. cell division, mitosis
4. organ, organ system
5. red blood cell, connective tissue
6. skin, epithelial tissue
7. organelles, cytoplasm
8. cell membrane, cell wall
9. ribosomes, protein
10. mitochondria, energy
11. chromosomes, daughter cell
12. theory, data

Content Challenges..

Multiple Choice *Write the letter of the term or phrase that best completes each statement.*

1. The first person to observe cells was
   a. Theodor Schwann. **b.** Matthais Schleiden. **c.** Anton van Leeuwenhoek. **d.** Robert Hooke.

2. The main parts of a cell are the cell membrane, the nucleus, and the
   a. plasma. **b.** cytoplasm. **c.** mitochondria. **d.** ribosomes.

3. The structures that act as "storage bins" for a cell are
   a. mitochondria. **b.** ribosomes. **c.** vacuoles. **d.** organelles.

4. Organelles float inside a cell's
   a. cytoplasm. **b.** cell membrane. **c.** nucleus. **d.** chromosomes.

5. One of the functions of the cell wall is to
   a. get rid of oxygen.   b. give the cell its shape.   c. give a cell its color.   d. move water through the cell.

6. Molecules are made up of
   a. atoms.   b. water.   c. oxygen.   d. cells.

7. When a cell divides, each chromosome makes a copy that is
   a. identical to the original.   b. slightly different from the original.   c. very different from the original.   d. a mutation of the original.

8. Cells that carry messages from one part of the body to another are
   a. guard cells.   b. blood cells.   c. nerve cells.   d. muscle cells.

9. An example of connective tissue is
   a. muscle.   b. skin.   c. ligaments.   d. nerves.

10. The kind of tissue that keeps germs out of your body is
    a. muscle tissue.   b. epithelial tissue.   c. connective tissue.   d. nerve tissue.

11. The number of organ systems inside the human body is
    a. 8.   b. 9.   c. 10.   d. 11.

True/False *Write true if the statement is true. If the statement is false, change the underlined term to make the statement true.*

1. The first person to see living cells was Anton van Leeuwenhoek.
2. Water, food, and oxygen enter a cell through its nucleus.
3. Most of a cell is made up of plasma.
4. The organelles responsible for breaking down food inside a cell are the mitochondria.
5. The cell wall of a plant is made up of living material.
6. Chlorophyll is the material that makes food in animal cells.
7. The vacuoles in plant cells usually are very large.
8. The even distribution of ink molecules in a beaker of water is an example of diffusion.
9. In plant cell reproduction, a new vacuole forms down the middle of the cell.
10. The messages carried by nerve cells are called stomates.
11. Glands are a part of the nervous system.

Understanding the Features...

Reading Critically *Use the feature reading selections to answer the following. Page numbers for the features follow each question in parentheses.*

1. What are Schwann cells? (79)
2. What is one disadvantage of a transmission electron microscope? What other electron microscope can make up for this disadvantage? (83)
3. **Infer:** Why is it important that cytology lab technicians keep careful records of test results from their work? (91)
4. How is a cornea transplant done? (93)
5. **Describe:** Describe the work done by a worker at an eye bank. (93)
6. **Compare:** How are multicellular organisms like compound machines. (95)

Critical Thinking *Answer each of the following in complete sentences.*

1. **Infer:** Why do you think little was known about cells before the invention of the microscope?
2. What are the three principles that make up the cell theory?
3. Why do you think that organelles can be compared to a factory?

4. **Contrast:** List three differences between plant and animal cells.
5. What makes osmosis a special kind of diffusion?
6. **Contrast:** How does plant cell reproduction differ from animal cell reproduction?

Interpreting a Diagram *Use the diagram to complete the following.*

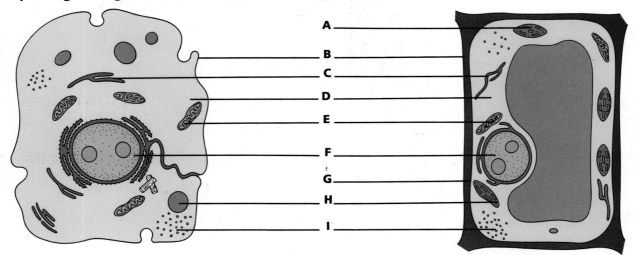

1. What structure is represented by A: Why is this structure missing from Figure 1?
2. What structures are represented by C?
3. What structure are represented by D?
4. What structure are represented by E?
5. What structure is represented by G? Why is this structure missing from Figure 1?
6. What structure are represented by H?
7. What structures are represented by I? What is their function?

Finding Out More..

1. **Observe:** Find and examine slides or photographs of cells magnified with a compound microscope or an electron microscope. Be sure to use photographs of both plant and animal cells. Identify as many different cell parts as you can in each photograph or slide.
2. **Predict:** Perform a simple experiment on diffusion by spraying some perfume at the front of the classroom. Ask your classmates to raise their hands then they smell the perfume. Predict beforehand how your classmates will respond. Will all students eventually raise their hands? In what order will the hands be raised? Write a summary of the experiment. Tell whether your prediction was correct.

3. **Observe:** Using a compound microscope, examine plant cells from the tip of a leaf. Follow the instructions for mounting the leaf given by your teacher. Make a sketch of one cell and try to identify the location of the vacuoles, the chloroplasts, the nucleus, the cell wall, and the cytoplasm.
4. Using an encyclopedia or other reference book, research organ transplants. Write a report on which organs can be successfully transplanted. Explain briefly how medical experts go about performing one of the transplants you researched.

CLASSIFICATION

CONTENTS

STUDY HINT As you read each lesson in Unit 5, write the topic sentence for each paragraph in the lesson on a sheet of paper. After you complete each lesson, compare your list of topic sentences to the Lesson Summary.

What is classification?

Objective ▶ Explain why it is necessary to classify things.

TechTerms

▶ **classification** (klas-uh-fi-KAY-shun): grouping things according to similarities
▶ **taxonomy** (tak-SAHN-uh-mee): science of classifying living things

Classification Imagine that you are in a bookstore. You are looking for a mystery novel. You look at the signs above each group of books. Each sign names the kinds of books in each area. In no time at all, you find your book and are on your way. In a library you are able to find books quickly because they are classified, or grouped, by subject. Grouping things according to similarities, or how they are alike, is called **classification** (klas-uh-fi-KAY-shun).

📁 *Classify:* Name one other way you could classify books.

Taxonomy About 1 1/2 million different kinds of organisms have been discovered so far. Each year the list of living things grows longer. Imagine having to keep track of so many living things. To help scientists keep track of living things, scientists use a system of classification. The science of classifying living things is called **taxonomy** (tak-SAHN-uh-mee). Scientists who classify living things are taxonomists.

Why do scientists classify living things? Classification is a way of organizing information about different kinds of living things. Classification also makes it easier for scientists to identify a newly discovered organism.

▶ *Name:* What is the science of classifying living things called?

Classifying Living Things Living things are classified based on how they are alike. Taxonomists do not group organisms together simply because they look alike. Taxonomists use many features to classify organisms. For example, taxonomists study the cells of an organism, and the substances that make up the cell. They study the way the organism grows and develops before it is born. They also study the blood of animals. Taxonomy is a very complex subject.

▶ *State:* What are two reasons scientists classify organisms?

LESSON SUMMARY

▶ Grouping things according to how they are alike is called classification.

▶ Taxonomy is the science of classifying living things.

▶ Classification is a way of organizing information about living things.

▶ Taxonomists use many features to classify living things.

CHECK *Complete the following.*

1. When you classify, you put things into _____ according to similarities.

2. The science of classifying living things is _____ .

3. Taxonomists study the way organisms grow and develop before they are _____ .

4. About _____ million different kinds of organisms have been discovered so far.

APPLY *Answer the following.*

▶ 5. **Infer:** Why is taxonomy an ongoing science?

▶ 6. **Analyze:** Food is classified in a supermarket. Books are classified in a library. Name two other classification systems that are used in daily life.

7. Why do you think taxonomists do not classify living things only according to appearance?

Skill Builder......................................

Comparing When you compare things, you look at how the things are alike and how they are different. Libraries classify books using either the Dewey Decimal System or the Library of Congress classification system. Find out how these two classification systems compare. What are some similarities and differences between these two systems?

ACTIVITY

CLASSIFYING OBJECTS

You will need a sheet of paper and a pencil.

1. Carefully examine the vehicles shown in the diagram.
2. On a sheet of paper, list as many features of the vehicles as you can.
3. Classify the vehicles according to one of the features.
4. Repeat Step 3 using another feature.

Questions

1. What feature did you use to classify the vehicles?
2. How many different ways were you able to classify the vehicles?
3. Which classification system had the most groups?
4. Which classification system had the fewest groups?

5-2 How are living things classified?

Objective ▶ Explain the different levels of classification.

TechTerms

- ▶ **genus** (JEE-nus): classification group made up of related species
- ▶ **kingdom:** classification group made up of related phyla
- ▶ **phylum** (FY-luhm): classification group made up of related classes
- ▶ **species** (SPEE-sheez): group of organisms that look alike and can reproduce among themselves

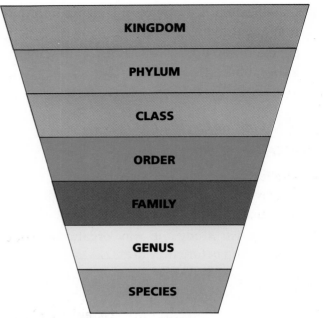

Classification Groups Today, organisms are classified into a series of groups. Organisms that are classified in the same group are alike in some ways. The more alike organisms are, the more classification groups they share. The largest classification group is the **kingdom.** Each kingdom is divided into smaller groups called phyla (FY-luh). One group is a **phylum** (FY-luhm). Each phylum is divided into still smaller groups. These groups are called classes.

Altogether, there are seven major classification groups. The number of different kinds of organisms in each group decreases as you move from the kingdom level to each of the next levels. From largest to smallest, the classification groups are: kingdom, phylum, class, order, family, genus, and species.

A **species** (SPEE-sheez) is the smallest classification group. A species is a group of organisms that look alike and can reproduce among themselves. Each species includes only one kind of organism. All dogs belong to the same species. A **genus** (JEE-nus) is made up of two or more species that are very much alike. For example, dogs and wolves belong to different species, but they belong to the same genus.

▶ *Sequence:* List the seven classification groups from largest to smallest.

Naming Organisms In the eighteenth century, a Swedish botanist, Carolus Linnaeus (luh-NAY-us), developed a system of naming organisms that is still used today. In this system, each kind of organism is identified by a two-part scientific name. A scientific name is made up of the genus and species names of an organism. For example, the scientific name for pet dogs is *Canis familiaris.*

▶ *Name:* What are the two parts of a scientific name?

Table 1	The Classification of Four Organisms						
	KINGDOM	PHYLUM	CLASS	ORDER	FAMILY	GENUS	SPECIES
Dog	Animal	Chordata	Mammalia	Carnivora	Canidae	Canis	familiaris
Wolf	Animal	Chordata	Mammalia	Carnivora	Canidae	Canis	lupus
Human	Animal	Chordata	Mammalia	Primates	Hominidae	Homo	sapiens
Chimpanzee	Animal	Chordata	Mammalia	Primates	Pongidae	Pan	troglodytes

LESSON SUMMARY

▶ Organisms are classified into a series of groups.

▶ The seven classification groups from largest to smallest are: kingdom, phylum, class, order, family, genus, and species.

▶ A species is a group of organisms that look alike and can reproduce among themselves.

▶ The two parts of a scientific name include the genus and species names of an organism.

CHECK *Write the letter of the term that best completes each statement.*

1. The largest classification groups are
   a. species. b. classes. c. orders.
   d. kingdoms.
2. The classification group that contains only one kind of organism is the
   a. phylum. b. genus. c. species.
   d. order.
3. A group of organisms that look alike and can reproduce among themselves is a
   a. species. b. class. c. order.
   d. division.

APPLY *Use Table 1 on page 102 to answer the following questions.*

4. How many classification groups do humans and chimpanzees have in common?
5. What is the scientific name for a human?
6. Which two animals are the most closely related?

InfoSearch

Read the passage. Ask two questions that you cannot answer from the information in the passage.

John Ray In the seventeenth century, an English botanist named John Ray identified and classified more than 18,000 different kinds of plants. Many of his ideas are used in modern plant taxonomy. Ray also classified animals, and was the first to use the term "species" for each different kind of living thing. Ray defined a species as a group of organisms that looked alike and could reproduce among themselves.

SEARCH: Use library references to find answers to your questions.

LOOKING BACK IN SCIENCE

HISTORY OF CLASSIFICATION

One of the first known classification systems was developed more than 2000 years ago by the Greek philosopher Aristotle (AR-is-taht-ul). Aristotle classified organisms as either plants or animals. Animals were classified into smaller groups based upon where the animals lived. Aristotle had three groups of animals: land animals, water animals, and air animals.

One of Aristotle's students, the Greek philosopher Theophrastus (thee-uh-FRAS-tus) classified plants according to their sizes and kinds of stem. Small plants with soft stems were called herbs. Medium-sized plants with many woody stems were called shrubs. Large plants with one woody stem were called trees.

In the eighteenth century, Carolus Linnaeus developed a new way to classify organisms. Linnaeus classified organisms according to their physical characteristics. Organisms that looked alike were grouped together. Linnaeus is known as the founder of modern taxonomy.

Land Animals

Water Animals

Air Animals

5-3 What are the five kingdoms?

Objective ▶ Name and describe the five kingdoms of living things.

TechTerms

- **fungi** (FUN-gy): plantlike organisms that lack chlorophyll
- **monerans** (muh-NER-uns): single-celled organisms that do not have a nucleus
- **protists** (PROHT-ists): single-celled organisms that have a nucleus

Five Kingdoms At one time, all organisms were classified as either plants or animals. With the development of the microscope, new organisms were discovered. These microscopic organisms were placed into a third kingdom. Using the electron microscope, scientists discovered that microscopic organisms are not all alike. Some microscopic organisms do not have a nucleus or other cell organelles. These organisms were placed into a kingdom of their own. Studies also showed that fungi and plants were not closely related. Scientists placed fungi in a kingdom of their own.

Today, most scientists accept the five kingdom classification system. Some scientists would like to make a sixth kingdom for viruses. As new information becomes available, the five kingdom classification system may change.

📄 *Hypothesize:* Why may the five kingdom classification system change over time?

- **Kingdom *Monera*** The **monerans** (muh-NER-uns) are single-celled organisms. Unlike members of the other four kingdoms, monerans do not have a nucleus. In the monerans, nuclear material is scattered throughout the cell. Monerans also lack most of the organelles found in other kinds of cells. Bacteria are examples of monerans.

- **Kingdom *Protista*** Most **protists** (PROHT-ists) are single-celled organisms. Some protists are simple, many-celled organisms. Unlike monerans, the cells of protists have a nucleus. Protist cells also have organelles. The protist kingdom includes both plantlike and animal-like organisms. Algae are one example of protists.

- **Kingdom *Fungi*** Most **fungi** (FUN-gy) are made up of many cells. Some fungi, such as yeast, are made up of only one cell. Like plants, the cells of most fungi have a cell wall. However, fungi do not have chlorophyll. Fungi do not make their own food. They absorb food from their environment.

- **Kingdom *Plantae*** Plants are many-celled organisms. Plant cells have a cell wall. Plant cells also have rounded structures that contain chlorophyll. Plants use chlorophyll to make food.

- **Kingdom *Animalia*** Animals are made up of many cells. Most animals have organs that form organ systems. Unlike plants, animal cells do not have a cell wall or chlorophyll. Animals cannot make their own food. Animals obtain food by eating plants and other animals.

KINGDOMS

| Monera | Protista | Fungi | Plantae | Animalia |

LESSON SUMMARY

- ▶ There are five kingdoms of classification.
- ▶ As new information becomes available, the five kingdom classification system may change.
- ▶ Organisms are classified as either monerans, protists, fungi, plants, or animals.

CHECK *Complete the following.*

1. Today, organisms are classified into _____ kingdoms.

2. The cell of a _____ does not have a nucleus.

3. Algae are members of the _____ kingdom.

4. Some scientists would like to make a sixth kingdom for _____ .

5. The yeasts belong to the _____ kingdom.

6. At one time, all organisms were classified as either plants or _____ .

APPLY *Complete the following.*

▬ 7. **Analyze:** Why are protists and monerans placed in separate kingdoms?

▬ 8. **Analyze:** Why are fungi and plants placed in separate kingdoms?

9. How has technology affected classification?

Skill Builder

▲ *Organizing Information* When you organize information, you put the information in some kind of order. A table is one way to organize information. Make a table that describes the five kingdoms. Place the names of the kingdoms across the top. For each kingdom, list whether its members are one-celled or many-celled, if they have a nucleus and organelles, and how they get food.

PEOPLE IN SCIENCE

ROBERT H. WHITTAKER (1924–1980)

The five kingdom classification system that is widely used today was developed by Robert Whittaker. Whittaker was an American ecologist and botanist. In 1969, he presented his ideas for classifying organisms into five kingdoms. In Whittaker's system, organisms are classified according to whether their cells contain a nucleus surrounded by a membrane, whether they are one-celled or many-celled, and how they obtain food.

While Whittaker is best known for his five kingdom classification system, he made many other contributions to science. As a graduate student in botany at the University of Illinois, Whittaker developed a method of recording how moisture, temperature, and altitude affect how and where plants grow. Whittaker also used radioactive tracers to learn how some substances move in a food chain.

Robert Whittaker died of cancer in 1980. In 1984, the Ecological Society of America established the R. H. Whittaker Travel Fellowship in Whittaker's honor. This fellowship enables young ecologists from other countries to study in the United States.

Are viruses living?

Objectives ▶ Describe the structure of a virus.
▶ Explain why viruses are hard to classify.

TechTerms

▶ **bacteriophage** (bak-TIR-ee-uh-fayj): virus that infects bacteria

▶ **capsid** (KAP-sid): protein covering of a virus

▶ **virus** (VY-rus): piece of nucleic acid covered with a protein

Structure of a Virus The structure of viruses is very different from that of living cells. Viruses do not have any cell parts. A **virus** (VY-rus) is just a piece of nucleic (noo-KLEE-ik) acid covered with an outer coat of protein. This outer coat is called a **capsid** (KAP-sid). The capsid makes up most of the virus.

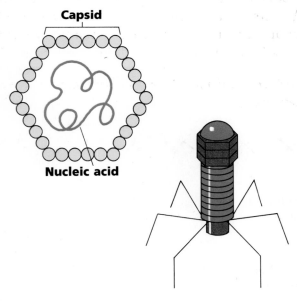

Figure 1

Capsids give viruses their shapes. Some viruses are round. Others look like long rods. The virus shown in Figure 2 attacks and destroys bacteria. A virus that infects bacterial cells is called a **bacteriophage** (bak-TIR-ee-uh-fayj). Notice the unusual shape of a bacteriophage.

▐▶ *Define:* What is a capsid?

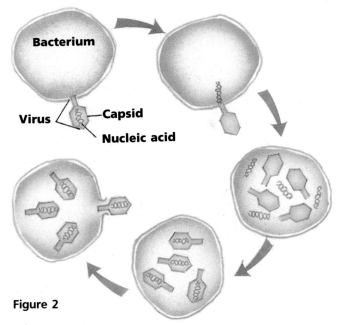

Figure 2

Classifying Viruses Scientists classify viruses according to the living things they infect. The three main groups of viruses are plant viruses, animal viruses, and bacterial viruses.

Viruses are like living cells in one important way. Viruses are able to produce more viruses. However, viruses can produce more viruses only inside a living cell. Outside a cell, viruses appear to be completely nonliving. They have no cytoplasm. They do not carry on life processes.

▐▶ *Contrast:* How do viruses differ from living cells?

Reproduction in Viruses Scientists first learned about reproduction in viruses by studying bacteriophages. When attacking a bacterial cell, a bacteriophage first attaches itself to the surface of the cell. The virus then gives off enzymes. The enzymes make a small hole in the cell wall of the bacterium. The nucleic acid of the virus passes through this hole and enters the cell. Once inside, the virus takes control of the cell and causes the cell to make new viruses. These new viruses may then leave the cell and attack other cells. When viruses leave the cell, it is destroyed.

▐▶ *Sequence:* What happens once a virus enters a living cell?

LESSON SUMMARY

▶ Viruses are made up of a piece of nucleic acid covered by a protein coat.

▶ Capsids give viruses their shape.

▶ Viruses can reproduce only inside a living cell.

▶ Once inside a living cell, viruses take control of the cell.

▶ Viruses are classified according to the living things they infect.

CHECK *Write true if the statement is true. If the statement is false, change the underlined term to make the statement true.*

1. A virus is a piece of <u>cytoplasm</u> covered with a protein coat.

2. A virus that infects a bacterial cell is called a <u>bacteriophage</u>.

3. Viruses can produce more viruses only inside <u>nonliving</u> cells.

4. Scientists first learned about reproduction in viruses by studying <u>bacteria</u>.

5. The outer coat of a virus is called a <u>capsid</u>.

6. The three main groups of viruses are plant, <u>fungal</u>, and animal viruses.

APPLY *Complete the following.*

7. **Sequence:** List the steps of viral reproduction.

8. **Contrast:** How does a virus differ from a living cell?

9. **Hypothesize:** Why is it difficult to classify viruses?

Ideas in Action

IDEA: Many human diseases are caused by viruses, including chicken pox, measles, hepatitis, the flu, and the common cold.

ACTION: Find out how each of these diseases is spread and what you can do to protect yourself from viral diseases. Organize your findings in a table.

InfoSearch

Read the passage. Ask two questions that you cannot answer from the information in the passage.

Naming Viruses A virus often is named for the disease it causes. For example, the virus that causes polio is called polio virus. Code letters and numbers also are used instead of actual names. T2, T7, and P1 are names given to the viruses that attack bacteria found in the human intestines.

SEARCH: Use library references to find answers to your questions.

ACTIVITY

MODELING A BACTERIOPHAGE

You will need a machine screw, 1/4 inch by 1 and 1/2 inches, 2 acorn nuts that fit the screw, pliers, 3 pieces of floral wire, and a pipe cleaner.

1. Screw the acorn nuts onto the top of the screw.
2. Twist 3 pieces of floral wire around the opposite end of the screw. Use the pliers to bend the ends of the wires down.
3. Wrap the pipe cleaner around the middle of the screw.

Questions

1. **Identify:** What organism does the model represent?
2. What kinds of organisms can be infected by the organism represented by the model?
3. **Apply:** In which part of your model would the substance that enters living cells be contained?

How are plants classified?

Objective ▶ Identify the characteristics of plants.

TechTerms

▶ **bryophytes** (BRY-uh-fyts): plant division that includes mosses, liverworts, and hornworts

▶ **photosynthesis** (foht-uh-SIN-thuh-sis): food-making process in plants

▶ **tracheophytes** (TRAY-kee-uh-fyts): division of plants that have conducting tubes

Characteristics of Plants You probably recognize a plant when you see one. However, some organisms, such as algae, look like plants. How can you tell if an organism is a plant? Plants have many cells. In plants, the cells are organized into tissues and organs. The cells of plants also have cell walls and contain chlorophyll.

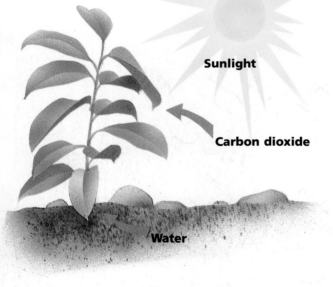

Sunlight

Carbon dioxide

Water

Plants use chlorophyll to trap the energy in sunlight. Plants use the sun's energy to combine carbon dioxide from the air and water from the soil to make food. The food-making process is called **photosynthesis** (foht-uh-SIN-thuh-sis). Photosynthesis provides plants with the food energy they need for growth and development.

▶ *State:* What are three plant characteristics?

Plant Divisions Botanists have classified more than 350,000 organisms in the plant kingdom. They have divided the plant kingdom into two large groups called divisions. The divisions are based on how water and dissolved nutrients (NOO-tree-unts) are moved throughout the plant. The two divisions are the bryophytes (BRY-uh-fyts) and the tracheophytes (TRAY-kee-uh-fyts).

▶ *Identify:* What are the two divisions of the plant kingdom?

Tracheophytes In **tracheophytes,** water and nutrients move through special tubelike cells to all parts of the plant. These tubelike cells make up the transport tissue of the plant. Transport tissue also helps support a plant. The presence of transport tissue allows some tracheophytes to grow to great heights.

Most of the plants you are familiar with probably are tracheophytes. Among tracheophytes, plant tissues are organized into roots, stems, and leaves. The roots of a tracheophyte anchor the plant in the soil. Roots also take in water and dissolved nutrients from the soil.

▶ *Describe:* What are two jobs of roots?

Bryophytes Unlike tracheophytes, **bryophytes** do not have true roots, stems, or leaves. They have rootlike and leaflike structures. Bryophytes also lack transport tissue. In bryophytes, water and nutrients seep from one cell to another. Bryophytes are small plants. They are usually only a few centimeters high.

Bryophytes are land plants. However, they need water to reproduce. For this reason, botanists believe that bryophytes once lived in water.

▶ *Infer:* Why do you think that bryophytes cannot grow to great heights?

LESSON SUMMARY

▶ Plant cells have a cell wall and chlorophyll, and are organized into tissues and organs.

▶ Photosynthesis is the food-making process in plants.

▶ The plant kingdom is divided into two divisions: bryophytes and tracheophytes.

▶ Tracheophytes have tubelike cells used to transport materials to all plant parts.

▶ Tracheophytes have roots, stems, and leaves.

▶ Bryophytes are small plants that lack transport tissues.

▶ Bryophytes need water to reproduce.

CHECK *Complete the following.*

1. The two plant divisions are the tracheophytes and the _____ .

2. The food-making process in plants is called _____ .

3. Tubelike cells make up the _____ tissue of tracheophytes.

4. Among tracheophytes, plant tissues are organized into roots, stems, and _____ .

5. Bryophytes need _____ to reproduce.

APPLY *Complete the following.*

6. **Contrast:** How do bryophytes differ from tracheophytes?

7. **Hypothesize:** Why do you think tracheophytes are successful land plants?

InfoSearch..

Read the passage. Ask two questions that you cannot answer from the information in the passage.

Pioneer Organisms Bryophytes are called pioneer organisms because they are often among the first plants to grow in bare, rocky places. As bryophytes grow, they help to break down rocks to form soil. When bryophytes die and decay, they add to the richness of the forming soil. As a soil layer builds up, other plants, such as trees and shrubs, may grow. In time, a forest may grow in an area that was once bare.

SEARCH: Use library references to find answers to your questions.

◆◆●● CAREER IN LIFE SCIENCE ●◆◆◆◆◆◆◆◆◆◆◆◆◆◆◆◆◆◆◆◆◆◆◆◆◆◆◆◆◆

FLORIST

Do you like growing plants at home? Do you enjoy gardening? If so, you may want to become a florist. A florist is a person who arranges and sells fresh flowers and plants. Flower arrangements are often in the form of corsages, bouquets, and table centerpieces.

Customers often ask a florist questions about caring for plants, the best flowers to buy, color schemes, and so on. A good florist is able to answer these questions and provide customers with helpful tips. A florist's success also depends upon ordering enough cut flowers and houseplants for times of heavy demand, such as Valentine's Day and Mother's Day. A successful florist always is well-stocked for special occasions.

If you are interested in becoming a florist, you need a high school diploma. Florists also need training in flower arranging, preserving cut flowers, and caring for houseplants. Courses in small business management also may be helpful. For more information, write to the Society of American Florists, 901 North Washington Street, Alexandria, VA 22314.

How are animals classified?

Objective ► Identify characteristics used to classify animals.

TechTerms

- ► **endoskeleton** (en-doh-SKEL-uh-tun): internal skeleton
- ► **exoskeleton** (ek-so-SKEL-uh-tun): external skeleton
- ► **invertebrates** (in-VUR-tuh-brayts): animals without backbones
- ► **vertebrates** (VUR-tuh-brayts): animals with backbones

Two Large Groups The animal kingdom is made up of more kinds of organisms than the other four kingdoms. Scientists classify animals into two large groups. One group is made up of animals with backbones. The other group is made up of animals without backbones.

▬▶ *Name:* What are the two groups into which the animal kingdom is divided?

Vertebrates Animals that have backbones are called **vertebrates** (VUR-tuh-brayts). Vertebrates live in water as well as on land. They are the most complex organisms in the animal kingdom. They are also the most widely recognized and familiar of all animals. Vertebrates include fishes, frogs, snakes, birds, cats, and many other animals. You are a vertebrate too.

Look at Figure 1. It shows a model of the largest animal on the earth. The largest animals on Earth are vertebrates. Vertebrates may grow very large because they have an **endoskeleton** (en-doh-SKEL-uh-tun), or a skeleton inside their bodies. The endoskeleton does not limit the growth and size of an animal. The endoskeleton covers and protects soft body parts. It also helps give shape and support to an organism.

▬▶ *Define:* What is an endoskeleton?

Invertebrates Animals without backbones are called **invertebrates** (in-VUR-tuh-brayts). Most animals are invertebrates. In fact, more than 97 percent of all animals are invertebrates. Sponges and jellyfish are invertebrates. Other invertebrates include worms, snails, clams, sea stars, and insects.

Some invertebrates, such as worms, do not have any skeleton at all. They are soft-bodied animals. Other invertebrates have an **exoskeleton** (ek-so-SKEL-uh-tun). An exoskeleton is a skeleton on the outside of the body. It is made up of a hard, waterproof substance. The exoskeleton protects and supports the body. Spiders, lobsters, and insects have an exoskeleton.

▬▶ *List:* What are three kinds of animals that are invertebrates?

Figure 1

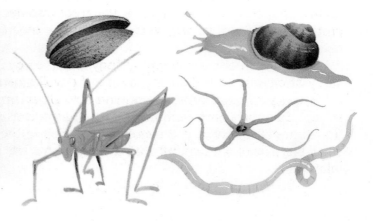

LESSON SUMMARY

▶ The animal kingdom is divided into two groups, animals with backbones and animals without backbones.

▶ Vertebrates are animals with backbones.

▶ Vertebrates have an endoskeleton.

▶ Invertebrates are animals without backbones.

▶ Some invertebrates are soft-bodied; others have an exoskeleton.

CHECK *Complete the following.*

1. Animals with backbones are classified as _____ .

2. An endoskeleton is a skeleton found _____ the body.

3. More than 97 percent of all animals are _____ .

4. Invertebrates are animals without _____ .

5. The largest animals on the earth are _____ .

APPLY *Complete the following.*

6. **Classify:** Write "I" if the organism has an endoskeleton and "E" if the organism has an exoskeleton.
   a. lobster   d. horse
   b. rattlesnake   e. grasshopper
   c. robin   f. frog

7. **Classify:** Classify each of the following organisms as a vertebrate or an invertebrate.
   a. worm   d. squirrel
   b. house fly   e. sea star
   c. robin   f. clam

Skill Builder.....................................

Building Vocabulary A prefix is a word part placed at the beginning of a word. The meaning of a prefix usually remains the same. Look up the meanings of the prefixes "endo-" and "exo-" in a dictionary. Write their meanings. List and define five words that begin with the prefix "endo-" and five words that begin with the prefix "exo-" in your own words. Circle the part of each definition that relates to the meaning of the prefix.

SCIENCE CONNECTION ◆○◆○◆○◆○◆○◆○◆○◆○◆○◆○◆○◆○◆○◆

TAXONOMY TODAY

The founder of modern taxonomy, Carolus Linnaeus, classified organisms according to their physical characteristics. In Linnaeus' system, organisms that looked alike were grouped together. Modern-day taxonomists still compare the physical appearances of organisms. Organisms that look the same often are related. However, classification is not based on physical traits alone.

Taxonomists also compare the chromosomes and blood proteins of different kinds of organisms. Scientists have discovered that the chromosomes and proteins in the blood of certain animals are quite similar. If the chromosomes and blood proteins are similar, taxonomists infer that the organisms are related.

Taxonomists also study embryology. Embryology is the study of organisms in the early stages of development. Embryology helps scientists determine how organisms are related. Scientists have found that certain organisms show similarities as embryos. If the embryos develop in the same way, taxonomists infer that the organisms are related.

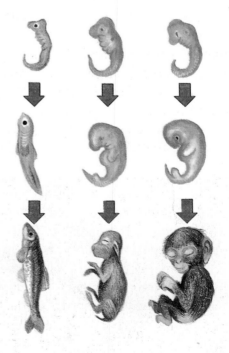

STUDY HINT Before you begin the Unit Challenges, review the TechTerms and Lesson Summary for each lesson in the unit.

TechTerms..

bacteriophage (106)
bryophytes (108)
capsid (106)
classification (100)
endoskeleton (110)
exoskeleton (110)
fungi (104)

genus (102)
invertebrates (110)
kingdom (102)
monerans (104)
photosynthesis (108)
phylum (102)
protists (104)

species (102)
taxonomy (100)
tracheophytes (108)
vertebrates (110)
virus (106)

TechTerm Challenges...

Matching *Write the TechTerm that best matches each description.*
1. plantlike organisms that lack chlorophyll
2. virus that infects bacteria
3. internal skeleton
4. single-celled organisms that have a nucleus
5. animals with backbones
6. food-making process in plants
7. animals without backbones
8. external skeleton
9. single-celled organisms that do not have a nucleus

Identifying Word Relationships *Explain how the words in each pair are related. Write your answers in complete sentences.*
1. genus, species
2. classification, taxonomy
3. capsid, virus
4. kingdom, phylum
5. bryophytes, tracheophytes

Content Challenges..

Multiple Choice *Write the letter of the term that best completes each statement.*

1. Tubelike cells make up the transport tissue of
   a. tracheophytes. **b.** bryophytes. **c.** algae. **d.** viruses.
2. Nuclear material is scattered throughout the cell in
   a. protists. **b.** fungi. **c.** monerans. **d.** yeasts.
3. Scientists first learned about viral reproduction by studying
   a. fungi. **b.** algae. **c.** bacteriophages. **d.** bacteria.
4. The largest classification group is the
   a. order. **b.** class. **c.** phylum. **d.** kingdom.
5. Plants trap the energy in sunlight using
   a. transport tissue. **b.** chlorophyll. **c.** carbon dioxide. **d.** the cell wall
6. Grouping things according to how they are alike is
   a. taxonomy. **b.** classification. **c.** photosynthesis. **d.** organization.
7. Two kinds of invertebrates are
   a. frogs and fishes. **b.** snakes and birds. **c.** birds and worms. **d.** worms and insects.

8. Today, most scientists accept a classification system with
   a. two kingdoms.   b. three kingdoms.   c. five kingdoms.   d. six kingdoms.
9. Organisms that look alike and can reproduce among themselves make up a
   a. species.   b. genus.   c. family.   d. kingdom.
10. The kingdom with the most kinds of organisms is the
    a. plant kingdom.   b. animal kingdom.   c. fungi kingdom.   d. protist kingdom.
11. Botanists have divided plants into two large groups called
    a. kingdoms.   b. phyla.   c. divisions.   d. tracheophytes.
12. The three main groups of viruses are plant viruses, animal viruses, and
    a. bacterial viruses.   b. protist viruses.   c. fungal viruses.   d. yeast viruses.
13. Bacteria are
    a. monerans.   b. protists.   c. fungi.   d. animals.
14. The two parts of a scientific name include the
    a. family and order names.   b. genus and species names.   c. kingdom and phylum names.
    d. class and order names.

True/False *Write true if the statement is true. If the statement is false, changes the underlined term to make the statement true.*
1. Some scientists would like to make a sixth kingdom for <u>fungi.</u>
2. <u>Bryophytes</u> need water to reproduce.
3. Vertebrates have an <u>endoskeleton.</u>
4. There are <u>ten</u> major classification groups for living things.
5. Dogs and wolves belong to the same <u>genus.</u>
6. Scientists who classify living things are <u>botanists.</u>
7. Algae are one example of <u>protists.</u>
8. The protein covering of a virus is called a <u>capsid.</u>
9. More than 97 percent of all animals are <u>vertebrates.</u>
10. About 1 1/2 <u>billion</u> different kinds of organisms have been discovered so far.

Understanding the Features...

Reading Critically *Use the feature reading selections to answer the following. Page numbers for the features are in parentheses.*
1. In what branches of biology did Robert Whittaker work? (105)
2. What do modern-day taxonomists study to classify organisms? (111)
3. **Identify:** Who is known as the founder of modern taxonomy? (103)
4. **Hypothesize:** Why might courses in small business management be helpful to a florist? (109)

Concept Challenges...

Critical Thinking *Answer each of the following in complete sentences.*
1. **Analyze:** If you were to discover a new species, how would you determine its classification?
2. Why is taxonomy a complex subject?
3. Do you think it is difficult to treat viral infections? Explain.
4. **List:** What four characteristics are used to classify an organism as a plant?
5. Fungi were once called the nongreen plants. Why do you think fungi were described in this way?

Interpreting a Diagram *Use the diagram to answer the following questions.*

1. What is shown in the diagram?
2. What is the structure labeled A called?
3. What is the structure labeled B called?
4. **Observe:** What is happening in step 2?
5. **Observe:** What is happening in step 5?

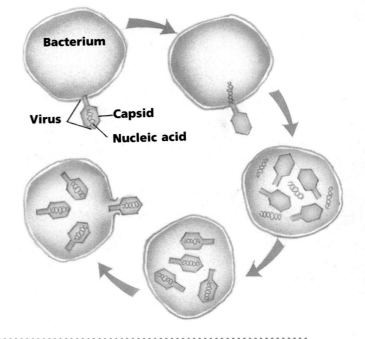

Finding Out More..

1. **Research:** Using library references, find out how the theory of evolution influenced the classification of living things. Write your findings in a report. Include the definition of the term phylogeny in your report.
2. Find and collect pictures of organisms from each of the five kingdoms. Attach the pictures to a posterboard. Label each picture with its proper kingdom.
3. **Model:** Using common materials, such as clay, styrofoam, and pipe cleaners, make models of a round virus and a bacteriophage. Label each part of the virus on your model.
4. **Experiment:** Design an experiment to show how liquid is transported in a tracheophyte. Fill a cup halfway with water and add a few drops of food coloring to the water. Cut off about 1 cm of the base of a celery stalk. Place the freshly cut base into the food-coloring solution. Observe the celery for the next 1/2 hour and after 24 hours. Write a laboratory report describing your observations.

SIMPLE ORGANISMS

CONTENTS

STUDY HINT As you read each lesson in Unit 6, write the lesson title and lesson objective on a sheet of paper. After you complete each lesson, write the sentence or sentences that answer each objective.

6-1 What are bacteria?

Objectives ▶ Describe a bacterium. ▶ Classify bacteria by shape.

TechTerms

- ▶ **bacilli** (buh-SIL-y): rod-shaped bacteria
- ▶ **cocci** (KAK-sy): spherical bacteria
- ▶ **endospore:** inactive cell surrounded by a thick wall
- ▶ **flagella** (fluh-JEL-uh): whiplike structures on a cell
- ▶ **spirilla** (spy-RIL-uh): spiral-shaped bacteria

Kingdom *Monera* Bacteria are simple, one-celled living things. They are microscopic. A bacterial cell is made up of cytoplasm, a cell membrane, and a cell wall. Bacteria do not have a nucleus. The material that makes up the nucleus of a bacterial cell is scattered in the cytoplasm of the bacterium. The Kingdom *Monera* includes all bacteria.

▶ *Compare:* How is a bacterium cell different from an animal cell?

Grouping Bacteria Bacteria are grouped according to their shapes. Some bacteria are round and look like tiny beads. Bacteria with a round, or spherical, shape are called **cocci** (KAK-sy). Some

Cocci

cocci often grow in pairs, chains, or large clusters that look like a bunch of grapes. Other types of bacteria are shaped like a spiral or corkscrew. These bacteria are called **spirilla** (spy-RIL-uh). The most common kind of bacteria are shaped like rods. These bacteria are called **bacilli** (buh-SIL-y). Bacilli often grow in pairs or in chains.

Bacilli

Spirilla

▶ *Identify:* What are the three shapes of bacteria?

Blue-green Bacteria Blue-green bacteria are unusual bacteria. They use sunlight to make their own food. Blue-green bacteria also contain the green pigment chlorophyll (KLAWR-uh-fil) in their cytoplasm. Chlorophyll is needed by blue-green bacteria and green plants to make food.

▶ *Explain:* Why can blue-green bacteria make food?

Movement in Bacteria Some kinds of bacteria can move on their own. Some bacilli and spirilla have **flagella** (fluh-JEL-uh). Flagella are whiplike structures than enable bacteria to move in liquids. Cocci cannot move on their own because they do not have flagella.

▶ *Describe:* How do some bacteria move?

Needs of Bacteria All bacteria need water and a proper temperature. Most live best in darkness. Some bacteria need oxygen. Others can live without oxygen. Some bacteria get their food by living inside plants or animals. Most bacteria feed on the remains of dead plants and animals. The bacteria break down the body of the dead plant or animal. This process is called decay.

Many bacteria can live through periods of extreme heat or cold. When their environments are not right, these bacteria form protective walls around themselves. A bacterium with a protective wall is called an **endospore.** When the environment is favorable, the endospore breaks open. The cell becomes active again.

▶ *Identify:* What is an endospore?

LESSON SUMMARY

▶ All bacteria are classified in the Kingdom *Monera.*

▶ Bacteria cells have cytoplasm, a cell membrane, and a cell wall. They do not have a nucleus.

▶ Bacteria are grouped according to their shapes.

▶ Blue-green bacteria contain chlorophyll and can make their own food.

▶ Some bacteria have flagella that help them move.

▶ An endospore is a bacterium covered by a protective outer wall.

CHECK *Write true if the statement is true. If the statement is false, change the underlined term to make the statement true.*

1. Bacteria are simple, <u>many-celled</u> organisms.
2. A bacterium cell does not contain a <u>nucleus</u>.
3. Bacteria with a round shape are called <u>bacilli</u>.
4. Blue-green bacteria have <u>chlorophyll</u> scattered through their cytoplasm.
5. The most common bacteria have a <u>spiral</u> shape.
6. Whiplike structures that help a bacterium move in liquids are called <u>flagella</u>.

7. A bacterium with a protective wall is called an <u>endospore</u>.

APPLY *Complete the following.*

▶ **8. Infer:** A kind of bacteria is discovered feeding on the remains of a dead plant. Are these monerans likely blue-green bacteria? Explain your answer.

9. Classify: Classify each of the bacteria shown.

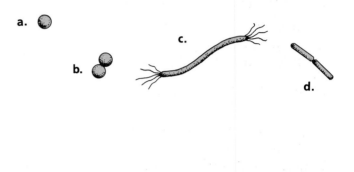

a.

b.

c.

d.

Building Vocabulary Look up the prefixes "diplo-", "staphylo-", and "strepto-" in the dictionary. Write the meaning of each prefix on a piece of paper. Then add each prefix to the root word "cocci" and define the new words. Describe a diplobacillus and a streptobacillus. Draw a model of these bacteria.

▼▼▼ LOOKING BACK IN SCIENCE ▼▼▼▼▼▼▼▼▼▼▼▼▼▼▼▼▼

CLASSIFICATION OF BACTERIA

At different times during scientific history, bacteria have been classified as plants and protists. Today, bacteria are classified in a separate kingdom. The kingdom is called *Monera.*

When organisms were classified as plants or animals, the bacteria and blue-green bacteria were classified as plants. In fact, blue-green bacteria were called blue-green algae. With the invention of the microscope, scientists discovered a world of microscopic, single-celled organisms. Scientists added the Protist kingdom to the classification system. Bacteria and blue-green bacteria were classified with the protists because they were simple, one-celled organisms.

With the invention of the electron microscope, scientists could see more detailed views of microscopic organisms. They discovered that the cells of bacteria and blue-green bacteria were basically different from those of protists, plants, and animals. So, in recent years, scientists classified bacteria and blue-green bacteria in a separate kingdom of their own.

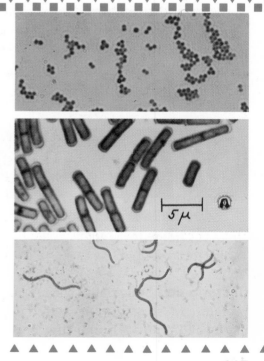

How are bacteria helpful and harmful?

Objective ▶ Explain how some bacteria are useful and others are harmful.

TechTerm

▶ **bacteriology** (bak-tir-ee-AHL-uh-jee): study of bacteria

Bacteriology In the mid-1800s the French scientist Louis Pasteur showed how important bacteria are. Pasteur proved that bacteria can cause many diseases. He started the new science of bacteriology (bak-tir-ee-AHL-uh-jee) or the study of bacteria. Pasteur is often called the "father of bacteriology".

▷ *Identify:* Who discovered the importance of bacteria?

Bacteria and Foods Many food products are made with the help of bacteria. For example, the action of certain bacteria gives flavor to butter. Other types of bacteria are used to make buttermilk and cheese. Bacteria also are used to make sauerkraut, coffee, and cocoa.

Bacteria play an important part in the digestion of the food you eat. Your digestive tract contains billions of bacteria that help digest proteins. Some digestive bacteria even help form important vitamins.

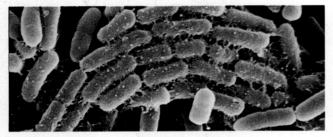

Figure 1 Digestive bacteria

▷ *List:* Name three foods made by using bacteria.

Bacteria and Soil Plants need nitrogen to make proteins. Most of the air is nitrogen. However, plants and animals cannot use nitrogen directly

Figure 2 Nodules of nitrogen-fixing bacteria

from the air. Plants can only use nitrogen that has been combined with other elements. Bacteria in soil and in the roots of some plants change nitrogen into compounds plants can use. These bacteria are called nitrogen-fixing bacteria. The "fixed" nitrogen is taken in and used by plants. Animals get the nitrogen they need by eating plants.

▷ *Explain:* Why do living things need nitrogen?

Plant Diseases Some types of bacteria can be harmful to plants. Blight is a plant disease caused by bacteria. Blight makes the flowers, young leaves, and stem of a plant die quickly. Rot is another plant disease caused by bacteria. Rot destroys the cell walls of plant tissues. Millions of dollars are lost each year from crop damage caused by these bacterial diseases.

▷ *Explain:* What effect does rot have on a plant?

Bacteria and Food Decay Bacteria can cause foods to spoil or decay. Some bacteria produce poisons while they are acting on food. It is important that food be protected from the action of bacteria. Cooking foods thoroughly at high temperatures kills harmful bacteria. Canning, pickling, and freezing food are other ways of preventing or slowing down the action of harmful bacteria.

▶ *Infer:* Why are many foods refrigerated?

LESSON SUMMARY

▶ Bacteriology is the study of bacteria.

▶ Bacteria are used to make many foods.

▶ Bacteria in your digestive tract help you digest foods.

▶ Nitrogen-fixing bacteria change nitrogen from the air into nitrogen compounds that can be used by plants to make proteins.

▶ Plant diseases, such as blight and rot, are caused by bacteria.

▶ Bacteria can cause foods to decay and spoil.

CHECK *Complete the following.*

1. The study of bacteria is called _____ .

2. Bacteria in the digestive tract can help digest _____ .

3. Bacteria that change nitrogen from the air into compounds that plants can use are called _____ .

4. The plant disease that causes flowers, young leaves, and stems to die quickly is called _____ .

5. Some bacteria produce _____ while acting on foods.

APPLY *Complete the following.*

▶ 6. **Infer:** Why do you think you should not buy canned foods if the can is bulging?

7. Foods that are canned are cooked at a high temperature and then placed in airtight containers. What effect do these actions have on the food?

8. **Apply:** Identify three ways that bacteria affect your life.

InfoSearch

Read the passage. Ask two questions that you cannot answer from the information in the passage.

Pasteurization Pasteurization is a process used to slow down the spoiling of milk and other dairy products. In this process, the milk is quickly heated to kill most of the bacteria. The milk is then quickly cooled. As long as the milk is kept cold, the few living bacteria in it grow very slowly. Pasteurized milk that is kept refrigerated is safe to drink for many days.

SEARCH: Use library references to find answers to your questions.

TECHNOLOGY AND SOCIETY

MAP: MODIFIED ATMOSPHERE PACKAGING

Fruits and vegetables spoil from airborne bacteria. As a result, keeping fruits and vegetables fresh has always been a problem. Chemical sprays have been used on produce to kill bacteria. However, these chemicals remain on the fruit. Some may be harmful to people.

A new method of keeping fresh foods from spoiling is based on controlling the food's environment. The method is called modified atmosphere packaging (MAP). The foodstuff is enclosed by special plastic coverings. This plastic covering only lets in certain gases in the air and lets out other gases. By reducing the amount of oxygen and other gases around the foodstuff, decay is slowed down. The types of packaging used before stopped gases from entering the package. However, they also kept gases from leaving the package. Gases released by the fruits and vegetables were trapped in the package. These gases sped up decay. M.A.P. is unique in that it permits such gases to leave the package. Scientist estimate that the technology of M.A.P. may double the shelf life of fresh produce.

Objective ▶ Identify some common protists and how they move.

TechTerms

- **cilia** (SIL-ee-uh): tiny, hairlike structures
- **protozoans** (proh-tuh-ZOH-uns): one-celled, animallike protists
- **pseudopod** (SOO-duh-pod): fingerlike projections of cytoplasm used for movement and food-getting

Kingdom *Protista* The protists are a kingdom of very simple organisms. Most protists have only one cell. Some are made up of many cells. All protist cells have a nucleus surrounded by a membrane.

The Protist kingdom is divided into three large groups. The three groups are the **protozoans** (proh-tuh-ZOH-uns), the algae, and the slime molds. The protozoans are one-celled organisms. They often are called the animallike protists because they cannot make their own food. Most protozoans also can move about on their own.

Some algae are one-celled. Others are multicellular (MUL-ti-sel-you-lur), or many-celled. All algae contain chlorophyll. Because they have chlorophyll, algae can make their own food. For this reason, algae are sometimes called plantlike protists.

📁 *Classify:* Seaweeds are protists that can make their own food. What kind of protist are they?

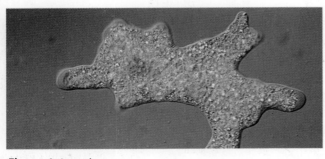

Figure 1 Amoeba

Amoeba The amoeba is a protozoan. Most amoeba live in fresh water. An amoeba uses its pseudopods (SOO-duh-pods) to move through wa-

ter. Pseudopods are fingerlike projections of cytoplasm. The pseudopods of an amoeba also are used to trap and take in food.

◀▶ *Define:* What are pseudopods?

Paramecium The paramecium is a slipper-shaped protozoan. It is part of a group of protozoans that use **cilia** (SIL-ee-uh) for movement. Cilia are tiny, hairlike structures around the edge of the organism. The cilia move back and forth like tiny oars. The beating of cilia moves a paramecium through water.

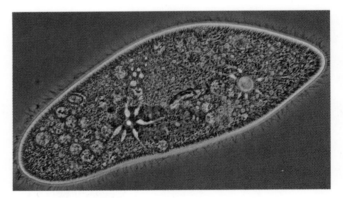

Figure 2 Paramecium

◀▶ *Identify:* What structure does a paramecium use to move?

Trypanosomes The trypanosome is a disease-causing protozoan. In humans, it causes the disease African sleeping sickness. Trypanosomes are part of a group of protozoans called the flagellates. They have flagella (fluh-JEL-uh). Flagella are whiplike structures used for movement. Some flagellates have one flagellum. Others have two or more.

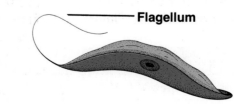

Flagellum

Figure 3 Trypanosomes

◀▶ *Name:* How do flagellates move?

LESSON SUMMARY

▶ The protists are a kingdom of very simple animals.

▶ Protozoans, algae, and slime molds are classified in the Protist kingdom.

▶ Algae are plantlike protists that can make their own food.

▶ Pseudopods are used by an amoeba to help it move and get food.

▶ A paramecium moves by the beating of cilia.

▶ Flagella are whiplike structures that push flagellates through water.

CHECK *Complete the following.*

1. Protozoa, algae, and slime molds all are classified in the _____ kingdom.

2. Amoeba move by using the _____ that extend from their cytoplasm.

3. A paramecium moves by beating the _____ that extend from its surface.

4. The whiplike structure that pushes some protozoans through water is called a _____ .

5. African sleeping sickness is caused by a _____ .

APPLY *Complete the following.*

6. **List:** What are the three groups of protists?

7. **Compare:** How are flagella and cilia alike?

InfoSearch

Read the passage. Ask two questions that cannot be answered from the information in the passage.

Slime Molds Slime molds are simple organisms that have two life stages. A slime mold changes as it goes through each of these stages. In its reproductive stage, a slime mold has features similar to a fungus. However, during its feeding stage, a slime mold acts like an amoeba. At one time, scientists classified slime molds as fungi. Today, most scientists classify slime molds with the protists.

SEARCH: Use library references to find answers to your questions.

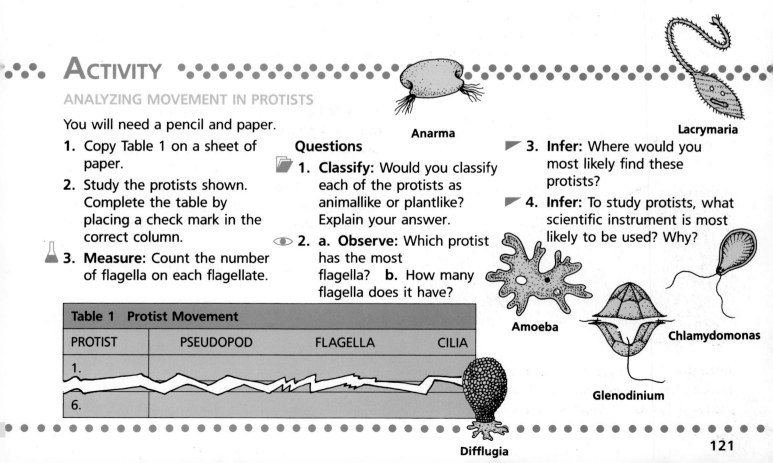

ACTIVITY

ANALYZING MOVEMENT IN PROTISTS

You will need a pencil and paper.

1. Copy Table 1 on a sheet of paper.

2. Study the protists shown. Complete the table by placing a check mark in the correct column.

3. **Measure:** Count the number of flagella on each flagellate.

Questions

1. **Classify:** Would you classify each of the protists as animallike or plantlike? Explain your answer.

2. **a. Observe:** Which protist has the most flagella? **b.** How many flagella does it have?

3. **Infer:** Where would you most likely find these protists?

4. **Infer:** To study protists, what scientific instrument is most likely to be used? Why?

Anarma

Lacrymaria

Amoeba

Chlamydomonas

Glenodinium

Difflugia

Table 1 Protist Movement			
PROTIST	PSEUDOPOD	FLAGELLA	CILIA
1.			
6.			

6-4 What are algae?

Objective ▶ Identify and describe different kinds of algae.

TechTerms

▶ **chlorophyll** (KLAWR-uh-fil): green pigment used by some organisms to make food

▶ **plankton** (PLANK-tun): organisms that float at the water's surface

Algae Algae are classified in the protist kingdom. All algae contain **chlorophyll** (KLAWR-uh-fil). Chlorophyll is a green pigment found in plant cells and algae. Plants and algae use chlorophyll to make food. The food-making process is called photosynthesis (foht-uh-SIN-thuh-sis).

▐▶ *Compare:* How are algae and plants alike?

One-Celled Algae Some kinds of algae are made up of only one cell. Most one-celled algae live in a watery environment. They float on the surface of the water. Algae are found in **plankton** (PLANK-tun). Plankton are tiny organisms that float on the surface of the oceans.

Although all algae have chlorophyll, all kinds of algae are not green. Fire algae are one-celled algae that are red in color. Golden-brown algae can have a color ranging from yellowish-green to golden-brown. These differences in color are due to other pigments, or colorings, in the algae.

▐▶ *Identify:* What is plankton?

Euglena The euglena is a unicellular, or one-celled, algae. It is both plantlike and animallike. Like all algae, the euglena contains chlorophyll. It uses the chlorophyll to make food from sunlight. When light is not available, the euglena also can take in food. A structure in the euglena called the eyespot can detect light. The euglena uses its eyespot to find light in the water. The euglena moves to the light by beating the flagellum that extends from its body.

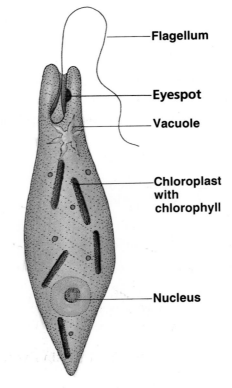

- Flagellum
- Eyespot
- Vacuole
- Chloroplast with chlorophyll
- Nucleus

▐▶ *Explain:* How is the euglena animallike?

Many-Celled Algae Some kinds of algae are made up of many cells. These are the multicellular algae. Pigments in multicellular algae may give them a green, red, or brown coloring. Most green algae live in fresh water.

Red and brown algae live in the ocean. Brown algae often are called seaweed or kelp. Brown algae are the largest and most complex kind of algae. If you have visited a beach, you may have seen brown algae washed up along the shoreline.

▶ *Infer:* Why are all types of algae not green?

LESSON SUMMARY

▶ All algae contain chlorophyll and make their own food by photosynthesis.

▶ Some kinds of algae are one-celled.

▶ All algae are not green, but still have chlorophyll.

▶ The euglena is a single-celled algae that is both plantlike and animallike.

▶ Some algae are multicellular.

▶ Red and brown algae are multicellular algae.

CHECK *Complete the following.*

1. Algae _____ make their own food.
2. Plants and algae both contain the green pigment _____ .
3. The euglena uses its _____ to detect light.
4. Fire algae are _____ in color.
5. The largest kind of algae is _____ .

APPLY *Complete the following.*

6. **Compare:** How is the Euglena similar to the trypanosome?

7. Suppose you needed to obtain a water sample that contained both algae and protozoans. Where would you go to obtain the sample?

Skill Builder..

▲ *Organizing* When you organize information, you put the information into some kind of order. A table is one way to organize information. Make a table with the following headings: "Type of Algae" and "Product." Use library references to determine the kinds of algae used to make the following products: toothpaste, marshmallows, ice cream, detergent, cheese, and silver polish.

◆○◆ SCIENCE CONNECTION ◆○◆○◆○◆○◆○◆○◆○◆○◆○◆○◆○◆○◆○◆○◆○◆○

ALGAL BLOOMS

Algae make their own food by photosynthesis. For photosynthesis to occur, three things must be present -- chlorophyll, sunlight, and carbon dioxide. However, algae also must have water and other nutrients, such as nitrogen and phosphorous, to carry out their life processes. When all of these elements are present, algae can reproduce quickly and in large numbers. A sudden, large growth of algae is called an algal bloom.

Algal blooms often occur when wastes containing phosphates are dumped into ponds or lakes. Phosphates are commonly found in detergents. When an algal bloom occurs, algae can be seen living at the surface of a lake or pond in great numbers. However, an algal bloom can be a problem for organisms that live in the lake or pond. As the algae die, they are broken down by bacteria. The bacteria use oxygen from the water when they break down the algae. If the bacteria use too much oxygen, fishes and other organisms living in the water do not have enough oxygen to carry out their life processes. As a result, the organisms may die.

6-5 What are fungi?

Objective ▶ Describe the different kinds of fungi.

TechTerms

▶ **cap:** umbrella-shaped top of a mushroom

▶ **gills:** produce mushroom spores

▶ **rhizoids** (RY-zoidz): rootlike structures anchor fungi

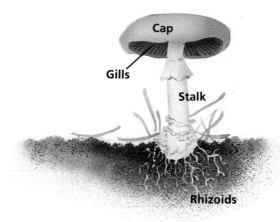

Classifying Fungi Fungi are like plants in some ways. The cells of fungi have cell walls. Many fungi also are multicellular, or many-celled, living things. Fungi often grow well in soil like most plants. For these reasons, scientists once classified fungi in the plant kingdom.

Scientists discovered that fungi and plants are not really very much alike. The cells of fungi do not have chloroplasts or chlorophyll. Therefore, fungi cannot make their own food as plants do. Fungi must get food from their surroundings. Fungi get the food they need from other living organisms or from dead organisms. Fungi often have very large cells with many nuclei. They grow well in dark, warm, and wet places. Today, scientists classify fungi in a kingdom of their own. The fungi kingdom includes yeasts, molds, and mushrooms.

▶ *Describe:* Where do fungi usually grow?

Yeasts Yeasts are colorless, one-celled fungi. Yeast cells are surrounded by a cell membrane and a cell wall. Within the cell is cytoplasm and a nucleus. Yeasts grow well where sugar is present. They use the sugar for food.

▶ *Describe:* What structures are found in a yeast cell?

Molds Molds are common kinds of fungi. They grow on bread, fruits, vegetables, and even leath-er. Most molds look like a mass of threads. If you look closely at a bread mold, you will see long threads growing along the surface of the bread. These threads are long, thin strands of cytoplasm. The threads are not divided into separate cells. Instead, each strand has many nuclei.

Some strands produce **spores.** Spores are reproductive cells of fungi. Short threads grow downward into the bread. These rootlike structures are called **rhizoids** (RY-zoidz). The rhizoids anchor the mold to the bread. The rhizoids also release chemicals that break down nutrients in the bread. The nutrients move up through the rhizoids and into the other parts of the mold.

▶ *Identify:* What are rhizoids?

Mushrooms You probably know a mushroom when you see one because of its shape. The stemlike part of a mushroom is called the **stalk.** At the top of the stalk is an umbrella-shaped top or **cap.** Mushrooms and molds do not look alike, but mushrooms also are made up of threads. In mushrooms, the threads are packed closely together. The underside of the cap is lined with **gills.** The gills produce spores. Spores are the reproductive structures of mushrooms. They are light in weight and easily carried by wind, water, or on the bodies of insects.

▶ *Infer:* What is the job of the stalk of a mushroom?

LESSON SUMMARY

▶ The fungi have cell walls, are usually many-celled, and grow in soil like plants.

▶ Fungi are classified in the kingdom Fungi because they cannot make their own food, have large cells with many nuclei, and grow well in dark, warm, wet places.

▶ Yeasts are one-celled fungi.

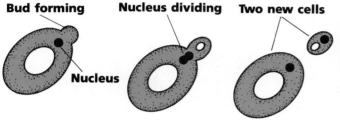

Bud forming   Nucleus dividing   Two new cells

Nucleus

▶ Molds are common fungi that are made up of many threads of cytoplasm.

▶ Mushrooms have a stalk, cap, and gills where spores are produced.

CHECK *Answer the following.*

1. What are three kinds of fungi?

2. Why can fungi not make their own food?

3. Which fungi is colorless and unicellular?

4. What structure anchors a mold to its food source?

5. What are the reproductive structures of a mushroom?

6. Where are spores produced in mushrooms?

APPLY *Complete the following.*

7. **Infer:** A mushroom may develop hundreds of kilometers away from the parent mushroom. How might this happen?

8. **Hypothesize:** Mushrooms form a huge number of spores. Why is the ground not covered with mushrooms?

9. **Infer:** Based on the color of a mushroom, how can you tell it does not have chlorophyll?

Health & Safety Tip..........................

Never eat mushrooms that you find growing outdoors. Poisonous mushrooms and edible mushrooms look very similar. Use library references to find out the names of three kinds of poisonous mushrooms. Then draw a picture of each type. Identify any unique trait for each poisonous mushroom you choose.

SCIENCE CONNECTION ◆○◆○◆○◆○◆○◆○◆○◆○◆○◆○◆○◆○◆○◆○◆○◆○◆○◆○

FUNGICIDES

Some fungi can cause infections and diseases in people, animals, and plants. Chemical preparations called fungicides are used to prevent or cure diseases caused by fungi. Athlete's foot is a common infection caused by fungi. It is corrected by using a fungicide cream or powder on the infected area. Some nail infections, and scalp diseases in young children are caused by fungi. A mouth infection called thrush is also a fungal infection. Fungal infections are very hard to get rid of. Usually it takes between ten to fourteen days to get ride of a fungal infection. Some infections can take as long as six months to cure. Fungal infections often are treated with medications that can be taken orally. Sulfa drugs and antibiotics are used to treat many fungal infections in people.

Avoiding Athlete's Foot
▶ Wear rubber thongs in public showers.
▶ Always change your socks or stockings each day.
▶ After bathing, be sure to dry the skin between your toes.
▶ Never share socks, shoes, or other footwear.

How do yeasts and molds reproduce?

Objective ▶ Compare and contrast reproduction of yeasts and molds.

TechTerms

▶ **asexual** (ay-SEK-shoo-wuhl) **reproduction:** reproduction needing only one parent

▶ **budding:** kind of asexual reproduction in which a new organism forms from a bud on a parent

▶ **spore cases:** structures that contains spores

▶ **sporulation** (spowr-yoo-LAY-shun): kind of asexual reproduction in which a new organism forms from spores released from a parent

Asexual Reproduction In many organisms, only one parent is needed for reproduction. Reproduction that needs only one parent is called **asexual** (ay-SEK-shoo-wuhl) **reproduction.** There are many kinds of asexual reproduction.

▶ *Define:* What is asexual reproduction?

Budding One kind of asexual reproduction is **budding.** In budding, a new cell is formed from a tiny bud on a parent cell. Yeast reproduce by budding. During budding, the cell wall of the parent cell pushes outwards. This is the beginning of the bud. The cell nucleus moves toward the bud. The nucleus divides. One nucleus moves into the bud. The other nucleus stays in the parent cell. The bud remains attached to the parent cell and grows larger. In time, a cell wall forms between

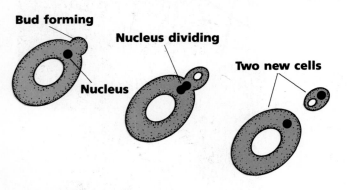

Figure 1 Budding in Yeast

the parent cell and bud. The bud breaks away from the parent cell. It develops into a mature yeast cell.

▶ *Classify:* By what kind of asexual reproduction do yeast cells reproduce?

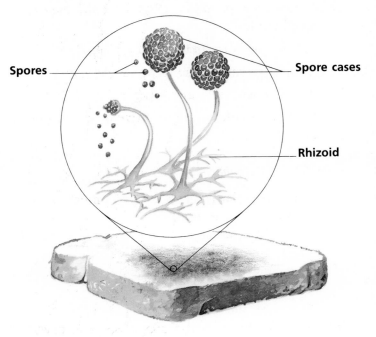

Figure 2 Sporulation in Bread Mold

Sporulation Have you ever seen mold growing on bread? It looks like tangled cotton. Many parts of it are black or gray. Under the low power of a microscope, bread mold looks like many tiny threads. On the top of some of the threads is a ball. This is the **spore case.** Each spore case holds thousands of spores. Spores are the reproductive cells of molds.

When a spore case breaks, thousands of microscopic spores are released. Each spore can grow into a new plant. This kind of reproduction is called **sporulation** (spowr-yoo-LAY-shun). Sporulation is another kind of asexual reproduction. It produces many more offspring than budding.

▶ *Classify:* What kind of reproduction is sporulation?

▶ Reproduction that needs only one parent is called asexual reproduction.

▶ Budding is a kind of asexual reproduction in which a new organism forms from a bud on a parent organism.

▶ Spore cases contain thousands of spores.

▶ Sporulation is a kind of asexual reproduction.

CHECK *Find the sentence in the lesson that answers each question. Then, write the sentence.*

1. How many parents are needed for asexual reproduction?

2. What fungus reproduces by budding?

3. How many spores are inside a spore case?

4. Where does a bud grow larger?

5. Does sporulation produce more offspring than budding?

APPLY *Complete the following.*

6. **Compare:** How is reproduction of mushrooms similar to that of mold?

7. **Hypothesize:** Why do you think molds are not growing all over everything around you?

InfoSearch

Read the passage. Ask two questions that you cannot answer from the information in the passage.

Alexander Fleming There are many different kinds of molds. Some are harmful. Some types cause disease. Others grow on foods and destroy them. However, there are also molds that are helpful. They are used to make medicines. In 1929, a British scientist named Alexander Fleming was studying bacteria. He discovered that a certain mold could destroy many different kinds of bacteria. This mold is called penicillium. Today penicillin is made from this mold. Penicillin is one of the most important medicines used today.

SEARCH: Use library references to find answers to your questions.

ACTIVITY

GROWING MOLD

You will need three plastic containers with covers, damp paper towels, a magnifying glass, a slice of bread made without preservatives, an orange rind, and a slice of potato.

1. Line the bottom of each container with damp paper towels. Place the bread in one container, the orange in another container, and the potato slice in the third container.

2. Store the containers in a dark place for a few days.

3. Use the magnifying glass to observe the changes in each food sample. Draw a diagram to show the structures you see.

Questions

1. On which food samples did mold grow?

2. **Observe:** Are the molds all the same color? Describe the color.

3. **Infer:** What do the differences in color tell you about the molds?

4. **Infer:** What conditions were needed to grow molds?

5. **Hypothesize:** How could the food samples be protected from mold growth?

UNIT 6 Challenges

STUDY HINT Before you begin the Unit Challenges, review the TechTerms and Lesson Summary for each lesson in this unit.

TechTerms

asexual reproduction (126)
bacilli (116)
bacteriology (118)
budding (126)
cap (124)
chlorophyll (122)

cilia (120)
cocci (116)
endospore (116)
flagella (116)
gills (124)
plankton (122)

protozoans (120)
pseudopod (120)
rhizoids (124)
spirilla (116)
spore cases (126)
sporulation (126)

TechTerm Challenges

Matching *Write the TechTerm that matches each description.*

1. spiral-shaped bacteria
2. one-celled animallike protists
3. rootlike structures that anchor fungi
4. inactive cell surrounded by a thick wall
5. fingerlike projections of cytoplasm used for movement and food-getting
6. spherical bacteria
7. organisms that float at the water's surface
8. material found in chloroplasts that is needed for plants to make food
9. asexual reproduction in which a new organism forms from a bud on a parent

Identifying Word Relationships *Explain how the words in each pair are related. Write your answers in complete sentences.*

1. spore cases, sporulation
2. bacteriology, bacilli
3. cilia, flagella
4. cap, gills
5. asexual reproduction, budding
6. plankton, chlorophyll
7. pseudopod, protozoan

Content Challenges

Multiple Choice *Write the letter of the term that best completes each statement.*

1. Fire algae are
   a. white.   **b.** red.   **c.** green.   **d.** brown.

2. The *Fungi* kingdom includes yeasts, molds, and
   a. slime molds.   **b.** algae.   **c.** mushrooms.   **d.** protists.

3. Blight is caused by
   a. bacteria.   **b.** fungi.   **c.** trypanosomes.   **d.** protists.

4. A paramecium moves by means of
   a. pseudopods.   **b.** flagella.   **c.** whiplike structures.   **d.** cilia.

5. Yeasts reproduce by
   a. budding.   **b.** sporulation.   **c.** photosynthesis.   **d.** pasteurization.

6. Bacteria with a round shape are called
   a. cocci.   b. spirilla.   c. bacilli.   d. flagella.

7. African sleeping sickness is caused by
   a. a paramecium.   b. a slime mold.   c. an amoeba.   d. a trypanosome.

8. The reproductive cells of fungi are called
   a. rhizoids.   b. gills.   c. spores.   d. threads.

9. The largest and most complex kind of algae are
   a. red algae.   b. brown algae.   c. green algae.   d. golden-brown algae.

10. Nitrogen is changed into compounds that plants can use by
    a. fungi.   b. bacteria.   c. algae.   d. protists.

11. Fungi grow best in places that are
    a. wet.   b. dry.   c. bright.   d. cold.

12. The eyespot of the *Euglena* can detect
    a. sound.   b. water.   c. heat.   d. light.

True/False *Write true if the statement is true. If the statement is false, change the underlined term to make the statement true.*

1. Most amoeba live in <u>salt</u> water.
2. The stemlike part of a mushroom is called the <u>cap</u>.
3. All algae contain <u>chlorophyll</u>.
4. Budding is one kind of <u>asexual</u> reproduction.
5. Brown algae often are called <u>seaweed</u>.
6. Bacteria are grouped according to their <u>colors</u>.
7. The study of bacteria was started by <u>Pasteur</u>.
8. Most protists are <u>many-celled</u> organisms.
9. A bread mold reproduces by <u>budding</u>.
10. The <u>paramecium</u> is a slipper-shaped protozoan.
11. The Kingdom *<u>Monera</u>* includes all bacteria.
12. Sporulation produces <u>less</u> offspring than budding.
13. The most common bacteria have a <u>spiral</u> shape.
14. Protozoans often are called <u>plantlike</u> protists.
15. Some bacteria can live without <u>oxygen</u>.

Understanding the Features...

Reading Critically *Use the feature reading selections to answer the following. Page numbers for the features follow each question in parentheses.*

1. **Identify:** Where do algal blooms often occur? (123)
2. **Define:** What is a fungicide? (125)
3. Why were bacteria and blue-green bacteria once classified with the protists? (117)
4. **Define:** What is modified atmosphere packaging? (119)

Concept Challenges...

Interpreting a Diagram *Use the diagram to answer the questions.*

1. What is the part labeled A called?
2. What are the parts labeled B called?
3. What is the function of the parts labeled B?
4. What is the part labeled C called?
5. What are the parts labeled D called?
6. What are the functions of the parts labeled D?

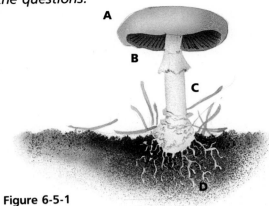

Figure 6-5-1

Critical Thinking *Answer each of the following in complete sentences.*

1. How do canning, pickling, and freezing foods prevent the growth of bacteria?
2. **Analyze:** Why do you think many-celled algae are not classified as plants?
3. **Compare:** How are fungi similar to plants? How are they different?
4. How do bacteria help animals in making proteins?
5. **Hypothesize:** If you wanted to find fungi in your area, where would you look?

Finding Out More...

1. Call the health department in your area and ask about careers in microbiology. Present your findings in an oral report.
2. Chemical preservatives are added to some foods to slow spoilage. Some preservatives are sodium or calcium proprionate, sorbic acid, and sulfur dioxide. Go to the grocery store and find two foods containing each preservative. In a brief report, describe the kinds of foods in which each preservative is used.
3. **Model:** A trypanosome has a complex life cycle that involves two hosts. Using library references, research the life cycle of a trypanosome. Draw a model illustrating the life cycle.
4. **Research:** Agar is a substance used to make foods such as pudding, ice cream, cheese, and marshmallows. Find out how red algae are used to make agar. Write your findings in a report.
5. Examine a mushroom with a hand lens. Identify the cap, the stalk, and the gills. Sketch a diagram of your mushroom.

PLANTS

CONTENTS

STUDY HINT Before beginning Unit 7, review the TechTerms for each lesson and read the Lesson Summary.

Why are plants important?

▶ Identify different plants used for food. ▶ Describe how plants are used in medicine and industry.

TechTerm

▶ **legumes** (LEG-yooms): group of plants that includes beans and peas

Plants for Food Early humans spent much of their time gathering plants to eat. About 10,000 years ago, people began to cultivate, or to plant and grow certain plants just for food. Evidence from prehistoric campsites shows that cereal grains, such as wheat and corn, were the first food crops. Later people began to plant **legumes** (LEG-yooms). Legumes are beans and peas. Fruit crops were cultivated next. The last food crops to be cultivated were roots. Today, about 100 different plants are grown as food crops.

👁 *Observe:* What are five plants cultivated for food?

Medicines from Plants Drugs used as medicines are made from many different plants. For instance, the leaves of the foxglove plant contain a substance that helps people who have had a heart attack. The drug helps the heart pump with more force. The bark of willow trees contains a chemical used to make aspirin. The bark of the cinchona tree is a source of quinine (KWI-nine). Quinine is used against malaria (muh-LER-ee-uh). Malaria is a disease caused by a single-celled organism that destroys red blood cells. When you go to the eye doctor, the doctor uses a drug that comes from the leaves and roots of the belladonna plant. When the drops of the drug are put in your eyes, the drug makes the pupil of your eye open up more. The eye doctor can examine your eyes easier.

▶ *Identify:* What are three plants that contain medicines?

Useful Plant Products Plants are used by people in many ways. Trees are used for building materials. Lumber and cork come from trees. Other products that come from trees are turpentine, waxes, and rubber. The oils used to make perfume come from flowers and other plant parts. Some plants provide us with fibers. The fibers can be spun into threads. The threads can be woven into cloth. For example, cotton is made from cotton plants. Linen is made from flax plants. Twine or string is made from linden plants.

The most useful plant product is cellulose (SEL-yoo-lohs). Cellulose is the hard, nonliving material in plant cell walls. It is used to make products you use or see every day. Cellulose is used to make paper. It also is used to make rayon cloth, paints, plastics, sponges, cellophane, camera film, and eyeglass frames.

▶ *List:* What are five common items obtained from plants?

LESSON SUMMARY

- Many different kinds of plants are grown for food, including cereal grains, legumes, and fruits and vegetables.
- Many different kinds of plants are the source of drugs and medicines.
- Plants are used for materials, to make perfume oils, and as fibers for cloth.
- Cellulose is a plant product that is used to make many different products.

CHECK *Complete the following.*

1. People began to cultivate food crops about _____ years ago.
2. The _____ plant contains a drug used to help people who have had a heart attack.
3. Beans and peas are examples of _____ .
4. Corn and wheat are examples of _____ .
5. Rayon and plastics can be made from the plant material called _____ .

APPLY *Complete the following.*

6. **Sequence:** List food plants in the order in which they were first cultivated.

7. **Infer:** What structure does the cotton plant have that makes it useful in making cloth?

Ideas in Action

IDEA: Many different plants are used for food.
ACTION: Work with a partner and list 15 different plants that are eaten by people. As a class, try to list more than 50 plants used for food.

PEOPLE IN SCIENCE

GEORGE WASHINGTON CARVER (1864–1943)

George Washington Carver was a botanist and plant chemist. He was born a slave, but as a young boy he worked hard to get an education. He graduated at the top of his college class.

Carver was invited by Booker T. Washington to join the staff of Tuskegee Institute. There he analyzed soil from local farms. He found that the main crops, cotton and tobacco, had taken most of the minerals from the soil. This poor soil kept the farmers from growing healthy, profitable crops. Carver experimented and found that growing legume crops, such as peanuts and soybeans, in rotation with cotton helped revive the soil.

The farmers soon had more peanuts than they could sell. To solve the problem of all those extra peanuts, Carver developed more than 300 uses for peanuts. These uses included cheese substitutes, paper, face cream, soap, ink, and dyes.

What are spore plants?

Objective ▶ Identify and describe the spore plants.

TechTerms

- ▶ **bryophyte** (BRY-uh-fyt): plant that reproduces by spores and has no transport tubes
- ▶ **rhizoid** (RY-zoid): fine hairlike structure that acts as a root
- ▶ **spore**: reproductive cell of bryophytes

Bryophytes Have you ever seen green, velvetlike plants growing on rocks and tree trunks? These plants are mosses. Mosses are classified in the phylum *Bryophyta* (BRY-uh-fuh-tuh). Figure 1 shows two other kinds of plants classified as **bryophytes** (BRY-uh-fyts). They are liverworts and hornworts. The bryophytes are thought to have been the first plants to live on land. Even though they live on land, bryophytes can live only where there is a good supply of water. They need water for reproduction.

Figure 1

The bryophytes are spore plants. Spore plants produce spores instead of seeds. A **spore** is a reproductive cell.

👁 *Observe:* What do the leaflike structures of the liverwort look like?

Structure of Bryophytes Bryophytes are small plants. They are usually only a few centimeters tall. Bryophytes differ from most other plants in a

Figure 2 Moss plants

few ways. Unlike most plants, bryophytes do not have true roots, stems, or leaves. Instead, they have fine hairlike structures called **rhizoids** (RY-zoids). Like roots, rhizoids grow down into the soil and anchor the plant. They also take in water and dissolved minerals.

Most plants have a system of tubes that carry water and dissolved materials throughout the plant. Bryophytes do not have these tubes. Instead, the individual cells get what they need as water and dissolved materials seep from one cell to another. The absence of transport tubes limits the size of these plants.

▍▶ *Identify:* What are two characteristics of bryophytes?

Mosses If you look closely at a moss plant you will see thin stalks with tiny leaflike parts. The leaflike parts make food for the plants. Rhizoids anchor the mosses. You also may see taller stalks. At the top of these stalks, there are spore cases. The spore cases are filled with spores.

▍▶ *Name:* What structure of a moss holds the spores?

Pioneer Plants Bryophytes are called pioneer plants. Pioneer plants are the first plants to grow in bare or rocky places. As they grow, their rhizoids help break down rocks to form soil. As the plants die and decay, they add richness to the forming soil. As a soil layer builds up, other plants, such as shrubs and trees, may grow. In time, a forest may grow.

▍▶ *Define:* What are pioneer plants?

LESSON SUMMARY

► The phylum *Bryophyta* includes mosses, liverworts, and hornworts.

► Bryophytes are spore plants.

► Bryophytes are small plants that do not have true roots, stems, or leaves.

► Bryophytes do not have tubes to carry water and dissolved materials throughout the plant.

► Mosses have thin stalks with tiny leaflike parts.

► Pioneer plants are the first plants to grow in bare or rocky places.

CHECK *Complete the following.*

1. What kind of plants are bryophytes?

2. What are the reproductive structures of bryophytes called?

3. Name three kinds of bryophytes.

4. What are the rootlike structures of bryophytes called?

5. In what structure are the spores of mosses produced?

APPLY *Complete the following.*

6. **Infer:** In what ways are bryophytes less suited to life on land than other plants?

7. **Compare:** In what ways are rhizoids like roots?

Skill Builder

Building Vocabulary "Bryophyte" is made up of the word parts "bryon" and "phyton." Look up the word parts in the dictionary. Write the meanings of these word parts. Find two other life science words that contain the word part "phyton." Define the two words. To what kind of living things do the word parts relate?

LOOKING BACK IN SCIENCE

SCOURING RUSHES

Horsetails are spore plants that have tubes for transporting materials. They are an ancient plant. Fossil horsetails have been found in rocks nearly 400 million years old. Tree-size horsetails once formed huge forests on the earth. Modern horsetails are smaller than the ancient horsetails. Modern horsetails grow in woodlands and swamps. They have hollow, jointed stems with small leaves around each joint. The stems and leaves look like the tail of a horse. Spores are located in spore cases at the top of some of the stems. The outer layer of the stem is made up of a single layer of cells. The cell walls contain silica (SIL-i-kuh). Silica is a compound in sand. Silica gives horsetails a rough, stiff and gritty texture. American colonists called horsetails "scouring rushes." They used horsetails to clean and scour pots and pans in the days before steel-wool pads and nonstick coatings.

What are ferns?

Fronds

Rhizome

Roots

Objective ▶ Name and describe the structure of a fern.

TechTerms

▶ **frond:** featherlike leaf of a fern
▶ **rhizome** (RY-zohm): underground stem

Ferns Ferns are common spore plants. They grow in great numbers in warm, moist areas, but they also live in cooler, drier areas. Most ferns can be recognized by their featherlike leaves. These leaves are called **fronds.** Each frond is made up of blades. The fronds are the part of the fern that can be seen above ground.

Ferns vary greatly in size and shape. Most ferns are about 1 m tall. Some ferns are only 3 cm tall. Some ferns, however, are giant fern trees of the South Pacific. They grow up to 15 m tall.

▶ *Identify:* What is the most recognizable part of a fern?

Prehistoric Ferns About 300 million years ago, giant tree ferns were the most common kind of land plant. They formed huge forests. The remains of these and other plants eventually became buried under thick layers of earth. The pressure of the overlying layers, along with chemical changes, gradually changed this material to coal. Most of the giant tree ferns died out about 225 million years ago.

▶ *Describe:* In what form did prehistoric ferns grow?

Structure of a Fern The structure of ferns differs from that of bryophytes. First, most modern ferns are about a meter or so tall. Second, ferns have true roots, stems, and leaves. Fern stems and roots grow underground. The underground stems are called **rhizomes.** They grow parallel to the surface. Roots grow downward from the rhizome, while the fronds grow upward.

Ferns are also similar to bryophytes. They reproduce by spores. If you look underneath a fern frond, you will see tiny brown spots. These are spore cases. Each case is filled with hundreds of spores.

▶ *Name:* What is the underground stem of a fern called?

LESSON SUMMARY

► Ferns are common plants with fronds.

► Ferns vary greatly in size and shape.

► Prehistoric ferns were giant trees that gradually were changed to coal.

► Ferns have true roots, stems, and leaves, and a transporting system of tubes.

► Ferns are spore plants.

CHECK *Complete the following.*

1. The featherlike leaves of ferns are the _____ .

2. The rhizome of a fern is an underground _____ .

3. Ferns have true stems, _____, and leaves.

4. In ferns, spore cases form on the underside of the _____ .

APPLY *Complete the following.*

▲ 5. **Model:** Draw a picture of a fern. Label the frond, rhizome, and blades.

◤ 6. **Infer:** What part of a fern makes food for the fern? How do you know?

InfoSearch..

Read the passage. Ask two questions about the topic that you cannot answer from the information in the passage.

Forming Coal Coal is a fossil fuel. It was formed from the remains of giant fern trees and other plants. The plants lived in swamps. When the plants died, they were covered by mud, water, and other sediments. Heat and pressure slowly caused chemical changes to take place. Bacteria also caused chemical changes. After millions of years, most of the compounds in the plants changed so that only carbon was left. The carbon formed coal.

SEARCH: Use library references to find answers to your questions.

Skill Builder..

Researching There are four kinds of coal. They are peat, lignite, bituminous coal, and anthracite. Find out about these types of coal. Which kind of coal is called soft coal and brown coal? Which kind of coal is the hardest? Write a report about your findings.

LEISURE ACTIVITY

INDOOR GARDENING

Indoor gardening is a popular activity. Most indoor gardeners like plants that look nice and are easy to take care of. For these reasons, many people grow ferns as house plants. The delicate, feathery fronds of ferns are very attractive. Ferns also grow well in shade. In fact, many kinds of ferns grow well in complete shade. They are perfect for indoor gardening in apartments.

All ferns need a very light, peat-filled soil in order to grow well. The soil should be kept damp at all times. Ferns also like moist air. You should mist ferns with a fine spray of water. It helps them stay healthy. A little liquid plant food during the growing season is useful. Most ferns like a cool place. They should not be put in direct sunlight. A north-facing window is ideal for many ferns.

What are gymnosperms?

Objective ▶ Name and describe three kinds of gymnosperms.

TechTerms

- ▶ **conifer** (KAHN-uh-fur): tree that produces cones and has needlelike leaves
- ▶ **gymnosperms** (JIM-nuh-spurms): group of land plants with uncovered seeds
- ▶ **seed:** structure that contains a tiny living plant and food for its growth

Gymnosperms Millions of years after the appearance of ferns, a group of woody land plants arose. These plants were the **gymnosperms** (JIM-nuh-spurms). The gymnosperms are a large group of land plants that have uncovered seeds. A **seed** is a reproductive structure. The seeds of most plants have a tough outer covering for protection. The seeds of gymnosperms do not have a protective covering. Gymnosperms also have stiff, woody stems. They have a system of tubes for carrying water and dissolved materials. They also have true roots, stems, and leaves.

▶ *Define:* What is a gymnosperm?

Conifers The most common and best known of the gymnosperms are the evergreens, or **conifers** (KAHN-uh-furs). Conifers are plants that produce cones. The seeds are in the cones. Conifers have special leaves called needles. The needles stay green throughout the year. Pines, cedars, spruces, and hemlock are conifers.

Conifers are grown in all climates and soil types. Many conifers grow in the forests of northern Europe and North America. The giant redwood trees, or sequoias (si-KWOI-uhs), of California also are conifers. Some of these trees are more than 100 m tall and 9 m in diameter. Other gymnosperms are palmlike trees. They grow in tropical areas.

▶ *Infer:* Why are conifers also called evergreens?

Importance of Conifers The conifers are an important group of plants. They are widely used as a source of lumber and fuel. They also are used in the production of paper. Turpentine, charcoal, tar, and alcohol are made from gymnosperms. Spruces and firs are planted along the edges of farms. The wall of trees slows down the wind, keeping rich soil from blowing away.

▶ *List:* What are three uses of gymnosperms?

Cycads and Ginkgoes Two other kinds of gymnosperms are cycads (SY-kads) and ginkgoes (GIN-kohs). Cycads grow mainly in tropical areas. Most look like small palm trees. They have large, feathery, fernlike leaves. Female trees have a very large conelike structure in the center. The ginkgoes have fan-shaped leaves. Unlike the evergreens, ginkgoes shed their leaves in the fall. Ginkgoes often are grown along city streets because they grow well even in polluted air.

▶ *Name:* What are two other groups of gymnosperms?

LESSON SUMMARY

▶ Gymnosperms are woody land plants that have true roots, stems, and leaves, transporting tubes, and uncovered seeds.

▶ Conifers are a group of gymnosperms that produce seeds in cones and have needles.

▶ Conifers are used as a source for many different products such as lumber and fuel.

▶ Cycads and ginkgoes are two other groups of gymnosperms.

CHECK *Write true if the statement is true. If the statement is false, change the underlined term to make the statement true.*

1. The most common of the gymnosperms are the <u>ferns</u>.

2. An example of a conifer is a <u>pine</u> tree.

3. Conifers produce seeds in <u>cones</u>.

4. The gymnosperms include <u>evergreens</u>.

5. Turpentine, lumber, and tar are made from <u>cycads</u>.

6. <u>Ginkgoes</u> shed their leaves in autumn.

APPLY *Complete the following.*

▲ 7. **Model:** A giant redwood tree has a diameter of 9 m. Use the scale 1 mm equals 50 cm. Draw a circle to represent the 9-m diameter of the redwood tree's trunk.

👁 8. **Observe:** Which of the following shows a gymnosperm? How can you tell?

Ideas in Action..

IDEA: Wood from pine and cedar trees often is used to make furniture, shelving, and doors.
ACTION: Identify objects in your home that are made from wood from these conifers. Organize your information in a table.

ACTIVITY

OBSERVING PINE CONES

You will need several different pine cones, a field guide to trees, and a metric ruler.

1. Collect a pine cone from three different conifers. Take a few cones that have already fallen from each kind of conifer.

2. Use the field guide to help you identify which kind of tree the cones come from. Label the cones.

🧪 3. **Measure:** Look at the cones carefully. Each scale of the cone had a seed. Count the number of seeds that one of the cones could form.

🧪 4. **Measure:** Measure the length of each cone. Organize the data in a table.

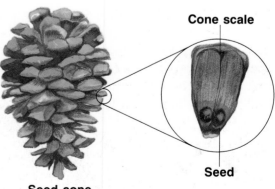

Cone scale

Seed

Seed cone

Questions

1. Are all the cones from the same tree identical?

2. **Compare and Contrast:** How are the cones from the same tree alike? How are they different?

3. How many seeds could form in the cone you chose?

7-5 What are angiosperms?

Objective ▶ Identify monocots and dicots as two kinds of angiosperms.

TechTerms

- **angiosperms** (AN-jee-uh-spurms): flowering plants
- **cotyledon** (kaht-LEED-on): seed structure that contains food for the developing plant
- **dicots:** seed plants with two cotyledons
- **monocots:** seed plants with one cotyledon

Angiosperms The **angiosperms** (AN-jee-uh-spurms) are the flowering plants. Scientists estimate that the angiosperm first appeared about 250 million years ago. They gradually replaced the gymnosperms as the major land plants. Today, scientists have classified about 250,000 kinds of angiosperms. They are the largest plant group.

▶ *Identify:* What is another name for angiosperms?

Characteristics of Angiosperms Most of the common plants you see every day are angiosperms. Like gymnosperms, angiosperms have true roots, stems, and leaves. They also have a highly-developed system of transport tubes.

All angiosperms have flowers. No other plant group has flowers. In some angiosperms, the flowers are not very noticeable. You probably have never seen the flowers of grasses, oak trees or corn. They have very small flowers. Other angiosperms have large, colorful flowers.

▶ *Name:* What is the main characteristic of angiosperms?

Seeds and Fruits Like gymnosperms, angiosperms are seed plants. The flowers of angiosperms produce seeds and fruits. The seeds of angiosperms are inside the fruit. Thus, the seeds of angiosperms are covered and protected. The next time you eat a piece of fruit look for the seeds. You will see one seed or many seeds. Most angiosperms produce a lot of seeds.

▶ *Name:* What do the flowers of angiosperms produce?

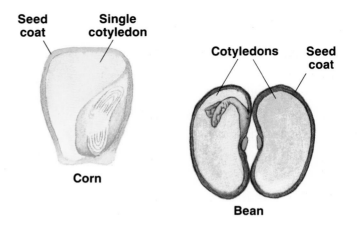

Corn — Seed coat, Single cotyledon

Bean — Cotyledons, Seed coat

Monocots and Dicots Angiosperms are classified into two groups based on their seed structure. Within the seeds of angiosperms are structures that contain food for the developing plant. These structures are called **cotyledons** (kaht-LEED-ons). The seeds of one group contain a single cotyledon. Plants of this group are **monocots.** In plants of the second group, the seeds have two cotyledons. These plants are **dicots.**

▶ *Identify:* What is a cotyledon?

Identifying Monocots and Dicots You can tell if a plant is a monocot or a dicot by looking at its flowers and leaves. Monocots have flowers with petals arranged in groups of three. Petals are the colorful parts of flowers. The veins in the leaves of monocots are parallel. Dicots have flowers with petals arranged in groups of four or five. The veins in the leaves of dicots are branched.

▶ *Contrast:* In what way are monocot flowers and dicot flowers different?

LESSON SUMMARY

▶ Angiosperms are the flowering plants.

▶ Angiosperms have true leaves, roots, and stems.

▶ The flowers of angiosperms produce seeds inside fruits.

▶ Angiosperms are classified as monocots and dicots based upon the number of cotyledons in their seeds.

▶ Monocots and dicots can be identified by examining their leaf vein pattern and the number of petals that make up the flowers.

CHECK *Complete the following.*

1. Monocots have leaves with _____ veins.

2. The seeds of angiosperms are enclosed in _____ .

3. Angiosperms have a transport system made up of _____ .

4. Fruits are produced by the plant parts called _____ .

5. The number of cotyledons in a dicot seed is _____ .

6. Monocot flowers have petals in groups of _____ .

APPLY *Complete the following.*

▲ 7. **Model:** Draw a model of an imaginary monocot and an imaginary dicot. Label the number of petals and the pattern of leaf veins.

Skill Builder

Using Prefixes Use a dictionary to find out the meanings of the prefixes "mono-" and "di-." Monocots also are called monocotyledons. Dicots are called dicotyledons. What do the prefixes "mono-" and "di-" tell you about the difference between monocots and dicots? Use the meaning of the prefix "mono-" to define the words *monorail* and *monosyllable*. Check your definitions with a dictionary.

ACTIVITY

CLASSIFYING MONOCOTS AND DICOTS

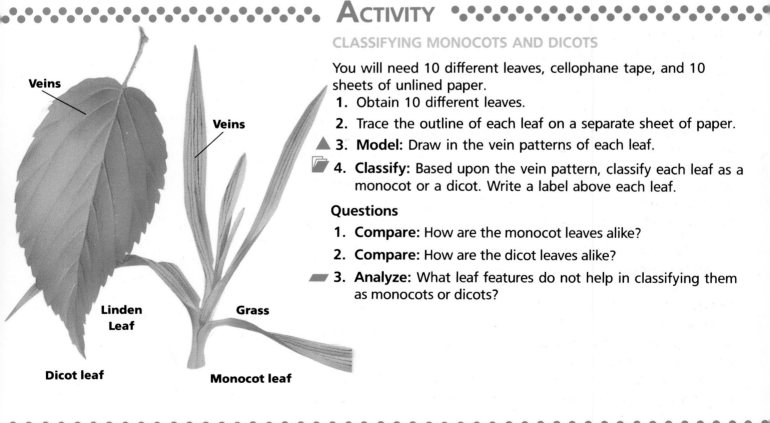

Veins

Veins

Linden Leaf

Grass

Dicot leaf

Monocot leaf

You will need 10 different leaves, cellophane tape, and 10 sheets of unlined paper.

1. Obtain 10 different leaves.

2. Trace the outline of each leaf on a separate sheet of paper.

▲ 3. **Model:** Draw in the vein patterns of each leaf.

4. **Classify:** Based upon the vein pattern, classify each leaf as a monocot or a dicot. Write a label above each leaf.

Questions

1. **Compare:** How are the monocot leaves alike?

2. **Compare:** How are the dicot leaves alike?

3. **Analyze:** What leaf features do not help in classifying them as monocots or dicots?

Challenges

STUDY HINT Before you begin the Unit Challenges, review the TechTerms and Lesson Summary for each lesson in this unit.

TechTerms..

angiosperms (140)
bryophyte (134)
conifer (138)
cotyledon (140)
dicots (140)

frond (136)
gymnosperms (138)
legumes (132)
monocots (140)
rhizoid (134)

rhizome (136)
seed (138)
spore (134)

TechTerm Challenges..

Matching *Write the TechTerm that best matches each description.*

1. fine hairlike structure that acts as a root
2. tree with cones and needlelike leaves
3. flowering plants
4. group of plants that includes beans and peas
5. reproductive cell of bryophytes
6. seed plants with two cotyledons
7. plant that reproduces by spores and has no transport tubes

Identifying Word Relationships *Explain how the words in each pair are related. Write your answers in complete sentences.*

1. seed, angiosperm
2. bryophyte, spore
3. conifers, gymnosperms
4. stem, rhizome
5. cotyledons, monocots
6. root, rhizoid
7. frond, leaf

Content Challenges..

Multiple Choice *Write the letter of the term or phrase that best completes each statement.*

1. The most useful plant product is
   a. rubber.   **b.** cellulose.   **c.** wax.   **d.** quinine.

2. The last food crops to be cultivated were
   a. fruit crops.   **b.** cereal grains.   **c.** roots.   **d.** legumes.

3. Plants that do not have true roots, stems, or leaves are classified as
   a. bryophytes.   **b.** gymnosperms.   **c.** ferns.   **d.** angiosperms.

4. Bryophytes can live only where there is a good supply of
   a. minerals.   **b.** soil.   **c.** water.   **d.** unpolluted air.

5. The underground stem of a fern is a
   a. root.   **b.** rhizome.   **c.** frond.   **d.** rhizoid.

6. The leaves of conifers are called
   a. needles.   **b.** fronds.   **c.** cones.   **d.** blades.

7. Plant structures common to all angiosperms are
   a. spore cases.   **b.** flowers.   **c.** rhizomes.   **d.** monocots.

8. The number of cotyledons in a dicot is
   a. one.   **b.** two.   **c.** three.   **d.** four.

9. The leaves of a foxglove plant contain a substance used to treat people with
   a. heart problems   b. kidney problems   c. cancer   d. asthma

10. Linen is made from
    a. cotton plants   b. belladonna plants   c. flax plants   d. cinchona trees

11. Pioneer plants are
    a. angiosperms   b. gymnosperms   c. bryophytes   d. tracheophytes

12. Pines, cedars, spruces, and hemlock are
    a. cycads   b. gingkos   c. angiosperms   d. conifers

13. The largest plant group is made up of
    a. gymnosperms   b. angiosperms   c. bryophytes   d. rhizoids

True/False *Write true if the statement is true. If the statement is false, change the underlined term to make the statement true.*

1. Plants are used to make many useful products such as <u>turpentine</u>, cork, and lumber.
2. Legumes are beans and <u>wheat</u>.
3. The bark of the <u>cinchona</u> tree is a source of a chemical used to make aspirin.
4. Three plants classified as bryophytes are mosses, <u>liverworts</u>, and hornworts.
5. Spore plants produce <u>flowers</u> instead of seeds.
6. Cedars and cycads are classified as <u>gymnosperms</u>.
7. Conifers are plants that produce <u>cones</u>.
8. Conifers also are called <u>evergreens</u>.
9. Flowering plants are classified as <u>gymnosperms</u>.
10. Both gymnosperms and bryophytes produce seeds.

Understanding the Selections

Reading Critically *Use the feature reading selections to answer the following. Page number for the features are in parentheses.*

1. **List:** What are some of the products that George Washington Carver made from peanuts? (133)
2. **Explain:** How did George Washington Carver revive soil that had been depleted through cotton and tobacco planting? (133)
3. **Identify:** What compound gives horsetails their rough gritty texture? (135)
4. What are two things the owner of a fern can do to keep the plant healthy? (137)
5. **Infer:** Why might direct sunlight harm a fern? (137)

Concept Challenges

Critical Thinking *Answer each of the following in complete sentences.*

1. **Contrast:** Explain two ways bryophytes differ from other groups of plants.
2. **Compare:** Identify one way in which ferns are similar to bryophytes.
3. **Identify:** What is the reproductive structure of a gymnosperm?
4. **Contrast:** How does a monocot differ from a dicot?
5. **Predict:** How would your life be different if there were no plants?

Interpreting a Diagram *Answer the questions about the diagram.*

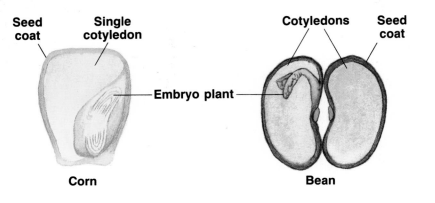

Corn

Bean

1. **Identify:** Both corn and beans are members of a group of plants called angiosperms. What features of the seeds from these plants identify them as angiosperms?

2. Which seed is a monocot? How do you know?

3. Which seed is a dicot? How do you know?

4. **Describe:** Where is the cotyledon located?

5. **Observe:** What structure encases and protects the cotyledon in both monocots and dicots?

Finding Out More..

1. **Communicate:** Some careers open to plant lovers include working as a florist, a gardener, or a landscape architect. Use library references to find out about the duties and job requirements of one of these careers. Present your findings to the class in an oral report. Be sure to include information about the skills and educational requirements for job as well as information about what the job involves.

2. Make a terrarium in a large bottle using mosses, ferns, and other spore plants. Be sure that your terrarium has the proper growth conditions for each kind of plant. Check your terrari-

um daily to note any changes in the appearance of your plants. Keep a written record of these changes. Be sure to look for changes in color, size, and growth rate.

3. **Organize:** Make a list of twenty different kinds of plants that are common in your community. Use a field guide to identify each plant. In a table, classify each of the plants you identified as a gymnosperm or an angiosperm.

4. **Classify:** Cut pictures of ten different kinds of flowers from magazines. Make a poster that classifies each kind of flower as a dicot or a monocot.

PLANT STRUCTURE AND FUNCTION

CONTENTS

STUDY HINT After you read each lesson in Unit 8, write a brief summary on a sheet of paper explaining how the information in each lesson applies to your everyday life.

Objectives ► Describe the structure of a root. ► Explain the jobs of roots.

TechTerms

- **fibrous** (FY-brus) **root system:** root system made up of many thin, branched roots
- **root cap:** cup-shaped mass of cells that covers and protects a root tip
- **root hair:** thin, hairlike structure on the outer layer of the root tip
- **taproot system:** root system made up of one large root and many small, thin roots

Kinds of Roots The two main kinds of root systems are **fibrous** (FY-brus) **root systems** and **taproot systems.** Fibrous root systems are made up of many thin, branched roots. Grass, wheat, and barley have fibrous roots. A taproot system has one large root. Many small, thin roots grow out from the large root. Some taproots can store food. Carrots, radishes, and dandelions have taproots.

||||▶ *Name:* What are two kinds of root systems?

Figure 1 Root systems

Parts of a Root Roots are tubelike structures made up of three layers. The outer part of the root is made up of one layer of cells. This layer mainly absorbs water. Many tiny, hairlike structures called **root hairs** extend from the outer layer. Root hairs increase the surface area of the root,

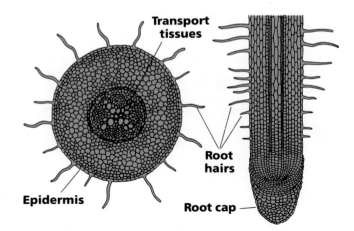

Figure 2 Root structure

allowing the root to absorb more water. The second root layer has soft, loose tissue. Water and food for the root are stored here.

The inner part of a root is made up of the tubes of the transport system. These tubes extend up through the root and into the stems and leaves. Some tubes carry water and dissolved minerals upward. Others carry food made in the leaves to all parts of the plant.

👁 *Observe:* Look at Figure 2. What is the outer part of the root tip called?

The Root Tip Roots grow from the tip. The tip of a root is covered by a **root cap.** The root cap is a cup-shaped mass of cells. It protects the root tip from damage as the root grows into the soil. As root cap cells are worn off, new cells are produced to take their place. Behind the root cap is an area where new root cells are formed. These new cells gradually change into the different kinds of cells that make up the root.

◣ *Analyze:* Why is the root cap at the end of the root?

Functions of Roots Most roots grow underground. Roots anchor, or hold, a plant firmly in the soil. Roots take in water and dissolved minerals from the soil. In some plants, food is stored in the roots.

||||▶ *List:* What are some functions of roots?

LESSON SUMMARY

▶ Two kinds of plant root systems are fibrous root systems and taproot systems.

▶ The outer root has root hairs to increase surface area for water absorption.

▶ The inner root contains the transport tubes.

▶ The root tip is protected by a root cap.

▶ Roots hold plants in the soil, absorb water and minerals, and some store food.

CHECK *Complete the following.*

1. Roots take in water and _____ from soil.

2. A taproot system has _____ main root.

3. Structures that increase the surface area of roots are the _____ .

4. The _____ part of the root contains transport tubes.

5. A root cap serves to _____ a root tip.

APPLY *Complete the following.*

6. **Calculating and Measuring:** If all the roots and root hairs of a rye plant were laid end to end, the length would be about 606,000 m.

How many kilometers is this? The surface area of the roots and root hairs would total about 210 m². Is this more, less, or about the same as the area of your classroom floor?

7. **Infer:** Would a plant with a fibrous root system be easy or difficult to pull out of the ground? Explain.

8. **Classify:** Which plants shown have a taproot? Which have fibrous roots?

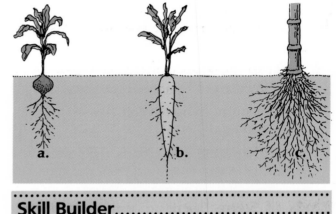

a.   b.   c.

Skill Builder...

△ *Organizing Information* A table is one way to organize information. Make a table with these headings: Name of Plant, Type of Root System, Depth of Roots. Use library references to find this information for the following plants: beet, oak tree, alfalfa, turnip, corn.

◆○◆ SCIENCE CONNECTION ◆○◆○◆○◆○◆○◆○◆○◆○◆○◆○◆○◆○◆○◆

SPECIALIZED ROOTS

Some roots have special functions. One kind of root gets water from the air. This is an aerial root. A plant with aerial roots is called an air plant. Orchids and English ivy are examples of plants with aerial roots.

Some roots grow out of other parts of a plant. Roots may grow from the stems above ground. Such roots are called prop roots. They help support a plant. Mangrove trees and corn are two plants that have prop roots. Look at the prop roots of the corn plant. They help keep the corn stem standing up.

Some plants do not grow in soil. For example, water lilies float along the surface of some ponds. These roots can absorb oxygen from water. Water lilies have aquatic (uh-KWAHT-ik) roots.

8-2 What are stems?

Objectives ▶ Distinguish between herbaceous and woody stems. ▶ Explain the jobs of stems.

TechTerms

- ▶ **herbaceous** (hur-BAY-shus) **stem:** stem that is soft and green
- ▶ **phloem** (FLOH-em): tissue that carries food from the leaves throughout the plant
- ▶ **woody stem:** stem that contains wood and is thick and hard
- ▶ **xylem** (ZY-lum): tissue that carries water and dissolved minerals upward from the roots

Kinds of Stems There are two kinds of plant stems. They are **herbaceous** (hur-BAY-shus) **stems** and **woody stems.** Herbaceous stems are soft, smooth, and green. Plants with herbaceous stems do not grow taller than 2 m. Plants with these stems grow during spring and summer and then die. Tomato plants and bean plants have herbaceous stems. Woody stems are thick, hard, and rough. The rough outer layer of the stem is the bark. Woody stems are not green. Woody stems may live for many growing seasons. They grow taller and wider each year. All trees, such as oaks and maples, have woody stems.

▷ *Name:* What are two kinds of plant stems?

Figure 1 Herbaceous stems

Functions of Stems Most stems grow above ground. The main job of stems is to support the leaves. Stems also are organs of transport. Special

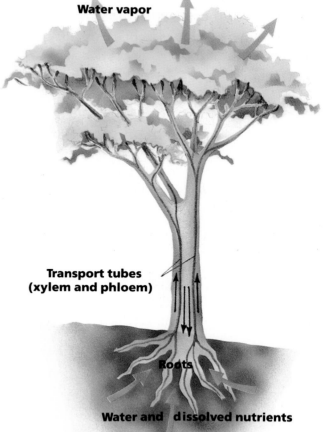

Water vapor

Transport tubes (xylem and phloem)

Roots

Water and dissolved nutrients

Figure 2 Stem structure

tubes within the stems carry materials between the roots and the leaves. In some plants, stems also store food. For example, the stems of sugar cane store large amounts of sugar.

▷ *List:* What are some functions of plant stems?

Stem Structure Both woody and herbaceous stems have similar tissues. The arrangement of the tissues is different in the two kinds of stems. Stems of both types contain tubes that carry water and dissolved minerals up from the roots. These tubes are called **xylem** (ZY-lum). Xylem carries water into the leaves. Extra water evaporates into the air through the leaves. Tubes that carry dissolved food made in the leaves downward throughout the plant are called **phloem** (FLOH-em). Xylem and phloem make up the conducting, or transport, system of the plant. In monocots and dicots, xylem and phloem are together in bundles.

▷ *Analyze:* Why does a plant need xylem and phloem?

LESSON SUMMARY

▶ Herbaceous stems are green and soft and live for a short time; woody stems are stiff, rough, not green, and live for many years.

▶ Stems support leaves and transport materials between the roots and the leaves.

▶ Xylem carries water and minerals up from the roots; phloem carries food down from the leaves to the rest of the plant.

CHECK *Write true if the statement is true. If the statement is false, change the underlined term to make the statement true.*

1. Plants with <u>herbaceous</u> stems live for many growing seasons.

2. Stems connect the <u>roots</u> and the leaves.

3. Oak trees have <u>woody</u> stems.

4. Xylem and phloem are found in <u>bundles</u>.

APPLY *Complete the following.*

5. **Sequencing:** List the plant structures in order to show how water gets from soil to leaves: leaves, phloem, roots, xylem.

6. **Hypothesize:** What do you think would happen if a herbaceous plant grew taller than 2 m?

7. **Classify:** Which plants listed have herbaceous stems? Which plants have woody stems?
   a. tulip   c. daisy   e. sunflower
   b. oak tree   d. apple tree f. willow

InfoSearch

Read the passage. Ask two questions that you cannot answer from the information in the passage.

Tree Rings In woody stems the layers of xylem form a pattern of rings. Xylem formed in spring has larger cells than xylem formed in summer. The pattern of small and large cells forms light and dark bands called annual rings. A tree's age may be determined by counting its rings.

SEARCH: *Use library references to find answers to your questions.*

ACTIVITY

TRANSPORT IN PLANTS

You will need a glass, water, ink, a celery stalk, a knife, and a metric ruler.

1. Put water in the glass. Add 3 drops of ink.

2. Use the knife to cut a slice off the bottom of the celery stalk. Discard the slice. **Caution: Be careful when using a knife.**

3. Put the cut end of the stalk into the water.

4. Copy the data table.

5. **Measure:** Record the height of the ink in the celery after 5, 10, and 15 minutes.

6. **Predict:** What do you think will happen to the ink in the celery after 24 hours?

7. Keep the celery in the ink for 24 hours. Measure and record the height of ink.

Questions

1. What happened to the ink in the celery?

2. Was your prediction correct? Explain.

3. What life process does this activity show?

Data Table	
TIME	HEIGHT OF INK (mm)
5 minutes	
10 minutes	
15 minutes	
24 hours	

8-3 What are leaves?

Objectives ▶ Describe the structure of leaves. ▶ Classify leaves as simple or compound.

TechTerms

- ▶ **blade:** wide, flat part of a leaf
- ▶ **epidermis** (ep-uh-DUR-mis): outer, protective layer of the leaf
- ▶ **mesophyll** (MES-uh-fil): middle layer of leaf tissue in which food-making occurs
- ▶ **stomata** (STOH-muh-tuh): tiny openings in the upper and lower epidermis of the leaf
- ▶ **vein:** bundle of tubes that contain the xylem and phloem in a leaf

Leaf Structure Most leaves have of a stalk and a wide, flat part called the **blade.** The blade is the most important part of the leaf. Food-making takes place in the blade. The stalk supports the blade and attaches it to the stem of the plant. Throughout the leaf there is a system of tubes called **veins.** They are made up of xylem and phloem. They connect to the xylem and phloem in the stem. The veins carry water and dissolved materials into and out of the leaf. Veins also support the leaf blade.

▶ *Identify:* What are the two main parts of a leaf?

Kinds of Leaves In some plants, the leaf blades are in one piece. This kind of leaf is called a simple

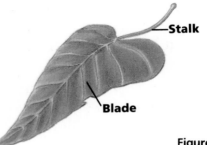

Figure 1 Leaf parts

leaf. Maple, oak, and elm trees have simple leaves. In other plants, leaf blades are divided into pieces. This kind of leaf is called a compound leaf. The pieces that make up a compound leaf are called leaflets. Each leaflet looks like a small leaf. Poison ivy and roses have compound leaves.

👁 *Observe:* Is the leaf in Figure 1 a simple or a compound leaf?

Leaf Tissues Leaves are covered by a protective layer called the **epidermis** (ep-uh-DUR-mis). This layer prevents the loss of water from the leaf. Scattered throughout the upper and lower epidermis are many tiny openings called **stomata** (STOH-muh-tuh). Two guard cells control the size of each stomate. They open and close the stomate, controlling water loss. They also control the exchange of oxygen and carbon dioxide between the inner tissues of the leaf and the surrounding air.

Beneath the epidermis is a layer of tissue called **mesophyll** (MES-uh-fil). Most of the food-making in the plant occurs in the mesophyll. Veins extend throughout the mesophyll.

▶ *Identify:* In what tissue layer does most food-making take place in a leaf?

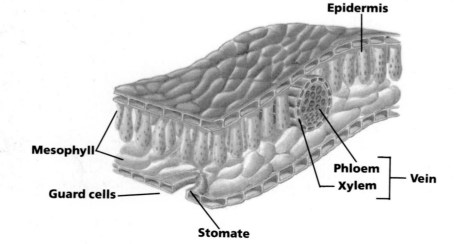

Figure 2 Leaf tissues

LESSON SUMMARY

▶ Most leaves have a blade, a stalk, and veins.

▶ The two kinds of leaves are simple and compound.

▶ The epidermis and the mesophyll are leaf tissues.

▶ Stomates control the gases that go into and out of a leaf and control water loss.

CHECK *Complete the following.*

1. The flat part of a leaf is called a _____ .

2. Leaves that are made up of leaflets are _____ leaves.

3. The outer, protective layer of a leaf is the _____ .

4. Water and gases pass into and out of leaves through tiny holes called _____ .

APPLY *Complete the following.*

5. How is the function of leaf epidermis similar to the function of your skin?

6. **Hypothesize:** Would the stomata of a leaf be opened or closed on a hot, dry day? Explain your answer.

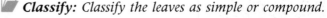

7. **Infer:** Why is air in a forest usually humid?

Classify: *Classify the leaves as simple or compound.*

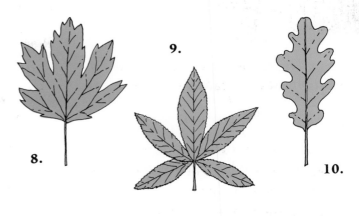

9.

8.

10.

Health & Safety Tip

Some plants are poisonous to the touch. They contain chemicals that can irritate the skin. Poison ivy, poison oak, and poison sumac are three common examples. Avoid touching any part of these plants. Poison ivy and poison oak have white berries. A useful rhyme to remember is: "Berries white, poisonous sight." Find pictures of poison ivy, oak, and sumac. Make sketches of each plant. How can you identify these plants? Are their leaves simple or compound?

ACTIVITY

CLASSIFYING LEAVES

You will need 10 different leaves and a field guide to plants.

1. Collect 10 different leaves. **CAUTION: Before collecting any leaves, learn how to identify poison ivy, poison oak, and poison sumac. Be sure not to touch the leaves of these plants.** Choose only those that have fallen to the ground.

2. **Model:** Trace an outline of each leaf. Draw in its vein pattern. Describe the vein patterns.

3. Use a field guide to help you identify the plants.

4. **Classify:** Identify each leaf as simple or compound.

Questions

1. **Analyze:** In what way is a field guide helpful in the identification of leaves?

2. **Measure:** Are more of the leaves you collected simple or compound?

3. How many leaflets make up each compound leaf?

4. **Observe:** How many basic vein patterns do you see?

What is photosynthesis?

Objective ▶ Explain the importance of photosynthesis in plants.

TechTerms

▶ **chlorophyll** (KLAWR-uh-fil): chemical pigment needed for photosynthesis

▶ **chloroplast:** organelle of green plant cells where photosynthesis takes place

▶ **photosynthesis** (foht-uh-SIN-thuh-sis): food-making process that uses sunlight

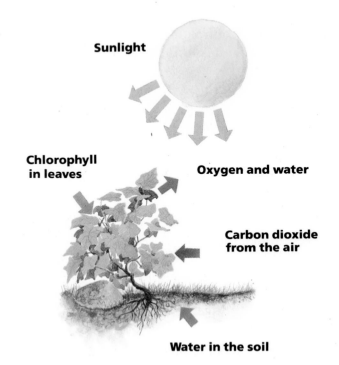

Sunlight

Chlorophyll in leaves

Oxygen and water

Carbon dioxide from the air

Water in the soil

Food Factories Plants do not have to eat. Green plants can make their own food in their leaves. The leaves are like "food factories." Some green stems also can make food. Food-making in plants is called **photosynthesis** (foht-uh-SIN-thuh-sis). The food that plants make is a sugar called glucose. Plants can change this sugar into starch, fats, and proteins. These nutrients are stored in the plants. They can be used at a later time.

▶ *Name:* What food do green plants make?

Photosynthesis In photosynthesis, water and carbon dioxide are used to make sugar. Roots absorb the needed water from the soil. Veins carry the water up into the leaves. Carbon dioxide enters the plant through the stomata. Sunlight supplies the energy the plant needs to make the sugar. During photosynthesis, oxygen and water are given off as waste materials. Look at the word and chemical equation for the process of photosynthesis at the bottom of the page.

◉ *Observe:* On which side of the equation are the waste products shown?

Chlorophyll Photosynthesis cannot occur without **chlorophyll** (KLAWR-uh-fil). Chlorophyll is a chemical pigment, or coloring, needed for photosynthesis. Other pigments in leaves are covered by chlorophyll. You see the colors of these pigments during the fall. Leaves change color. The different colors are caused by the other pigments.

Wherever there is chlorophyll in a plant, photosynthesis can occur. The chlorophyll is inside special parts of plant cells. The cell parts, or organelles, that contain chlorophyll are called **chloroplasts.** Leaf mesophyll cells contain many chloroplasts. Therefore, most photosynthesis takes place in this layer.

▶ *Name:* In what cell parts of a plant leaf is food made?

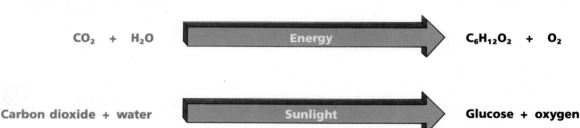

$$CO_2 \ + \ H_2O \xrightarrow{\text{Energy}} C_6H_{12}O_2 \ + \ O_2$$

$$\text{Carbon dioxide} + \text{water} \xrightarrow{\text{Sunlight}} \text{Glucose} + \text{oxygen}$$

LESSON SUMMARY

▶ Green plants can make their own food.

▶ In photosynthesis, carbon dioxide and water react to form sugar, and oxygen and water are given off as wastes.

▶ Chlorophyll is the chemical pigment needed for photosynthesis to take place.

▶ Photosynthesis occurs in the chloroplasts of plant cells.

CHECK *Write the letter of the term that best completes each statement.*

1. The food plants make is
   a. sugar. **b.** chlorophyll. **c.** oxygen. **d.** xylem.

2. Chloroplasts contain the green material
   a. mesophyll. **b.** chlorophyll. **c.** stomates. **d.** phloem.

3. Materials a plant uses to make sugar are
   a. oxygen and starch. **b.** carbon dioxide and water. **c.** sugar and protein. **d.** starch and mesophyll.

4. Water enters the plant through the
   a. roots. **b.** leaves. **c.** stomata. **d.** veins.

APPLY *Use the diagram to answer the following.*

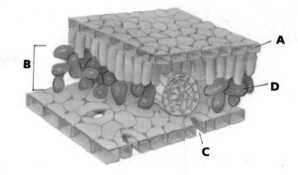

5. In which layer does photosynthesis occur?

6. Does water for photosynthesis enter or leave the leaf through C?

▶ 7. **Infer:** Does photosynthesis occur in layer A? How do you know?

Skill Builder

Building Vocabulary Look up the meaning of the word parts "photo-," "synthesis," "chlor-," "-plast," and "-phyll." Write their meanings. Explain the relationship of these word parts to the TechTerms.

ACTIVITY

SEPARATING "PLANT PIGMENTS"

You will need a paper towel, scissors, a metric ruler, a black felt pen, two colored felt pens, isopropyl alcohol, and a glass.

1. Cut a strip of paper towel 3 cm wide and 12 cm long.

2. Use one color felt pen to make a dot about 5 cm from the bottom of the strip. On top of this dot, place a dot of the second color.

3. After the dots dry, use the black felt pen to cover the dots.

4. Pour 1 cm of alcohol into the glass. **Caution: Do not breathe the fumes.** Stand the strip of paper towel in the glass.

▶ 5. **Predict:** What will happen to the dot on the strip?

👁 6. **Observe:** After 5 minutes, look at the strip.

Questions

1. What happened to the dot after 5 minutes?

👁 2. **Observe:** How many colors do you see?

▲ 3. **Model:** What would the ink dot represent in a green plant?

📄 4. **Hypothesize:** What would happen if plant pigment were used instead of ink?

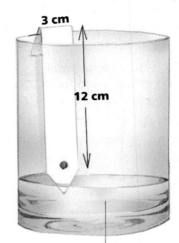

Bend strip over rim of glass

3 cm

12 cm

Isopropyl alcohol

What are flowers?

Objectives ► Recognize the flower as the reproductive organ of a plant. ► Identify the parts of a flower.

TechTerms

- ► **petal:** kind of leaf that often is brightly colored
- ► **pistil:** female reproductive organ in a flower
- ► **sepal** (SEE-pul): special kind of leaf that protects the flower bud
- ► **stamen** (STAY-mun): male reproductive organ in a flower

Flowers Not all plants have flowers. However, in plants with flowers, the flower is the organ of sexual reproduction. Flowers contain the male and female parts of a flowering plant.

▌▌▶ *Define:* What are flowers?

Parts of a Flower A flower is made up of several parts. The bottom of a flower is surrounded by **sepals** (SEE-puls). Sepals are a special kind of leaf. They protect the flower bud. In some flowers the sepals look like small green leaves. In others they are large and brightly colored. The **petals** of a flower are just inside the sepals. The petals are another kind of leaf. They may be white or brightly colored. Petals protect the reproductive organs of the plant. Inside the petals are the organs of reproduction for the plant.

▌▌▶ *Identify:* What are two special kinds of leaves in a flower?

Reproductive Organs The reproductive organs in a flower are the **pistil** and the **stamen** (STAY-mun). The stamen is the male reproductive organ. The pistil is the female reproductive organ. A flower usually has one pistil, with several stamens around it. These organs are located inside the circle of petals.

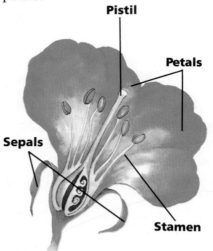

Some flowers have both male and female reproductive organs. These flowers are called perfect flowers. Other flowers may have only male or only female reproductive organs. These flowers are called imperfect flowers.

▶ *Classify:* What system of the plant are the pistil and the stamen part of?

LESSON SUMMARY

▶ Flowers are organs of sexual reproduction.

▶ The parts of a flower are the sepals, petals, pistil, and stamen.

▶ The reproductive organs in a flower are the stamens and pistils.

▶ Some flowers have male and female reproductive organs. Others have only male or only female reproductive organs.

CHECK *Write the letter of the term that best completes each statement.*

1. The organs of sexual reproduction in plants are the
   a. petals. **b.** flowers. **c.** leaves. **d.** stalks.

2. Sepals and petals are special kinds of
   a. veins. **b.** flowers. **c.** leaves. **d.** plants.

3. The _____ is the female reproductive organ in a flower.
   a. stamen **b.** pistil **c.** sepal **d.** vein

4. The _____ is the male reproductive organ in a flower.
   a. stamen **b.** pistil **c.** sepal **d.** vein

APPLY *Complete the following.*

▲ 5. **Model:** Draw a model of a flower. Label the petals, sepals, stamen, and pistil.

📁 6. **Classify:** Label each flower as perfect or imperfect.

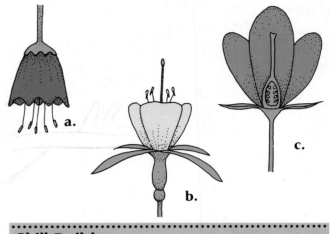

··
Skill Builder ···

📁 *Classifying* When you classify, you group things based upon how they are alike. Use a dictionary or library references to find out the meaning of the terms "pistillate" and "staminate." Classify each of the flowers shown above as perfect and imperfect. Then classify each imperfect flower as pistillate or staminate.

LEISURE ACTIVITY

GARDENING

Gardening is a popular hobby. People who live in the country or who have large back yards often plant gardens. A home garden may be large or small. It may be made up of flowers, herbs, vegetables, or fruits. Even people who live in cities or who have little or no land around their homes plant gardens. Their gardens may be made up of many different pots or other containers. A popular garden used by many people is a window box garden.

People like to garden for many reasons. Some people have gardens to grow flowers and other plants to beautify the grounds of their yards and homes. They cut the flowers to have fresh flowers for inside their homes. Many people like to garden because they enjoy raising their own fruits and vegetables.

No matter what the size or content of your garden, gardening can be fun. It also is good exercise. Gardening also can help lower the food bill. You can find out about amateur gardening by visiting local garden supply centers. Many free gardening booklets are available from companies that sell garden supplies and seeds.

How do flowering plants reproduce?

Objectives ▶ Describe pollination. ▶ Explain how fertilization and pollination are related.

TechTerms

- ▶ **anther:** part of the stamen that produces pollen
- ▶ **fertilization** (fur-tul-i-ZAY-shun): joining of the male and female reproductive cells
- ▶ **filament:** stalk of the anther
- ▶ **pollen grains:** male plant reproductive cells
- ▶ **pollination** (pahl-uh-NAY-shun): movement of pollen from a stamen to a pistil

Stamens and Pollen Stamens have two main parts. These are the **filament** and the **anther.** The filament is the thin stalk that holds up the anther. The anther is the part that produces **pollen grains.** Pollen grains are the male reproductive cells.

▶ *Define:* What is pollen?

Fertilization In sexual reproduction, reproductive cells from the male and female must meet. The joining of the nuclei of male and female reproductive cells is called **fertilization** (fur-tul-i-ZAY-shun). In flowering plants, fertilization takes place inside the pistil.

▶ *Define:* What is fertilization?

Pollination For fertilization to take place, pollen grains must first move from a stamen to a pistil. The movement of pollen from a stamen to a pistil is called **pollination** (pahl-uh-NAY-shun). The pollen often is moved by wind, by insects, and even birds. Sometimes, pollen is carried by water.

▶ *Classify:* What activity of a plant is pollination part of?

Self-Pollination In many flowers, the stamens are taller than the pistil. Pollen can fall off the anther on top of a stamen and land on the pistil in the same flower. This kind of pollination is called self-pollination. Self-pollination also occurs when pollen from one flower is carried to the pistil of another flower on the same plant.

Self-Pollination

▶ *Explain:* In what two ways may self-pollination occur?

Cross-Pollination Sometimes, pollen is carried from a stamen of a flower on one plant to the pistil of a flower on another plant of the same kind. This kind of pollination is called cross-pollination. Some plants have separate male flowers and female flowers. The male flowers have only stamens. The female flowers have only pistils. Usually the male and female flowers are on different plants. These plants have to cross-pollinate. Cross-pollination also can occur between two plants that have flowers with both stamens and pistils.

▶ *Name:* What kind of pollination occurs between flowers on different plants?

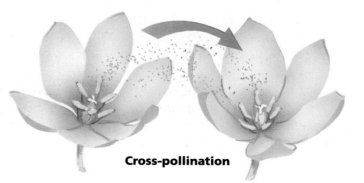

Cross-pollination

LESSON SUMMARY

▶ Stamens have a filament and an anther.

▶ Fertilization occurs when the nuclei of a male and a female reproductive cell join.

▶ Pollination is the movement of pollen from a stamen to a pistil.

▶ Self-pollination is pollination of flowers on the same plant.

▶ Cross-pollination is pollination between flowers on different plants.

CHECK *Complete the following.*

1. Nuclei of male and female reproductive cells join in the process of _____ .

2. Male reproductive cells are the _____ .

3. Movement of pollen from a stamen to a pistil is _____ .

4. Pollination between flowers on different plants is _____ .

5. Pollination between flowers on the same plant is _____ .

APPLY *Complete the following.*

➤ 6. **Infer:** Would flowers that depend on wind for pollination probably have flat, open flowers or tall, closed flowers? Explain your choice.

➤ 7. **Analyze:** How does an insect move pollen from one flower to another?

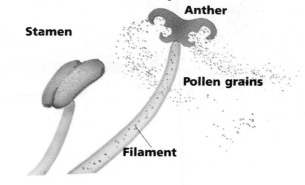

Stamen · Anther · Pollen grains · Filament

SCIENCE CONNECTION

MUTUALISM

Living things interact with each other in many different ways. Some of these interactions are helpful to both living things. These interactions help both living things survive. Other interactions make life more difficult for some living things. An example of a helpful interaction is mutualism (MYOO-choo-wul-izm). Mutualism is a relationship between two living things that is helpful to both living things.

Bees get food from plants. They gather nectar from flowers. When the insects gather the nectar, pollen sticks to their legs. The bees then may fly to the same kind of flower on a different plant. In this way the flowers of the plant are pollinated. The bees get food. Both the bees and flowers are helped by this interaction.

There are many examples of mutualism. Hummingbirds and flowers also are examples of mutualism. A bird called the oxpecker eats pests off rhinoceroses. The oxpecker gets food. The rhinos get rid of pests on their bodies.

What are seeds and fruits?

Objectives ► Identify the parts of the pistil. ► Explain how seeds and fruits form.

TechTerms

► **embryo:** undeveloped plant or animal

► **ovary** (OH-vuhr-ee): bottom part of the pistil

► **ovule** (OH-vyool): part of the ovary that develops into a seed after fertilization

► **stigma** (STIG-muh): top part of the pistil

Parts of a Pistil A pistil is made up of three parts. The top part is the **stigma** (STIG-muh). Below the stigma is a tube called the style. The style connects the stigma to the bottom of the pistil. The bottom of the pistil is called the **ovary** (OH-vuhr-ee). Inside the ovary is the **ovule** (OH-vyool). The ovule contains the female reproductive cells.

▐▐▐▶ *Identify:* What are the parts of a pistil?

Forming a Seed After a pollen grain lands on the stigma, it begins to change. The pollen cell begins to grow a tube. This tube is called a pollen tube. The pollen tube grows down into the stigma. It continues to grow down through the style and the ovary. Finally, the tip of the pollen tube enters the ovule.

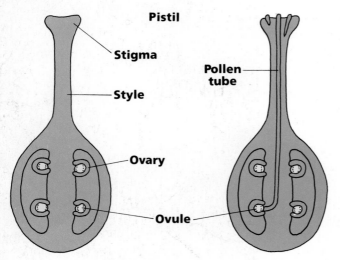

After the pollen tube enters the ovule, the tip of the tube dissolves. The nucleus of the pollen cell moves into the ovule. The nucleus of the pollen cell joins with the nucleus of the reproductive cell in the ovule. The joining of the two nuclei is called fertilization. After fertilization, the ovule develops into a seed. A new plant can grow from a seed.

◣ *Analyze:* Why does the pollen grain grow a tube after it lands on the stigma?

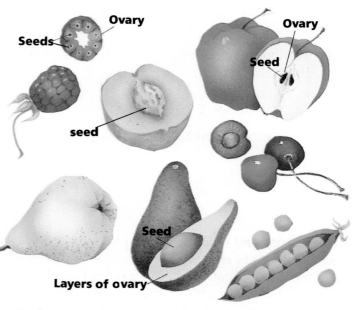

Fruits Ovules are found inside an ovary. An ovary may have only one ovule, or it may have more than one ovule. When the ovules are fertilized, they develop into seeds. Each seed contains a very young, or undeveloped, plant. Scientists call an undeveloped plant or animal an **embryo.** A seed contains a plant embryo.

While the seeds are forming, the ovary is growing. The ovary becomes very large. It surrounds and protects the seeds and the embryos inside the seeds. The large ovary and its seeds is called a fruit.

▐▐▐▶ *Distinguish:* What is the relationship between the ovary and a seed?

LESSON SUMMARY

► The parts of a pistil are the stigma, the style, and the ovary.

► A pollen grain that lands on a stigma forms a pollen tube that grows down into the ovary.

► Fertilization occurs in the ovules.

► A fertilized ovule becomes a seed that contains an embryo.

► The ovary surrounds and protects the seeds and becomes a fruit.

CHECK *Write true if the statement is true. If the statement is false, changed the underlined term to make the statement true.*

1. The female reproductive cells are in the <u>anther</u>.

2. A <u>seed</u> is formed after fertilization.

3. The undeveloped plant inside a seed is called an <u>ovary</u>.

4. A pollen grain that lands on a stigma grows a <u>pistil</u>.

APPLY *Complete the following.*

▲ 5. **Model:** Draw and label the parts of a pistil.

6. **Infer:** The top of a stigma is usually sticky. What might be an advantage of this stickiness?

Skill Builder

▲ *Organizing Information* Make a table showing the number of seeds in different fruits. List the names of 10 fruits in the first column. The relative number of seeds in the second column. Use the terms "one," "a few," and "many." In the third column, classify the seeds as small, medium, or large. What is the relationship between the number of seeds and their sizes?

Skill Builder

Classifying Scientists classify any plant part with seeds as a fruit. Vegetables are the leaves, stems, or roots of plants. People classify fruits and vegetables differently. Classify each of the plant parts listed as if you were a scientist and then as you would everyday.

a. tomato   d. carrot   g. lettuce
b. beets   e. peach   h. papaya
c. cucumber   f. celery   i. green pepper

CAREER IN LIFE SCIENCE

GREEN GROCER

Are you interested in food? Are you curious about the wide variety of fruits and vegetables in the world? Would you enjoy putting together attractive displays of fresh fruits and vegetables? If you answered yes to these questions, you may enjoy a career as a green grocer. A green grocer sells fresh fruits and vegetables to the public. They buy fruits and vegetables from growers. They choose items they think are high quality. Green grocers decide on fair prices. They arrange the fruits and vegetables in appealing displays. Green grocers also must be able to answer questions that their customers may have about cooking and storing fresh produce.

Many green grocers own their own stores. Some green grocers work in large supermarkets or in outdoor fresh produce markets. If you would like to work as a green grocer, you should take science and mathematics courses in high school. Courses at technical schools in managing a small business also are helpful.

For more information, you should talk to your guidance counselor. You also may want to talk to a green grocer.

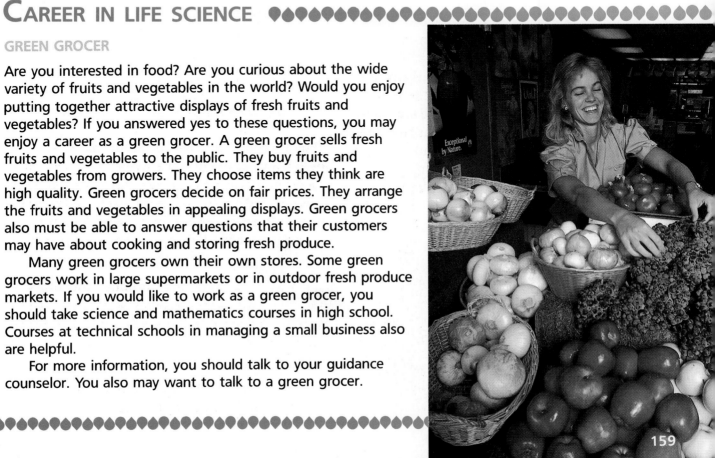

What are the parts of a seed?

Objective ▶ Identify the parts of a seed.

TechTerms

▶ **hilum** (HY-lum): mark on the seed coat where the seed was attached to the ovary

▶ **seed coat:** outside covering of a seed

Seed Coat All seeds have an outside covering. This outer covering is called the **seed coat.** Most seed coats are hard. The hard coat protects the embryo. Some seeds are protected so well that they can be kept for many years. They will still grow when they are planted.

▦▶ *Name:* What is the outside covering of a seed called?

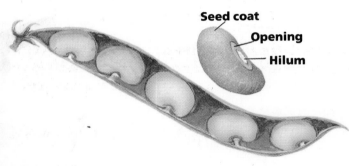

Figure 1 Outer Structure of a Seed

Hilum Look at Figure 1. Find the **hilum** (HY-lum). The hilum is a small mark, or scar on the seed. The hilum is where the seed was attached to the ovary. Near the hilum, there is a small opening. The opening is where the pollen tube entered the ovule.

▦▶ *Describe:* What is the hilum?

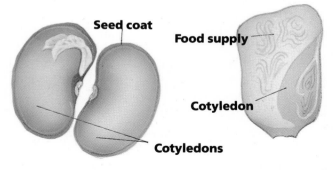

Figure 2 Inside a Seed

Inside a Seed If you soak a bean seed overnight, the seed coat gets soft. You can easily peel it off. Figure 2 shows the insides of two seeds. Notice that the one seed has two large halves, or cotyledons (kaht-ul-EED-uns). These halves store food as starch. Some seeds do not have two halves. Food is stored outside the cotyledon. The food is absorbed through the cotyledon. Food is still stored in the seed. The developing embryo uses this food as it grows.

Look again at Figure 2. You will see that the embryo is attached to one of the seed halves. The embryo has a tiny root, a stem, and leaves. It is only a small part of the seed. If you plant the seed, the embryo will begin to grow, or germinate (JUR-muh-nayt). As the seed germinates, the two halves, or cotyledons, are pushed above ground. When the first leaves unfold and begin to make food, the cotyledons dry up and die. In time, the embryo will develop into an adult plant.

▶ *Infer:* Why would the cotyledons no longer be needed once photosynthesis begins?

Figure 3 Seed Germination

LESSON SUMMARY

▶ The outer protective covering of a seed is the seed coat.

▶ The mark or scar on the seed coat where the seed was attached to the ovary is the hilum.

▶ Seeds contain food that the embryo uses as it grows.

▶ The embryo in a seed is a tiny plant that has a tiny root, stem, and leaves.

CHECK *Complete the following.*

1. What is the outside covering of a seed called?

2. What part of a seed develops into an adult plant?

3. What is the mark on the seed coat where the seed was attached to the ovary called?

4. What makes up most of a seed?

5. In what form is food stored in a seed?

6. What are the three parts of all seeds?

APPLY *Complete the following.*

▶ 7. **Infer:** All plants need water, sunlight, and nutrients to grow. Which of these needs does a seed not need to begin to germinate?

Ideas in Action

IDEA: People use many different kinds of seeds as food.

ACTION: Make a list of the kinds of seeds you eat.

Skill Builder

▶ *Inferring* When you infer, you form a conclusion based upon facts. Seeds are dispersed, or spread, by wind, animals, and water. Study the seeds shown. Decide how each seed is dispersed based upon how it looks. Give reasons for your answers.

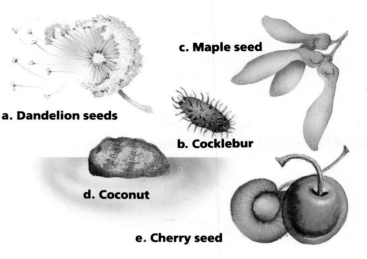

c. Maple seed

a. Dandelion seeds

b. Cocklebur

d. Coconut

e. Cherry seed

ACTIVITY

TESTING SEEDS FOR STARCH

You will need iodine, a medicine dropper, 5 different kinds of seeds, a small dish, water, and a knife.

1. Soak the seeds in a dish of water overnight.

2. The next day, remove the seed coats from the seeds. Observe the seed coats.

3. Break apart the seeds. You may need to cut some of the seeds in half with the knife. **Caution: Be careful when cutting with a knife.**

4. Place a drop of iodine on the side of each seed. Iodine changes color when starch is present.

Iodine

Starch present in bread

Questions

1. a. **Observe:** To what color did the iodine change when starch was present?   b. Which of the seeds contained starch?

2. a. What is the starch in the seeds?   b. What is the starch used for?

3. How many seeds had a seed coat?

How do plants reproduce asexually?

Objectives ▶ Recognize that vegetative propagation is a kind of asexual reproduction. ▶ Identify ways that plants reproduce asexually.

TechTerms

▶ **asexual reproduction:** reproduction without the joining of male and female cells
▶ **bulb:** underground stem covered with fleshy leaves
▶ **tuber:** underground stem
▶ **vegetative propagation** (VEJ-uh-tayt-iv prahp-uh-GAY-shun): kind of asexual reproduction that uses parts of plants to grow new plants

Reproduction Without Seeds Some plants can reproduce without male and female cells joining to form seeds. This method of reproduction is called **asexual reproduction.** Asexual reproduction in plants is called **vegetative propagation** (VEJ-uh-tayt-iv prahp-uh-GAY-shun). During vegetative propagation, the growing parts of plants develop into new plants. The growing parts are roots, stems, or leaves.

▶ *Define:* What is vegetative propagation?

Tubers Some plants, such as white potatoes, have underground stems. An underground stem is called a tuber. A white potato may have many small white buds growing on its skin. These buds are called eyes. The eyes of the potato are the organs of vegetative reproduction. When planted in soil, each eye may grow into a new potato plant. Each new plant is identical to the parent plant.

▶ *Define:* What is a tuber?

Bulbs A **bulb** is an organ of vegetative reproduction in some plants. A bulb is an underground stem. It differs from the underground stem called a tuber. A bulb is covered with thick leaves. An

onion is an example of a bulb. Each onion plant produces many bulbs. When planted, each bulb may grow into a new onion plant. Other plants that grow from bulbs are daffodils, lilies, and tulips.

▶ *Name:* What are two plants that grow from bulbs?

Cuttings Some plants can grow new plants from pieces of themselves. This method of vegetative propagation is called cuttings. Roots make good cuttings. For example, carrots are roots. It also is an organ of vegetative reproduction. A carrot may be cut in two. When the top part is planted, it will begin to grow new roots. In time, a new carrot plant will grow. The new plant will be identical to the parent.

Carrot

New plants may also grow from some leaf and stem cuttings. A leaf is cut off of the plant. The leaf stalk is placed in water. Roots will grow from the stalk, and a new plant will grow. The same thing can be done with the stems of some plants.

▶ *Classify:* What type of reproduction are cuttings?

LESSON SUMMARY

▶ Plants reproduce asexually by vegetative propagation.

▶ Underground stems called tubers are organs of vegetative propagation in some plants.

▶ Bulbs are organs of vegetative propagation.

▶ Plants may reproduce asexually by using root cuttings.

▶ Leaf and stem cuttings can be used to grow new plants.

CHECK *Write true if the statement is true. If the statement is false, change the underlined term to make the statement true.*

1. White potatoes are examples of <u>bulbs</u>.

2. A bulb of a daffodil is a <u>root</u>.

3. A carrot plant may be grown asexually from its <u>seed</u>.

Answer the following.

4. What kind of reproduction is vegetative reproduction?

5. What kind of organ of reproduction is an onion?

APPLY *Complete the following.*

6. **Classify:** Which part of the plant is used for reproduction in the following diagrams?

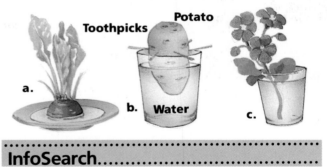

....
InfoSearch..

Read the passage. Ask two questions that you cannot answer from the information in the passage.

Grafting Grafting is a kind of vegetative propagation. It is used to produce new kinds of plants. Twigs from one tree are attached to another. For example, seedless oranges cannot reproduce by themselves. They can only reproduce by grafting. Twigs from a branch with seedless oranges are grafted onto a regular orange tree. Seedless oranges grow on the grafted branch.

SEARCH: Use library references to find answers to your questions.

ACTIVITY

GROWING PLANTS ASEXUALLY

You will need a paper cup, a glass, soil, water, 4 toothpicks, an onion, potato with eyes, and a knife.

1. Half-fill the cup with soil. Half-fill the glass with water.

2. Equally space 4 toothpicks around the onion. Rest the toothpicks and onion on the glass rim. Add water until the bulb bottom is underwater.

3. Cut a piece of potato that has an eye. **Caution: Be careful when using a knife.** Put it in the cup. Add some soil. Water the soil.

4. **Observe:** Check your plants every day. Keep the bulb bottom wet and the soil moist.

Questions

1. **Compare:** How is the reproduction of the onion and potato similar?

2. Which plant part grew roots first?

What are tropisms?

Objective ▶ Relate different kinds of stimuli to the tropisms they cause.

TechTerms

▶ **stimulus** (STIM-yuh-lus): something that causes a reaction to take place

▶ **tropism:** growth of a plant in response to something in the environment

Plant Responses All plants respond to changes in their environment. They respond to light, gravity, and water. Each response is caused by a **stimulus** (STIM-yuh-lus). A stimulus is a change that causes a response. The reaction of a plant to a stimulus is called a **tropism** (TROH-pizm). A plant responds to each stimulus by growing in a certain direction. Tropisms happen very slowly. You may not notice them.

🔲 *Hypothesize:* What stimulus do you think would make a plant part grow downward?

Light Green plants respond to the stimulus of light. Most stems and leaves grow toward light. A plant left near a sunny window will bend toward the light.

Some stems even change their response to light. The flower stem of a peanut plant grows up towards sunlight. After pollination, the flower stem grows away from light. It grows down into the soil, where its peanuts develop in the dark.

▥▶ *Explain:* How do most plant stems respond to light?

Gravity Plants respond to the stimulus of gravity. Roots grow down in response to gravity. A plant in a flowerpot could be tilted on its side. The roots will bend and grow down. If the plant is turned upright, the roots will bend again and grow down.

Most stems grow away from the pull of gravity. A plant in a flowerpot could be tilted on its side. The stem will respond by bending and growing up again. Rhizomes (RY-zohms), the underground stems of some plants, grow sideways in response to gravity. They grow horizontally just under the surface of the soil.

▥▶ *Describe:* How does a plant root respond to gravity?

Water Plants respond to the stimulus of water. Roots grow toward water. In most plants, this tropism is not very strong. It occurs only when water touches the roots. In some plants, this tropism is very strong. For example, the roots of a willow tree will grow into and clog sewer and water pipes.

▥▶ *Explain:* How do roots respond to water?

LESSON SUMMARY

▶ Tropisms are plant growth responses to stimuli.

▶ Leaves and stems grow toward light.

▶ Some stems change their response to light.

▶ Roots grow down in response to gravity.

▶ Stems grow up in response to gravity.

▶ Roots grow toward water.

CHECK *Write true if the statement is true. If the statement is false, change the underlined term to make the statement true.*

1. Stems grow toward the stimulus of <u>light</u>.

2. Roots grow <u>against</u> the stimulus of gravity.

3. Something in the environment that causes a reaction to occur is a <u>tropism</u>.

4. Plant responses to most stimuli occur <u>quickly</u>.

5. <u>Roots</u> grow toward water.

APPLY *Complete the following.*

6. **Analyze:** In what direction would plant stems grow in the dark grow? Explain your answer.

7. **Infer:** A drain pipe from a house seems to be clogged. A willow tree has been growing near the house for many years. What is one possible reason for the clogged drain?

Skill Builder.......................................

Building Vocabulary Each plant tropism has a name based on the stimulus involved. Look up the prefixes "geo-," "hydro-," and "photo-." Use the meanings of these prefixes to help you make new words. Add each prefix to the word "tropism" to make a word that names a plant's response to each stimulus. What are the words?

SCIENCE CONNECTION ◆○◆○◆○◆○◆○◆○◆○◆○◆○◆○◆○◆○◆○◆○◆

CARNIVOROUS PLANTS

Some plants respond immediately to stimuli. Growth is not involved. The Venus' flytrap is a plant that responds quickly to touch. The Venus' flytrap plant is a carnivorous (kahr-NIV-uh-rus) plant. It ingests small insects.

The leaves of the Venus' flytrap are hinged so that they can close like a book. The leaf edges have stiff spines on their edges. The surface of each leaf half has three stiff hairs. When an insect walks across the leaf, it may touch the hairs. Touching only one hair causes no reaction. Touching two hairs or jiggling one hair twice makes the leaf snap shut in 1 second or less. The spines interlock to form a cage. The insect is trapped inside the closed leaves. The leaf stays closed.

Special chemicals are given off by the leaf. The chemicals digest the soft parts of the insect. Digestion may take several weeks. After the insect has been digested, the trap opens again. After a leaf has caught several insects, it dries up and dies.

Challenges

STUDY HINT Before you begin the Unit Challenges, review the TechTerms and Lesson Summary for each lesson in this unit.

TechTerms...

anther (156)
asexual reproduction (162)
blade (150)
bulb (162)
chlorophyll (152)
chloroplast (152)
embryo (158)
epidermis (150)
fertilization (156)
fibrous root system (146)
filament (156)
herbaceous stem (148)
hilum (160)

mesophyll (150)
ovary (158)
ovule (158)
petal (152)
phloem (148)
photosynthesis (152)
pistil (152)
pollen grains (156)
pollination (156)
root cap (146)
root hair (146)
seed coat (160)

sepal (152)
stamen (152)
stigma (158)
stimulus (162)
stomata (150)
taproot system (146)
tropism (162)
tuber (162)
vegetative propagation (162)
vein (150)
woody stem (148)
xylem (148)

TechTerm Challenges...

Matching *Write the TechTerm that matches each description.*

1. root system made up of large root, and many small, thin roots
2. stem that contains wood and is thick and hard
3. kind of leaf that often is brightly colored
4. thin, hairlike structure on the outer layer of the root tip
5. stalk of the anther
6. middle layer of leaf tissue in which food-making occurs
7. cup-shaped mass of cells that covers and protects a root tip
8. underdeveloped plant or animal
9. top part of the pistil
10. special kind of leaf that protects the flower bud
11. stem that is soft and green
12. root system made up of many thin, branched roots
13. underground stem
14. asexual reproduction that uses parts of plants to grow new plants

Identifying Word Relationships *Explain how the words in each pair are related. Write your answer in complete sentences.*

1. asexual reproduction, vegetative propagation
2. chlorophyll, chloroplast
3. pollen grains, fertilization
4. stimulus, tropism
5. chlorophyll, photosynthesis
6. pistil, pollination
7. stamen, anther
8. ovary, ovule
9. hilum, seed coat
10. vein, blade
11. xylem, phloem
12. epidermis, stomata
13. bulb, tuber

Content Challenges..

Multiple Choice *Write the letter of the term that best completes each statement.*

1. In flowering plants, fertilization takes place inside the
   a. stamen. b. pistil. c. anther. d. sepal.

2. Most of the food-making in a plant occurs in the
   a. epidermis. b. mesophyll. c. stomata. d. guard cells.

3. Perfect flowers have
   a. only male reproductive organs. b. only female reproductive organs. c. both male and female reproductive organs. d. neither male nor female reproductive organs.

4. The rough outer layer of a woody stem is the
   a. root cap. b. xylem. c. mesophyll. d. bark.

5. The food that plants make is a
   a. starch. b. protein. c. sugar. d. fat.

6. Fibrous root systems are found in
   a. grass. b. carrots. c. radishes. d. dandelions.

7. Chloroplasts contain the green material
   a. chlorophyll. b. mesophyll. c. xylem. d. phloem.

8. The small white buds growing on a potato are called
   a. bulbs. b. eyes. c. tubers. d. cuttings.

9. A large ovary and its seeds is called
   a. an embryo. b. an ovule. c. a fruit. d. a seed coat.

True/False *Write true if the statement is true. If the statement is false, change the underlined term to make the statement true.*

1. The two main parts of a <u>pistil</u> are the filament and anther.

2. Asexual reproduction in plants is called vegetative <u>pollination</u>.

3. Most seed coats are <u>soft</u>.

4. Tropisms happen very <u>slowly</u>.

5. Large amounts of sugar are stored in the <u>roots</u> of sugar cane.

6. Roots grow from the <u>tip</u>.

7. An onion is an example of a <u>bulb</u>.

8. Petals are a kind of <u>leaf</u>.

9. The pieces that make up a <u>simple</u> leaf are called leaflets.

Understanding the Features...

Reading Critically *Use the feature reading selections to answer the following. Page numbers for the features follow each question in parentheses.*

1. **Define:** What is mutualism? (157)

2. What are three things a green grocer does on the job? (159)

3. **Infer:** Why do you think gardening is good exercise? (155)

4. How does a Venus' flytrap digest insects? (165)

5. **List:** What are three kinds of specialized roots? (147)

Content Challenges..

Interpreting a Diagram *Use the diagram to complete the following.*

1. Which structure is the male reproductive organ?
2. Which structure is the female reproductive organ?
3. Which structures protect the reproductive organs of a plant?
4. Which structures protect the flower bud?
5. Which structures are specialized leaves?
6. Is the flower a perfect flower or an imperfect flower? Explain.
7. Which structure produces pollen grains?
8. Which structure contains female reproductive cells?

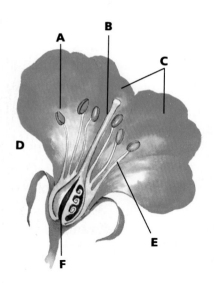

Critical Thinking *Answer each of the following in complete sentences.*

1. **Apply:** What would happen to the stem of a potted plant if you placed the plant on its side?
2. **Hypothesize:** What do you think would happen if a herbaceous plant grew taller than 2 m?
3. **Contrast:** What is the difference between pollination and fertilization?

4. **Analyze:** Fruit does not usually ripen until a seed is mature. Why is this important for the survival of the plant?
5. A slimy substance is produced by the root cap. How do you think this substance helps a plant?

Finding Out More...

1. **Research:** The xylem of a tree goes through two growth periods each year, one in the spring and one in the summer. These different growth periods result in annual rings on a tree's woody stem. Using library references, find out about annual rings, why they form, and how they can be used to determine the age of a tree that has been cut down. Write your findings in a report.
2. Collect leaves from a variety of trees. Find the blade and stalk of each leaf. Locate the veins in the blades. Make a sketch of each leaf and label its parts. Classify each leaf as a simple of compound leaf.

3. Obtain a flower. Identify as many parts of the flower as possible. Draw a sketch of your flower and label each part. Next to the label, write the function of each part. Classify your flower as a perfect or an imperfect flower.
4. Grafting and layering are two methods of vegetative propagation. Find out how plants can be grown using these methods. Present your findings in an oral report.

ANIMALS WITHOUT BACKBONES

CONTENTS

STUDY HINT Before beginning Unit 9, write the title of each lesson on a sheet of paper. Below each title, write a short paragraph explaining what you think each lesson is about.

Objective ▶ Describe the structure of a sponge.

TechTerms

▶ **flagellum** (fluh-JEL-uhm): whiplike structure on some cells

▶ **pores:** tiny openings

▶ **spicules** (SPIK-yools): small, hard needlelike structures of a sponge

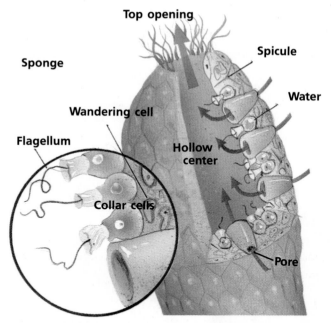

Porifera In the past, people thought that sponges were plants because they do not move from place to place like most other animals do. Sponges live attached to objects on the ocean floor. Today, scientists classify sponges in the Animal Kingdom in the phylum *Porifera* (paw-RIF-uhr-uh). The word *"Porifera"* means "pore-bearer." If you look at a sponge, you will see it has many **pores,** or tiny openings.

▶ *Identify:* In what phylum are sponges classified?

Structure of a Sponge Sponges are very simple animals. You can compare the body of a sponge to an empty sack. The sponge is closed at the bottom, and has a large opening at the top. The center of the sponge is hollow. The body of a sponge is made up of only two layers of cells. The outer layer is made up of thin, flat cells. The inner layer is made up of special cells called collar cells. The collar cells line the hollow center of the sponge. Each collar cell has a long whiplike structure called a **flagellum** (fluh-JEL-uhm). A jellylike substance fills the space between the two cell layers. In the jellylike substance, there are special cells that can move around like amoebas (uh-MEE-buhs). These wandering cells carry food to other cells. They also help a sponge to reproduce.

▶ *Describe:* What is the shape of a sponge?

Spicules Small, needlelike structures called **spicules** (SPIK-yools) are in the jellylike layer. They link together to form a simple skeleton. Spicules support the sponge. Some spicules are chalklike. Others are made up of the mineral that makes up sand. They are glasslike. Still other spicules are made up of a rubbery, flexible material called spongin (SPUN-jin). Spongin skeletons are dried and sold in stores as natural sponges. Most sponges you use are not natural sponges. They are factory-produced.

▶ *Define:* What are spicules?

Life Functions Sponges feed by filtering tiny organisms and bits of material from the water that passes through their bodies. Water flows through the pores into the hollow center. Water flows out through the large opening at the top of the sponge. The constant beating of each flagellum keeps the water moving into and out of the sponge. As the water flows by, tiny bits of food are trapped by the collar cells. They also absorb oxygen from the water. The wandering, amoeba-like cells carry food and oxygen to all cells. These cells also pick up wastes. The wastes are given off into the water passing through the sponge. The wastes are then carried out through the large opening.

▶ *Observe:* In how many directions does water move in a sponge?

LESSON SUMMARY

▶ Sponges are classified in the phylum *Porifera*.

▶ Sponges are shaped like empty sacks that are closed at the bottom and open at the top.

▶ The body of a sponge is made up of two cell layers with a jellylike substance between the layers.

▶ Spicules form a simple skeleton and support the sponge.

▶ Sponges feed by filtering food from the water that passes through them.

CHECK *Write true if the statement is true. If the statement is false, change the underlined term to make the statement true.*

1. Sponges have <u>three</u> cell layers.

2. The <u>collar cells</u> help move water through a sponge.

3. Water enters a sponge through its large <u>top opening</u>.

4. Sponges are classified in the phylum <u>*Animalia*</u>.

5. The rubbery material that makes up some spicules is called <u>spongin</u>.

APPLY *Complete the following.*

6. **Explain:** Think about the structure of a sponge. Why is *Porifera* a good name for this phylum?

7. **Infer:** Why can a sponge be considered a living filter?

8. **Hypothesize:** Suppose each flagellum was removed from the collar cells of a sponge. Do you think the sponge would be able to feed? Explain.

Ideas in Action...............................

IDEA: Most sponges sold in stores are not natural sponges.

ACTION: Find out what types of sponges are sold at a hardware store, drug store, or supermarket near you. Look at the labels on the sponges. What materials are the sponges made of? Try to find natural sponges. How do the prices of the natural sponges and the synthetic sponges compare?

LEISURE ACTIVITY

SNORKELING

Have you ever wondered what kinds of plants and animals live in the ocean? Snorkeling is a good way to find out. You can discover how fish and other water animals live together.

You only need a few pieces of equipment to go snorkeling. Fins, a face mask, and a snorkel, or breathing tube are needed. Fins help you move quickly and easily through the water. The face mask gives you a clear view through the water. The snorkel allows you to swim along the surface of the water without raising your head to breathe.

If you are snorkeling in cool water, you may want to wear a wet suit. A wet suit is a form-fitting covering that helps keep in body heat. It is called a wet suit because the swimmer gets wet while wearing it. Water seeps in between the suit and the body. Once your body warms this layer of water, your body stays warm.

Always snorkel with a buddy and under proper supervision. Be sure you know all the safety rules.

Objective ▶ Identify and describe the body forms of cnidarians.

TechTerms

▶ **cnidocytes** (NY-duh-syts): stinging cells

▶ **medusa** (muh-DOO-suh): umbrellalike form of a cnidarian

▶ **polyp** (PAL-ip): cuplike form of a cnidarian

▶ **tentacles:** long, armlike structures

Cnidarians If you have seen a jellyfish, you have seen a cnidarian (ni-DER-ee-un). Jellyfish are classified in the phylum *Cnidaria.* All cnidarians have **tentacles,** or long, thin armlike structures. The tentacles have special stinging cells, called **cnidocytes** (NY-duh-syts). The phylum gets its name from the cnidocytes.

All cnidarians live in water. Corals, jellyfish, and most other cnidarians live in the ocean. The hydra lives in lakes and ponds.

Classify: Name three cnidarians.

Body Forms Cnidarians have two body forms, or shapes. Some cnidarians have an umbrella shaped body called a **medusa** (muh-DOO-suh). Tentacles hang down from the edge of the umbrella. The mouth is in the center of the bottom surface. The medusa form can float on the surface of the water. It also can swim through the water. Other cnidarians have a tube-shaped body. It is called a **polyp** (PAL-ip). A polyp does not usually move from place to place. It lives attached to a surface in water. The mouth is at the top. It is surrounded by tentacles.

Classify: Is a jellyfish a medusa or a polyp?

Structure of Cnidarians Like sponges, cnidarians have two cell layers. There is a jellylike layer in between the two layers. In cnidarians, however, the cells are organized into tissues. Cnidarians have digestive, muscle, nerve, and sensory tissues. The tissues surround a central cavity (KAV-uh-tee), or hollow space. The mouth opens into the cavity. The mouth is the only body opening.

▶ *Infer:* Why are cnidarians considered more complex than sponges?

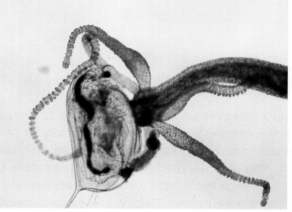

Figure 2 Food-getting in hydra

Food Getting Cnidarians use their tentacles to catch their food. Cnidocytes on the tentacles have coiled stingers inside. When a small animal gets close or touches the tentacles, the stingers shoot out. The poison in the stinger stuns or kills the animal. The tentacles wrap around the animal and pull it into the mouth and body cavity. Here the food is changed into a form that the cells can use.

Explain: How do cnidocytes get food?

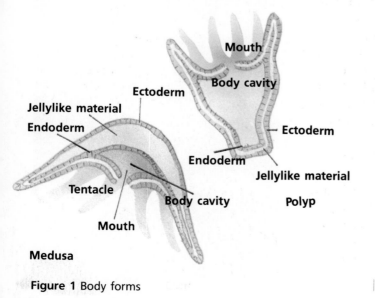

Medusa

Polyp

Mouth

Body cavity

Ectoderm

Jellylike material

Endoderm

Ectoderm

Endoderm

Jellylike material

Tentacle

Body cavity

Mouth

Figure 1 Body forms

LESSON SUMMARY

► All cnidarians have tentacles and cnidocytes.

► All cnidarians live in water.

► The two body forms of cnidarians are polyps and medusas.

► Cnidarians have two cell layers and tissues.

► Cnidarians use their tentacles to catch their food.

CHECK *Find the sentence in the lesson that answers each question. Then, write the sentence.*

1. How did the phylum *Cnidaria* get its name?

2. Where do cnidarians live?

3. What is the umbrella-shaped body form of a cnidarian called?

4. What types of tissue are found in cnidarians?

APPLY *Use Figure 1 on page 172 to answer the following.*

5. **Observe:** Does a jellyfish or a hydra have a larger jellylike layer?

6. **Compare:** How are the two body shapes of cnidarians alike?

7. **Analyze:** How are cnidarians more complex than sponges?

Health & Safety Tip

Never touch a jellyfish, even if it looks dead. It may still be able to sting you. The poison given off by some cnidarians can be harmful to humans. The sting of a jellyfish is painful. One type of jellyfish, the sea wasp, gives off a deadly poison. Its sting can kill a human in less than 3 minutes. If you do get stung by a jellyfish, get first aid as soon as you can. Use a first aid manual or other reference to find out how to treat a jellyfish sting.

SCIENCE CONNECTION

CORAL REEFS

Corals live together in colonies. As each coral polyp grows, it forms a hard rocklike external skeleton. When corals die, their skeletons remain in place. Young polyps attach themselves to these skeletons. As new layers of coral form on top of the skeletons, large mounds are built up. Billions of skeletons covered by a layer of living coral make up a coral reef.

A coral reef is one of the busiest places in the ocean. Sponges may live attached to the surface of the reef. Many different types of animals including sea urchins, marine worms, crustaceans, and fish, spend some of their time around the reef. The reef supplies nooks and crannies in which small animals can hide. The small animals serve as food for larger ones.

There are three kinds of coral reefs. A fringing reef forms close to shore. A barrier reef forms farther out in the ocean. One of these reefs is the Great Barrier Reef off the coast of Australia. It is about 6000 km long and 80 km wide. An atoll forms a ring around a lagoon when a fringing reef sinks.

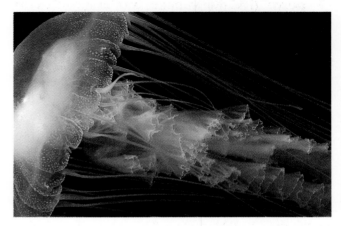

Barrier reef

Fringing reef

Atoll

What are flatworms?

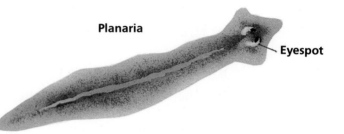

Eyespot

Objective ▶ Discuss the different kinds of worms and their features.

TechTerms

- ▶ **ectoderm** (EK-tuh-durm): outer tissue layer
- ▶ **endoderm** (EN-duh-durm): inner tissue layer
- ▶ **mesoderm** (MES-uh-durm): middle tissue layer

Classification of Worms When you think of worms, you probably think of earthworms. However, there are many other kinds of worms. They are so different from each other that they are classified in three different phyla. These three phyla are flatworms, roundworms, and segmented worms.

All common worms are invertebrates. They have three cell layers. The outer layer is the **ectoderm** (EK-tuh-durm). The inner layer is the **endoderm** (EN-duh-durm). The middle layer is the **mesoderm** (MES-uh-durm). Some organs and organ systems develop from the mesoderm.

All worms also have a top and bottom. They have a head end and a tail end. They also have a left side and a right side.

▶ *Identify:* What are the three cell layers in worms?

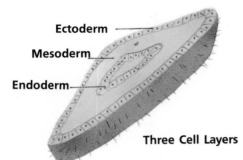

Ectoderm
Mesoderm
Endoderm

Three Cell Layers

Flatworms Flatworms are classified in the phylum *Platyhelminthes* (plat-ee-HEL-minths). They have flattened, ribbonlike bodies. Flatworms are the simplest worms. There are three groups of flatworms. One group lives in ponds and streams.

The most common flatworm is the planarian (pluh-NER-ee-uhn). Planaria are small flatworms between 5 and 25 mm long. They live attached to underwater twigs and branches. They have two small eyespots that can sense light.

The other two groups of flatworms are parasites (PAR-uh-syts). Parasites live inside or on other organisms. They feed on and harm the organisms they live in. Flukes and tapeworms are parasites. Special body parts help parasites live inside other organisms. Hooks and suckers hold the parasite in place. One kind of fluke lives in the human liver. Flukes range in length from 2 mm to 15 cm. Tapeworms live in the intestines of many kinds of animals, including humans. Tapeworms can grow to several meters in length. In the intestines, tapeworms absorb, or take in food that is already digested.

▶ *Classify:* Name three examples of flatworms.

Roundworms Roundworms make up the phylum *Nematoda* (NEM-uh-tohd-uh). Roundworms can live almost anywhere. They can live in soil or in water. Some are parasites of plants or animals. Hookworms and *ascaris* worms are two roundworms that live in the intestines of humans.

Roundworms have a threadlike body with pointed ends. It is covered by a smooth, tough skin. Roundworms are one of the simplest animals to have a complete digestive system. Food can pass through in only one direction. There is a mouth at one end and an anus (AY-nus) at the other end. An anus is an opening for getting rid of wastes. Roundworms also have simple excretory and nervous systems. They do not have a circulatory or a respiratory system. However, roundworms have a well-developed reproductive system.

▶ *Identify:* Name two kinds of roundworms.

LESSON SUMMARY

▶ Worms are classified in three different phyla.

▶ The body of a worm is made up of three tissue layers—the ectoderm, endoderm, and mesoderm.

▶ Flatworms, members of the phylum *Platyhelminthes,* have a flattened body.

▶ Parasites, such as flukes and tapeworms, feed on and harm the organisms they live in.

▶ Roundworms, members of the phylum *Nematoda,* can live almost anywhere.

▶ The body of a roundworm is more complex than that of a flatworm.

CHECK *Complete the following.*

1. There are _____ different phyla of worms.

2. The tissue layer between the ectoderm and the endoderm is the _____ .

3. The phylum *Platyhelminthes* is made up of the _____ .

4. Special body parts, such as hooks and _____, help some parasites live inside other organisms.

5. The phylum *nematoda* is made up of the _____ .

APPLY *Complete the following.*

▶ 6. **Infer:** The name *Platyhelminthes* is based on the Greek word "*platys*" meaning broad and flat. How does this definition help you remember an important trait of this phylum?

7. **Hypothesize:** A parasite often causes the organism it lives in to lose weight. Why?

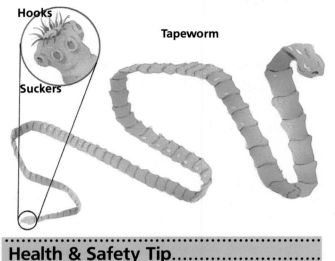

Hooks
Tapeworm
Suckers

Health & Safety Tip............................

Cooking pork at high temperatures kills the *trichina* (tri-KY-nuh) worm which is a parasite of pigs. Eating undercooked pork can cause a disease called trichinosis (trik-uh-NOH-sis). Find out how long pork must be cooked to kill the trichina worm.

SCIENCE CONNECTION ◆○◆○◆○◆○◆○◆○◆○◆○◆○◆○◆○◆○◆○◆○◆

PETS AND WORMS

Do you take your dog or cat to the veterinarian for check-ups? One thing that the veterinarian checks for is the presence of worms. Roundworms and hookworms are common parasites of both dogs and cats. Both parasites are found in the dirt. Pets can swallow roundworm eggs by eating scraps of food found in the dirt. Pets can get hookworms by walking on an infected area. Hookworms go through the skin on the animal's feet.

Heartworms are the most dangerous parasite of dogs. Heartworms are carried by mosquitos. Dogs can get heartworms by being bitten by an infected mosquito. Veterinarians supply pills that can prevent infections.

To prevent worms, keep your pet on a leash or in your yard. Clean its area regularly. Feed it a balanced diet. Also, have your pet checked by a veterinarian. If worms are found, the pet can be treated. It is important to start treatment as soon as possible. It can save your pet's life.

Female   Male

What are segmented worms?

Objective ▶ Describe the features of segmented worms.

TechTerms

- **closed circulatory system:** an organ system in which blood moves through blood vessels
- **setae** (SEET-ee): tiny, hairlike bristles

Segmented Worms The most complex worms are the segmented worms. They are classified in the phylum *Annelida* (AN-ul-id-uh). The word "annelid" means "little rings." If you look at the body of a segmented worm, you will see that it is made up of many ringlike sections, or segments.

There are more than 9000 species of segmented worms. Most live in the ocean. Others live in fresh water. The best-known segmented worm is the earthworm. It lives in the soil.

▶ *Name:* Where do most segmented worms live?

Earthworms Earthworms are the most complex worms. Look at Figure 1 as you read about the features of the earthworm.

- Like all worms, earthworms have a right side and a left side. They have a head end and a tail end. There is a mouth at the head end and an anus at the tail end.
- Each segment except the first and last has four pairs of small bristles. They are called **setae.**

Earthworms use their setae and tiny sets of muscles to move.

- Earthworms have a complex digestive system. Food passes into two organs where it is stored, and then ground up. Food is digested and absorbed in the small intestine.
- Earthworms have a **closed circulatory system.** In a closed circulatory system, blood moves through tubes in the body. In the head end, the two large vessels meet and form five pairs of hearts. The hearts pump blood through the blood vessels.
- Earthworms have a nervous system. Nerves run along the body and connect to a simple brain at the head end.
- Earthworms have male and female sex organs. A single worm, however, does not mate with itself. Earthworms reproduce sexually.
- Earthworms do not have a respiratory system. Gases pass into and out of the earthworm through its moist skin.

▶ *Observe:* What are the two organs that store and grind up food called?

Leeches The leech is a segmented worm that lives in ponds and streams. Most leeches are parasites. They live on the surface of other animals. The leech uses suckers to attach itself to the animal. The leech feeds on the blood of the animal.

▶ *Compare:* How do leeches differ from some other worms that are parasites?

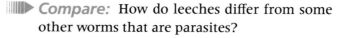

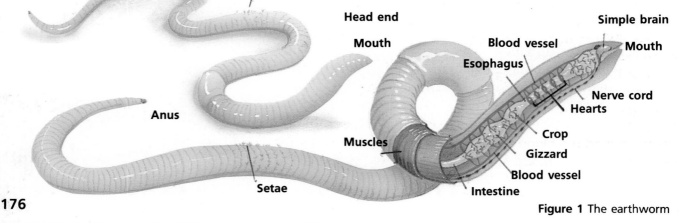

Tail end

Setae

Anus

Setae

Head end

Mouth

Muscles

Blood vessel

Esophagus

Simple brain

Mouth

Nerve cord

Hearts

Crop

Gizzard

Blood vessel

Intestine

Figure 1 The earthworm

LESSON SUMMARY

▶ Segmented worms belong to the phylum *Annelida*.

▶ There are more than 9000 species of segmented worms.

▶ An earthworm is divided into segments and has setae on its bottom surface.

▶ The earthworm is typical of segmented worms.

▶ Earthworms have a closed circulatory system.

▶ Earthworms have a complex digestive system.

▶ Earthworms have a reproductive system and a nervous system.

▶ Earthworms lack a respiratory system.

▶ Most leeches are parasites.

CHECK *Complete the following.*

1. Segmented worms are classified in the phylum _____ .

2. Segmented worms live in the _____, fresh water, and soil.

3. The earthworm uses small bristles called _____ for movement.

4. Gases pass in out of the earthworm through the _____ .

APPLY *Complete the following.*

6. **Explain:** The dictionary shows that the word *Annelida* means "little ring." Explain why this term is a good description for segmented worms.

▶ 7. **Infer:** Why do earthworms avoid direct sunlight?

InfoSearch

Read the passage. Ask two questions about the topic that you cannot answer from the information in the passage.

Importance of Earthworms Earthworms help to improve the soil. They make tunnels as they move through the soil. Air and water can then flow through the tunnels and reach the roots of plants. Earthworms get food from soil. Undigested food is passed out through the anus as wastes. These wastes help make rich soil.

SEARCH: Use library references to find answers to your questions.

◢◣◥ LOOKING BACK IN SCIENCE ▽▽▽▽▽▽▽▽▽▽▽▽▽▽▽▽▽▽

MEDICAL USES OF LEECHES

In the past, "bleeding" was a common medical treatment. Leeches would be placed on a patient's skin. The leeches would attach themselves and suck out blood. In a few cases, bleeding was helpful to the patient. In most cases, however, it was not a good treatment.

Barbers used to use leeches to treat black eyes. The blackness of the eye is caused by blood from broken capillaries. To speed up the healing, barbers would place a leech on the blackened part of the eye. The leech would take in the blood that collected. Today, leeches are again being used in medicine. Scientists have discovered special chemicals given off by leeches. One of these chemicals keeps blood from clotting. Another chemical widens blood vessels. Others help heal damaged tissue. One chemical even helps to prevent infection.

Using chemicals from leeches has important medical uses. In some types of surgery, tiny blood vessels are reconnected. Often the blood stops flowing through the vessels. As a result, tissues around the blood vessels die. Chemicals taken from the leech can keep the blood flowing and keep the tissues healthy.

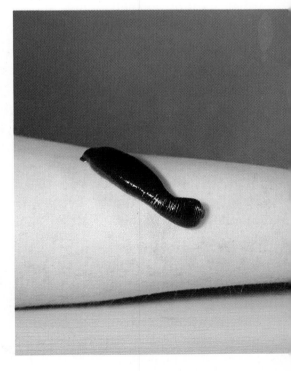

Objectives
▶ Describe the features of mollusks. ▶ Give examples of different kinds of mollusks.

TechTerms

▶ **mantle:** thin, membrane that covers a mollusk's organs

▶ **radula** (RAJ-oo-luh): rough, tonguelike organ of a snail

Mollusks The mollusks are the soft-bodied animals. They are classified in the phylum *Mollusca* (muh-LUS-kuh). All mollusks have a soft, fleshy body. Most mollusks are covered by hard shells. Mollusks live in salt water, in fresh water, and on land.

The mollusks are divided into three major classes. One class is made up of mollusks with one-shell. This class includes snails. Another class consists of mollusks made up of two-part shells. Clams, oysters, and mussels are mollusks with two-part shells. The third class of mollusks have a small shell or no shell. Squids and octopuses are classified in this group.

📄 *Classify:* Name four animals classified as mollusks.

Three Body Parts Mollusks have three body parts. They have a head, a foot, and a mass of tissue that contains well-developed organ systems, reproductive organs, and a heart. A thin membrane covers the soft fleshy body of a mollusk. The membrane is the **mantle.** In some mollusks, the mantle forms the shell.

▷ *Define:* What is the mantle?

Organ Systems Mollusks have a digestive system and a reproductive system. Mollusks also have a nervous system. The nervous system has many sense organs. It is well-developed in octopuses and squids.

Mollusks have an open circulatory system. The blood is not always inside blood vessels. Instead, the blood flows out of vessels into open spaces in the body. It bathes the body organs. At another point, the blood flows back into a blood vessel. Blood is moved by the beating of the heart.

▷ *Name:* What two kinds of mollusks have well-developed nervous systems?

Special Structures In mollusks with shells, the foot is a muscular organ. It is used for movement. In squids and octopuses, the foot is divided into tentacles. They are used for movement and feeding. Squids and octopuses swim by jet action. They take in water and squirt the water out through a tubelike structure in its head end.

Many mollusks, such as snails, have a **radula** (RAJ-oo-luh). The radula is a tongue-like organ. It is covered with hooked toothlike structures. The radula is used to scrape food, such as algae, from plants and rocks.

▷ *Define:* What is a radula?

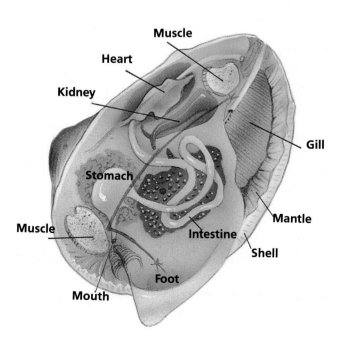

178

LESSON SUMMARY

▶ Mollusks are soft-bodied animals classified in the phylum *Mollusca*.

▶ Mollusks are classified into three major classes.

▶ Mollusks have a head, a foot, and a fleshy mass of tissue that contains well-developed organs and organ systems.

▶ Mollusks have a digestive system, reproductive system, nervous system, and an open circulatory system.

▶ In mollusks with shells, the foot is muscular; in squids and octopuses, the foot is divided into tentacles.

▶ Many mollusks have a radula used for getting food.

CHECK *Write true if the statement is true. If the statement is false, changed the underlined term to make the statement true.*

1. The foot is a fleshy tissue that covers the internal organs of a mollusk.

2. The radula is a tongue-like organ covered with hooked, toothlike structures.

3. Snails are one-shelled mollusks.

4. Octopuses and clams have tentacles.

APPLY *Complete the following.*

5. Newton's Third Law of Motion states that for every action there is an equal and opposite reaction. Explain what this means. Compare this law of motion to a squid's movement.

InfoSearch

Read the passage. Ask two questions that you cannot answer from the information in the passage.

Chambered Nautilus The chambered nautilus is classified as a head-foot mollusk. Unlike the octopus and squid, the chambered nautilus forms a shell outside its body. When the nautilus outgrows its shell, it adds on a new chamber, or section. In time, the nautilus shell has many chambers. The nautilus lives in the outer chamber. It can move up and down in the water by filling or emptying its chambers with gases.

SEARCH: Use library references to find answers to your questions.

ACTIVITY

MODELING SQUID JET PROPULSION

You will need a plastic straw, scissors, 4 meters of string, cellophane tape, a balloon, and 2 desk chairs.

1. Use the scissors to cut a 2-cm long piece of the straw. **CAUTION: Be careful when using scissors.**
2. Thread the string through the short piece of straw. Tape the straw to the middle of the uninflated balloon.
3. Tie the string to the backs of two chairs placed about 4 m apart.
4. Blow up the balloon. Hold the end closed with your fingers.
5. Release the balloon. Observe the direction it moves.

Questions

1. In which direction did the balloon move?
2. How does the movement of the balloon compare to the movement of the squid?
3. **Analyze:** How do you think the squid would move if it took in a large amount of water?

Objective ▶ List the main features of arthropods and give examples of some arthropods.

TechTerms

- ▶ **chitin** (KYT-in): hard material that makes up the exoskeleton of arthropods
- ▶ **exoskeleton** (ek-so-SKEL-uh-tun): hard covering that supports and protects the body
- ▶ **molting**: process by which an animal sheds its outer covering

Classifying Arthropods Arthropods (ar-thruh-PODS) are animals that have jointed legs. They are classified in the phylum *Arthropoda* (AHR-thruh-pahd-uh). Arthropods make up the largest group of animals. Scientists think that there are more than one million species.

Arthropods are divided into several classes. The largest classes are the insects, crustaceans (krus-TAY-shuns), and arachnids (uh-RAK-nidz). Insects include flies, butterflies, moths, ants, bees, mosquitoes, cockroaches, and many others. Crustaceans include lobsters, crabs, and shrimp. Arachnids include spiders, ticks, mites, and scorpions.

▶ *Identify:* Name the three largest classes of arthropods.

Arthropod Features All arthropods have a hard body covering and a segmented body. The hard covering is an **exoskeleton** (ek-so-SKEL-uh-tun). An exoskeleton is an external skeleton. It protects and supports the soft, inner body parts. The exoskeleton is made up of a material called **chitin** (KYT-in). Chitin is tough, but light in weight.

Arthropods have jointed legs. The legs are made up of several pieces. The pieces are connected by joints, which can bend. The legs are moved by muscles attached to the exoskeleton.

Arthropods have an open circulatory system. Blood moves through open spaces in their bodies. It does not flow through tubes.

▶ *Identify:* What material makes up the exoskeleton of an arthropod?

Molting When you grow, your bones grow. Your skeleton grows along with the rest of your body. An exoskeleton, however, cannot grow. It is not made of living material. As an arthropod grows, it becomes too big for its exoskeleton. At this time, the exoskeleton is shed. This process is called **molting.** The exoskeleton splits open and the animal works its way out. Gradually a new, larger exoskeleton grows.

▶ *Define:* What is molting?

Special Body Parts Depending upon where they live, what they eat, and how they move, arthropods have different body parts. Insects have special mouth parts that help them get food. These mouth parts are used for biting, chewing, or sucking, depending on how the insect feeds. Most insects also have wings. Some crustaceans, such as lobsters and crabs, have large front claws. They

use the claws to get food and to protect themselves. Spiders have silk glands that make silk for their webs. Many different types of arthropods also have antennae that are used for touch and taste.

▶ *List:* What are some special body parts of arthropods?

LESSON SUMMARY

► Arthropods are animals with jointed legs.

► The largest arthropod classes are the insects, crustaceans, and arachnids.

► Arthropods have jointed legs, an exoskeleton, and a segmented body.

► Arthropods have an open circulatory system.

► When an arthropod grows, it molts its old exoskeleton, and grows a new one.

► Arthropods have special body parts that are suited to the foods they eat, the places they live, and the way they move.

CHECK *Explain the relationship between the words in each pair.*

1. chitin, exoskeleton
2. arthropod, arachnid

Complete the following.

3. What is the name of the largest animal phylum?
4. What are the three arthropod features?
5. Describe molting.
6. How are large claws helpful for a crustacean?

APPLY *Complete the following.*

7. **Classify:** Identify the class of each arthropod shown on page 180.
8. **Analyze:** Can you tell which of these arthropods can fly? How?
9. **Infer:** How are the grasshopper's large back legs suited to how it moves?

....
Health & Safety Tip..........................

Watch out for ticks in heavily wooded areas. When you walk through the woods, tuck your pants into your socks. Wear long sleeves and a cap. Some ticks are carriers of Lyme disease. Lyme disease is a serious illness with many different symptoms.

Use reference books to find out about the rash that may form from Lyme disease. What are other symptoms of the disease? What is the role of the white-tailed deer in Lyme disease?

TECHNOLOGY AND SOCIETY

USES OF CHITIN

What do crab shells, some contact lenses, and certain types of varnish have in common? They are all made of chitin. Scientists are finding new uses for tough, protective chitin.

Chitin is especially good as a material for medical uses. It does not cause allergic reactions in most people. For this reason, chitin is used to make some kinds of contact lenses. Chitin is also used to make artificial skin. This skin can be used to protect badly burned parts of the body until new skin can form. A type of thread made from chitin is used for sewing up wounds or incisions. This type of thread dissolves in the body. It does not need to be removed.

Researchers have found many other uses for chitin. Varnish made of chitin makes an excellent finish for some musical instruments. Seeds coated with chitin are protected from infection. Chitin has also been used in water filters. Research on the uses of chitin continues. Chitin is now being tested as a coating for fabrics and paper. Who knows? Some day you may be wearing chitin-treated T shirts and reading science lessons printed on chitin-coated paper!

Objective ▶ Describe the features of insects.

TechTerms

- ▶ **abdomen** (AB-duh-mun): third section of an insect's body
- ▶ **spiracles** (SPIH-ruh-kuls): openings to air tubes of a grasshopper
- ▶ **thorax** (THOR-aks): middle section of an insect's body
- ▶ **tympanum** (TIM-puh-num): hearing organ in a grasshopper

Insects Insects are the largest class of arthropods. There are more different kinds of insects than any other kind of animals. In fact, more than 700,000 insects have been identified. Insects live almost everywhere, except the oceans. They live in the air and in the soil. They live inside the walls of buildings and on the leaves of plants. Beetles, fleas, termites, moths, and ants are all insects.

☛ *Classify:* What is the largest class of arthropods?

The Grasshopper The grasshopper is a good model to use when you study insects. Like all arthropods, a grasshopper has jointed legs, a segmented body, and an exoskeleton.

Like all insects, a grasshopper has three pairs of legs and antennae (an-TEN-ay). Antennae are sense organs. All insects have three body segments. They are the head, **thorax** (THOR-aks), and **abdomen** (AB-duh-mun). The thorax is the middle section of an insect's body. The three pairs of legs are attached to the thorax. The abdomen is the third section of an insect's body. Like most insects, the grasshopper also has wings.

Find the **spiracles** (SPIH-ruh-kuls). The spiracles are openings to air tubes inside the grasshopper. The grasshopper breathes by moving air in and out through the spiracles. Find the **tympanum** (TIM-puh-num). It is used for hearing.

▶ *State:* What are three features of all insects?

Social Insects Most insects live alone. Some live together in colonies. Insects that live in colonies are called social insects. Ants and honeybees are social insects.

In a honeybee hive, there are three kinds of honeybees. They are the queen, the drones, and the workers. Each one has its own job in the beehive. The queen bee lays eggs. The drones fertilize the eggs. The workers take care of the eggs. They feed the queen and the drones. Workers gather honey and pollen. Workers also build and protect the hive.

▶ *Name:* Name the three kinds of honeybees.

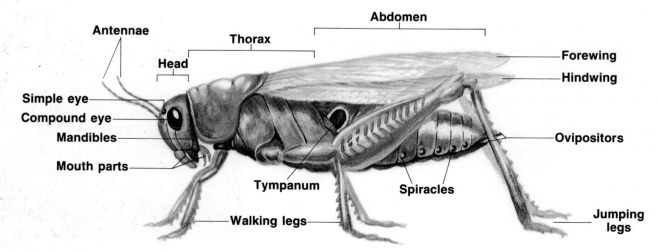

LESSON SUMMARY

- There are more species of insects than of all other animal species put together.

- A grasshopper is a good insect model to study.

- All insects have three pairs of legs, three body segments, and antennae.

- Spiracles and the tympanum are located on the abdomen of a grasshopper.

- Social insects, such as honeybees and ants, live together in colonies.

- In a beehive, each kind of bee has its own job.

CHECK *Answer the following.*

1. What animals make up the largest class of arthropods?

2. What are three body parts of an insect?

3. How many legs do insects have?

4. What are the openings to air tubes on a grasshopper called?

5. What do grasshoppers use for hearing?

6. What are insects that live in colonies called?

APPLY *Use the pictograph to answer the following.*

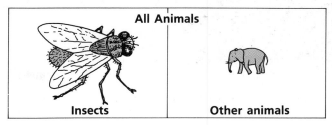

All Animals

Insects | Other animals

7. **Interpret:** How does the number of insect species compare to the number of all other animal species?

8. **Explain:** Why is the drawing of the fly larger than the drawing of the elephant?

InfoSearch

Read the passage. Ask two questions about the topic that you cannot answer from the information in the passage.

Helpful and Harmful Insects Some insects are helpful. Others are harmful. Bees and butterflies help pollinate flowers. Bees produce honey and wax that are used by people. Insects are food for many larger animals. Some insects, such as termites, eat wood and destroy buildings. Weevils ruin food crops. Other insects carry diseases.

SEARCH: Use library references to find answers to your questions.

SCIENCE CONNECTION

CAMOUFLAGE AND MIMICRY

Can you find the walking stick in the photograph? The walking stick looks like a twig. When it stands still, birds cannot see it. The walking stick is camouflaged, or hidden, by its body shape. Many insects and other animals are protected by camouflage. Their body shapes or coloring help them blend into their surroundings.

Some animals do not hide from their predators. They may have bright colors and bold patterns. Often these brightly colored animals mimic, or look like, other animals. The viceroy butterfly looks like a monarch butterfly. Birds avoid eating the monarch butterfly because it tastes bad to birds. Because the harmless viceroy butterfly looks like the monarch butterfly, birds avoid the viceroy butterfly. Mimicry can be found in other insects and in larger animals too. For example, some kinds of fish have a large dark circle on their tails. These circles mimic the eyes of a larger and more dangerous fish.

How do insects develop?

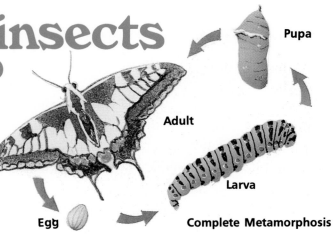

Pupa

Adult

Larva

Egg

Complete Metamorphosis

Objective ▶ Identify and describe the stages of metamorphosis.

TechTerms

▶ **cocoon** (kuh-KOON): protective covering around the pupa

▶ **larva** (LAHR-vuh): wormlike stage of insect development

▶ **metamorphosis** (met-uh-MOR-fuh-sis): changes during the stages of development of an insect

▶ **nymph** (NIMF): young insect that looks like the adult

▶ **pupa** (PYOU-puh): resting stage during complete metamorphosis

Metamorphosis All insects lay eggs. The developing insect, or embryo (EM-bree-oh), feeds on yolk stored in the egg. The eggs are laid on or near a food supply. After the eggs hatch, the young insect uses this food.

When an insect hatches from its egg, the insect usually does not look at all like the adult insect. For example, a small butterfly does not hatch from a butterfly egg. After the insect hatches from the egg, the insect goes through changes in form and size. The changes during the stages of development of an insect are called **metamorphosis** (met-uh-MOR-fuh-sis).

▶ *Define:* What is metamorphosis?

Incomplete Metamorphosis Some insects, such as grasshoppers, undergo incomplete meta-

morphosis. Incomplete metamorphosis has three stages. They are the egg, **nymph** (NIMF), and adult. The eggs hatch into nymphs. Nymphs look very much like a small adult. Nymphs have big heads and no wings. They also do not have reproductive organs. Nymphs gradually change and grow into an adult insect.

▶ *List:* What are the three stages of incomplete metamorphosis?

Complete Metamorphosis Some insects, such as butterflies and moths, undergo complete metamorphosis. Complete metamorphosis has four stages. There is a change in body form in each stage of development.

The egg hatches into a **larva** (LAHR-vuh). The larva is a wormlike stage of an insect. A caterpillar is the larva of a butterfly or moth. Larvas eat a lot of food. They grow very quickly. After a time, the larva goes into a resting stage called the **pupa** (PYOU-puh). During this stage, many insects spin a covering around themselves. The covering is a **cocoon** (kuh-KOON). In the pupa stage, the insect does not eat. Many body changes take place inside the cocoon. The organs of the adult form. When the adult has formed, the cocoon opens. An adult insect comes out.

▶ *Sequence:* Place the stages of complete metamorphosis in the correct order.

Incomplete Metamorphosis

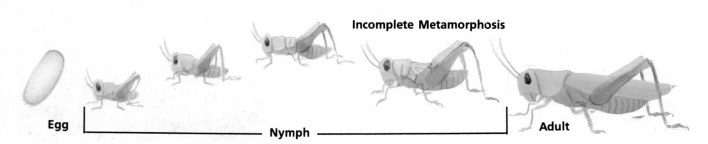

Egg

Nymph

Adult

LESSON SUMMARY

▶ All insects lay eggs on or near a food supply.

▶ The changes in form and size during the stages of development of an insect are called metamorphosis.

▶ The three stages of incomplete metamorphosis in insects are egg, nymph, and adult.

▶ The four stages of complete metamorphosis in insects are egg, larva, pupa, and adult.

CHECK *Complete the following.*

1. In insects, the third stage in complete metamorphosis is the _____ .

2. In insects, the first stage of metamorphosis is always an _____ .

3. A caterpillar is the _____ stage of a butterfly.

4. The stage of complete metamorphosis in which some insects look like a small adult is the _____ .

5. The larva stage of a moth is a _____ .

APPLY *Complete the following.*

6. **Compare:** In what way are the eggs of insects like the eggs of birds?

7. **Infer:** Why is the pupa called the resting stage?

8. **Sequence:** Place the stages in the development of the beetle shown below in the correct order.

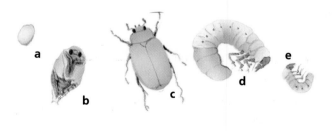

Skill Builder

📋 ***Classifying*** When you classify, you group things based upon how they are alike. Find out whether each of the insects listed below undergoes complete or incomplete metamorphosis. If the insect undergoes complete metamorphosis, write the name of its larval stage.

a. locust   e. fly
b. cricket   f. moth
c. beetle   g. wasp
d. cockroach

◆●◆ CAREER IN LIFE SCIENCE ◆◆◆◆◆◆◆◆◆◆◆◆◆◆◆◆◆◆◆◆◆◆◆◆◆◆◆◆◆◆◆

EXTERMINATOR

Many insects, such as bees, are helpful to people. They supply honey and pollinate flowers. Other insects, such as termites, can destroy property. Still others, such as grasshoppers and locusts, can destroy crops. An exterminator (ik-STUR-muh-nayt-ur) gets rid of unwanted pests in buildings. Exterminators usually are called to get rid of termites and cockroaches. Some exterminators also rid buildings of mice and rats.

Exterminators work for city or county governments. Some work for private exterminating companies. Others have their own companies.

A high school diploma is needed to become an exterminator. You also need to take vocational courses in pest control methods. Courses in biology and chemistry should be taken in high school.

For more information, you may want to call your local board of health or write a letter to a private exterminating company.

What are echinoderms?

TechTerms

- ► **echinoderms** (ee-KY-noh-durms): spiny-skinned animals
- ► **endoskeleton** (en-doh-SKEL-un-tun): internal skeleton
- ► **tube feet:** small suckerlike structures of echinoderms

Echinoderms The **echinoderms** (i-KY-noh-durms) are the spiny-sinned animals. They are classified in the phylum *Echinodermata* (i-KY-noh-dur-muh-toh). They live on the ocean bottom. Sea stars are the best known echinoderms. Other echinoderms are sea urchins, sand dollars, brittle stars, and sea cucumbers. Some echinoderms move about slowly. Others attach themselves to rocks or other objects.

▐▶ *List:* Name three echinoderms.

Anatomy of Echinoderms Echinoderms have an **endoskeleton** (en-doh-SKEL-un-tun), or internal skeleton. The endoskeleton is made up of spines. The most noticeable thing about echinoderms are their spiny skin.

In some echinoderms, such as the sea stars, the spines are hard, rounded lumps. In the sea urchins, the spines are like long needles. The skin of the sea cucumber is soft and leathery.

Echinoderms do not have a left side or a right side. Most have rays, or arms around a central point. Their body structures are arranged like the spokes of a wheel around a central body point. Echinoderms do not have circulatory, excretory, or respiratory systems. They have a nervous system, but echinoderms do not have a brain.

👁 *Observe:* What is the most noticeable thing about the echinoderms in the photographs?

Tube Feet Sea stars move using small, suckerlike structures called **tube feet.** The tube feet are part of a water-pumping system that works to pull the sea star forward. There are hundreds of tube feet. Each foot is like a suction cup. Using its tube feet, a sea star can move slowly across the ocean floor.

Tube feet also are used to get food. Sea stars eat mollusks. They use their tube feet to pull on mollusk shells, such as clams. The clam uses its muscles to keep its shell closed. Soon, the clam muscles tire and the shell opens a bit. The sea star then pushes its stomach out of its mouth into the clam. The clam is digested, and the sea star pulls its stomach back inside its body.

▐▶ *Name:* What two things does a sea star use its tube feet for?

LESSON SUMMARY

▶ Echinoderms are the spiny-skinned animals.

▶ Echinoderms have an endoskeleton made up of spines.

▶ Echinoderms have body structures arranged like the spokes of a wheel around a central body point.

▶ Tube feet are part of a water-pumping system used to move and to get food.

CHECK *Answer the following.*

1. Name four echinoderms.

2. What kind of skeleton do echinoderms have?

3. Which echinoderm has long needle-like spines?

4. What are tube feet?

Complete the following.

5. Sea stars eat oysters, mussels, and _____ .

6. Sea stars are classified in the phylum _____ .

APPLY *Complete the following.*

7. **Infer:** Why do you think a sea star would have trouble moving over sand or mud?

8. **Infer:** If all the sea stars in a clam bed were destroyed, what would happen?

Skill Builder

Building Vocabulary Animals have different kinds of symmetry. For example, sea stars have radial symmetry. Use a dictionary or library references to define bilateral symmetry, spherical symmetry, and radial symmetry. Then trace the outlines of the animals shown below. What kind of symmetry does each animal have? Draw a line or lines through each outline to show the animal's symmetry.

LEISURE ACTIVITY

SEASHORE COLLECTIONS

Making a seashore collection is a fun hobby. It also is an easy hobby to begin, especially if you live near an ocean. You can collect dried sea stars from the beach. Many different kinds of shells also can be collected. If you are very lucky, you may even find sand dollars on the beach. You should avoid collecting living animals.

If you do not live near a beach or if a trip to the ocean is not possible, you still can begin a seashore collection. Shells, sea stars, and sand dollars can be bought from shell dealers or other seashore collectors. There are mail-order catalogues, too. These catalogues offer many different seashore collectibles.

Once you begin gathering specimens for your seashore collection, you may want to buy a handbook or guide for collectors to help you identify your specimens. Most collectors label their collections. They mount their collections or keep them in small containers.

What is regeneration?

Objective ▶ Explain how some organisms can regrow lost parts.

TechTerm

▶ **regeneration** (ri-jen-uh-RAY-shun): ability to regrow lost parts

Regeneration New plants can grow from a part of a plant by vegetative propagation (VEJ-uh-tayt-iv prahp-uh-GAY-shun). Does this kind of asexual reproduction also take place in animals? Can a new animal grow from part of an animal? Can parts of an animal that are broken off grow back? Some animals have the ability to regrow lost parts. The ability of an animal to regrow lost parts is called **regeneration** (ri-jen-uh-RAY-shun). In other animals, a whole new animal can develop from just a part of an animal. If an entire animal develops from a part, regeneration is a kind of asexual reproduction.

▶ **Define:** What is regeneration?

Regeneration in Animals All animals do not have the same abilities to regenerate, or regrow lost body parts. A few kinds of animals can regenerate large body parts. Most animals, however, can regrow only a small part. Lobsters and crabs can regenerate claws that have broken off. The glass-tailed lizard can regenerate its tail. If the lizard is attacked and grabbed by its tail, its tail breaks off. The lizard escapes. Gradually, its tail grows back. Most sea stars have five arms, or rays. If a ray is cut off, a new ray grows back.

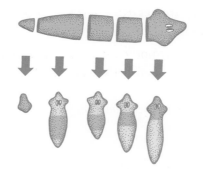

Some animals can reproduce asexually by regeneration. The flatworm, planaria can be cut in half. Each half can regenerate into a new planarian. Sea stars also can reproduce by regeneration. If a sea star is cut into pieces, each piece can develop into a new sea star. However, each piece must have part of the center of the sea star's body to regenerate into a new sea star.

▶ **Name:** What are two organisms that can reproduce by regeneration?

Regeneration in Humans The ability to regenerate body parts in humans is very small. The human body can regenerate some lost or damaged cells and tissues. If you cut yourself, new skin cells grow back to heal the cut. Broken bones grow back together again. The human body, however, cannot regenerate a whole organ, such as a kidney. The human body cannot regenerate large parts. A lost finger, for example, cannot grow back.

▶ **Describe:** What ability of regeneration does the human body have?

LESSON SUMMARY

▶ The ability of an animal to regrow lost parts is called regeneration.

▶ All animals do not have the same abilities to regenerate body parts.

▶ Some animals can reproduce asexually by regeneration.

▶ The human body only can regenerate some lost or damaged cells and tissues.

CHECK *Write true if the statement is true. If the statement is false, change the underlined term to make the statement true.*

1. Regeneration is a kind of <u>sexual</u> reproduction.

2. Regeneration of lost parts takes place in <u>some</u> animals.

3. The ability to regenerate in humans is very <u>small</u>.

4. A <u>sea star</u> can regenerate lost rays.

5. Crabs and lobsters can regenerate their <u>entire</u> bodies.

APPLY *Complete the following.*

6. How are vegetative propagation and regeneration similar?

7. **Infer:** Why is regeneration of a lobster's claw not a kind of asexual reproduction?

8. **Contrast:** How is regeneration in a crab different than regeneration in a sea star?

State the Problem

Sea stars feed on clams, oysters, and mussels. People who gather clams, oysters, and mussels to sell as food used to catch sea stars. They would chop up the sea stars and throw the pieces back into the shellfish beds. What problem do you see with doing this?

TECHNOLOGY AND SOCIETY

ORGAN TRANSPLANTS

Humans cannot regenerate lost or damaged organs. Doctors have found another way to help people who have lost certain organs or have damaged organs. Doctors have learned to transplant organs from one person to another. Transplanting organs makes it possible to replace some organs that are damaged or diseased.

Today, many different organs are transplanted. Some organs that can be transplanted in humans are the liver, kidneys, heart, pancreas, and lungs. Parts of the human digestive system also can be transplanted. Many lives can be saved by organ transplants.

Before an organ or organs can be transplanted, a person's tissue must be matched to the donor. A tissue match is important so that the person's body receiving the organ does not reject the organ. Transplants have become more successful since a new anti-rejection drug was discovered. The drug is called cyclosporine. The drug was discovered by a surgeon, Dr. Thomas L. Starzl. The drug helps keep the person's body receiving the organ from rejecting the transplanted organ.

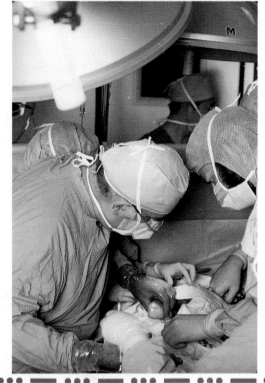

STUDY HINT Before you begin the Unit Challenges, review the TechTerms and Lesson Summary for each lesson in this unit.

TechTerms..

abdomen (182)
chitin (180)
closed circulatory system (176)
cnidocytes (172)
cocoon (184)
echinoderms (186)
ectoderm (174)
endoderm (174)
endoskeleton (186)
exoskeleton (180)

flagellum (170)
larva (184)
mantle (178)
medusa (172)
mesoderm (174)
metamorphosis (184)
molting (180)
nymph (184)
polyp (172)
pores (170)

pupa (184)
radula (178)
regeneration (188)
setae (176)
spicules (170)
spiracles (182)
tentacles (172)
thorax (182)
tube feet (186)
tympanum (182)

TechTerm Challenges..

Matching *Write the TechTerm that matches each description.*

1. whiplike structure on some cells
2. small, hard needle-like structures of a sponge
3. long, thin armlike structures of a cnidarian
4. organ system where blood moves through the body in tubes
5. thin membrane that covers the mollusk's body
6. tough, lightweight material that makes up an arthropod's exoskeleton
7. third section of an insect's body
8. section of insect where legs attach to
9. protective covering of a pupa
10. resting stage of metamorphosis
11. small, suckerlike structures that help echinoderms move
12. ability to regrow lost parts
13. spiny-skinned animals
14. changes during the stages of development of an insect
15. hearing organ of grasshopper
16. openings to the air tubes of a grasshopper
17. shedding of an exoskeleton
18. tongue-like organ of snails and other mollusks
19. bristles on an earthworm that help it move
20. middle layer of cells
21. stinging cells of a cnidarian
22. tiny openings

Applying Definitions *Explain the difference between the terms in each pair.*

1. ectoderm, endoderm
2. exoskeleton, endoskeleton
3. medusa, polyp
4. larva, nymph

Content Challenges..

Multiple Choice *Write the letter of the term that best completes each statement.*
1. Cnidarians can live either as a polyp or as a
   a. medusa. **b.** sponge. **c.** gemmule. **d.** echinoderm.

2. Spicules are found in a sponge's
   a. jellylike substance. b. mesoderm. c. endoskeleton. d. ectoderm.

3. All porifera and cnidarians have
   a. four cell layers. b. three cell layers. c. two cell layers. d. one cell layer.

4. The simplest worms are the
   a. segmented worms. b. roundworms. c. flatworms. d. tapeworms.

5. The worms that can live in the human body are
   a. planaria and flukes. b. flukes and tapeworms. c. tapeworms and segmented worms. d. planaria and hookworms.

6. Roundworms belong to the phylum
   a. *Platyhelminthes.* b. *Porifera.* c. *Nematoda.* d. *Annelida.*

7. Segmented worms include
   a. hookworms and trichina. b. hydra and tapeworms. c. roundworms and leeches.
   d. earthworms and leeches.

8. The body of segmented worms is
   a. ringlike. b. smooth. c. threadlike. d. rough.

9. Clams and oysters belong to the class of mollusks that have
   a. no shell. b. a one-part shell. c. a two-part shell. d. a three-part shell.

True/False *Write true if the statement is true. If the statement is false, change the underlined term to make the statement true.*
1. The outer layer of cells of a cnidarian is called the endoderm.
2. Invertebrates are animals with a backbone.
3. The umbrellalike form of a cnidarian is called a polyp.
4. An earthworm uses its setae and muscles to move.
5. Snails, clams, and squids all are examples of echinoderms.
6. A sea star uses flagella to move.
7. All echinoderms have an endoskeleton.
8. The development of a whole organism from one of its parts is called generation.
9. The largest group of animals on Earth is the mollusks.
10. Insects develop through a series of stages of development called mimicry.

Understanding the Features...

Reading Critically *Use the feature reading selections to answer the following. Page numbers for the features follow each question in parentheses.*

1. **Infer:** Why is it important to always snorkel with a buddy? (171)
2. **Name:** What are the three kinds of coral reef? (173)
3. **Hypothesize:** Do you think humans can get hookworms by walking in infected areas? (175)
4. How are leeches used in modern medicine? (177)
5. How is chitin used in surgery? (181)
6. **Explain:** How does the viceroy butterfly use mimicry? (183)
7. **Infer:** Why do you think biology and chemistry classes are important for an exterminator? (185)
8. Why should you avoid collecting live animals for your seashore collection? (187)
9. What are some organs that have been successfully transplanted? (189)

191

Interpreting a Diagram *Use the diagram of the grasshopper to answer the following.*

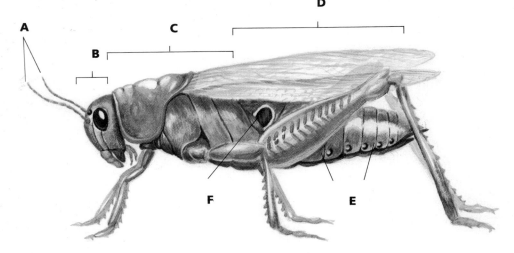

1. What letter represents a grasshopper's hearing organ?
2. What does letter B represent?
3. Which letter represents the openings to air tubes?
4. Which letter represents where an insect's legs attach to?
5. What does letter D represent?
6. Which letter represents the grasshopper's sense organs?

Critical Thinking *Answer each of the following in complete sentences.*

▶ 1. **Infer:** Until the mid-nineteenth century, most scientists thought that sponges were plants. Why might scientists have considered sponges to be part of the Plant kingdom?

▶ 2. **Infer:** Why do you think sponge farmers cut up sponges and throw them back in the water?

▶ 3. **Infer:** Why do you think earthworms avoid sunlight?

4. How does regeneration help some animal species survive?

5. **List:** What are some of the differences between insects and spiders? Why are spiders not classified as insects?

Finding Out More..

1. Due to increasing technology, scientists are now able to build artificial reefs. Use library references to find information about the advantages and disadvantages of artificial reefs. Present your findings in a report.

2. Research some of the methods of insect control. Include one natural method of insect control as well as chemical methods. Write your findings in a report.

3. Use library references to find out how a pearl is formed in an oyster. Present your findings in a report to the class.

4. Soft-shell crabs only are available during certain months of the year. Visit a local fish market and find out when these crabs are available. Also, find out why soft-shell crabs are not always available.

ANIMALS WITH BACKBONES

CONTENTS

STUDY HINT Before beginning Unit 10, scan through the lessons in the unit looking for words that you do not know. On a sheet of paper, list these words. Work with a classmate to try to define each word on your list.

What are chordates?

Objective ► Identify the features of chordates.

TechTerms

► **endoskeleton** (en-duh-SKEL-uh-tun): skeleton inside the body
► **notochord** (NOHT-uh-kowrd): strong, rod-like structure in chordates that can bend
► **vertebrate** (VUR-tuh-brit): animal with a backbone

Phylum Chordata Fishes, frogs, snakes, birds, cats, and many other animals are classified in the phylum *Chordata* (KOWR-dayt-uh). All animals in this phylum have a notochord (NOHT-uh-kowrd) at some time during their development. A notochord is a strong, rodlike structure that can bend. It is used for support. The notochord is located just below the nerve cord.

▌▶ *Define:* What is a chordate?

Other Chordate Features Besides a notochord, all chordates have two other features. All chordates have a hollow nerve cord. It runs down the back of the animal. Chordates also have paired gill slits at some time during their development. In fishes, the gill slits become gills. Gills are used to take in oxygen dissolved in water. In most other chordates, the gill slits disappear as the animal develops.

▌▶ *Name:* What are two chordate features besides a notochord?

Vertebrates Animals with a backbone are called **vertebrates** (VUR-tuh-brits). There are five major groups of vertebrates. They are fishes, amphibians, reptiles, birds, and mammals. Vertebrates have an **endoskeleton** (en-duh-SKEL-uh-tun). An endoskeleton is a skeleton that is inside the body. The endoskeleton protects the inside organs. It also gives support.

▶ *Infer:* What organ is protected by the hard, bony skull of a vertebrate?

Lancelets In most adult chordates, the notochord is replaced by a backbone. Only a few kinds of simple chordates do not have a backbone. The best-known of these is the lancelet. Lancelets are small, fishlike animals. They live buried in the sand on the ocean floor. A lancelet keeps its notochord throughout its life. It is never replaced by a backbone.

◢ *Analyze:* Look at the lancelet. Why is it classified as a chordate?

Body Plan of Chordates Chordates have a left side and a right side. They also have a head end and a tail end. During development there are three tissue layers. They are the endoderm, mesoderm, and ectoderm. Each tissue layer develops into different tissues and organs.

Chordates have many organ systems. One of these organ systems is a closed circulatory system. It is made up of a heart and many blood vessels. Chordates also have the most highly-developed nervous systems of all animals.

▶ *Infer:* Based upon the body plan of chordates, are you a chordate?

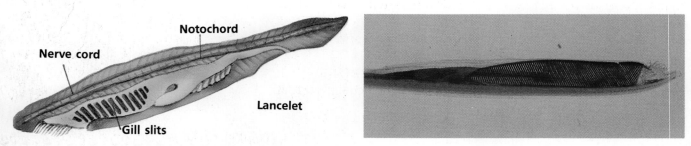

Notochord
Nerve cord
Gill slits
Lancelet

LESSON SUMMARY

▶ Chordates are animals that have a notochord at some time during their development.

▶ A hollow nerve cord and gill slits are other chordate features.

▶ Animals with a backbone are vertebrates.

▶ All chordates have a left and a right side, a head and tail end, and three tissue layers.

▶ Chordates have many organ systems.

CHECK *Complete the following.*

1. A notochord is a strong, flexible _____ structure that can bend.

2. The nerve cord of a chordate runs down the _____ of the animal.

3. A skeleton that is inside the body is an _____ .

4. Animals with backbones are called _____ .

5. An endoskeleton protects inside organs and gives the body _____ .

6. The tissue layers of chordates are the endoderm, the _____, and the ectoderm.

7. Chordates have a _____ circulatory system.

APPLY *Complete the following.*

8. **Identify:** What chordate characteristic makes a fish suited to life in the water?

▶ 9. **Infer:** All vertebrates are chordates. What structure do you think replaces the notochord in vertebrates?

10. **Contrast:** Identify two features of vertebrates that invertebrates do not have.

Skill Builder

Applying Definitions A prefix is a word part that is placed at the beginning of a word. Look up the meanings of the prefixes ''endo-,'' ''meso-,'' and ''ecto-'' in a dictionary. Write the definitions in your notebook. The diagram shows the endoderm, mesoderm, and ectoderm tissue layers.

Use the meanings of the prefixes to identify each of the lettered tissue layers in the diagram. Explain your choices.

CAREER IN LIFE SCIENCE

ANIMAL TECHNICIAN

Do you own a pet? Do you like to care for animals? If so, you may enjoy a career as an animal technician. Animal technicians help veterinarians treat sick animals. Animal technicians keep records for veterinarians. They do laboratory work, and dress wounds of animals. Animal technicians also help veterinarians with equipment. Technicians also may help animals during surgery.

Animal technicians need a high school diploma and must complete a two-year animal technology program. The program trains you in laboratory work and animal research. If you would like to become an animal technician, you should concentrate on doing very well in science courses. Summer or part-time job experience working with animals also is helpful.

For more information: Write to the American Veterinary Medical Association, 930 North Meacham Road, Schaumburg, IL 60196.

What are fishes?

PERCH

Fins

Tail fin

Eye

Gill cover

Mouth

Scales

Fins

Objectives
▶ List three characteristics of fishes.
▶ Name examples of fishes in each class.

TechTerms

▶ **cartilage** (KART-ul-idj): kind of strong, flexible tissue
▶ **coldblooded:** having a body temperature that changes with the temperature of the surroundings
▶ **gills:** organs that absorb dissolved oxygen from water

Fishes Fishes are the simplest and oldest group of vertebrates. Fishes first appeared more than 500 million years ago. Today, fishes live in most of the earth's waters.

Fishes are **coldblooded** animals. The body temperature of a coldblooded animal changes with the temperature of its surroundings. For example, when the water temperature drops, the body temperature of a fish also drops.

▶ *Predict:* What would happen to the body temperature of a fish if the water temperature rose?

Gills Fish breathe through gills. Gills are feathery organs that have many blood vessels. Dissolved oxygen in the water is absorbed by the gills. The oxygen passes into the blood in the gills. Blood carries the oxygen to all the cells in the fish's body. Carbon dioxide passes out of the blood in

the gills, and then into the water.

▷ *Identify:* What organs do fish breathe through?

Three Classes The fishes are the largest group of vertebrates. Fishes have a two-chambered heart. The top chamber is the atrium (AY-tree-um). The bottom chamber is the ventricle (VEN-tri-kul). Fishes are classified in three classes. They are the jawless fishes, the cartilaginous (kahrt-ul-AJ-uh-nus) fishes, and the bony fishes.

Lampreys (LAM-prees) and hagfish are the only jawless fishes. They look very different from most fishes. These fishes are long and snakelike. They do not have scales. They have slimy skin.

The cartilaginous fishes include sharks, rays, and skates. These fish have skeletons made of **cartilage** (KART-ul-idj). Cartilage is a strong, flexible tissue. The tip of your nose and your ears are made of cartilage.

Bony fishes have skeletons made of bone. You probably are most familiar with these fish. Bony fishes include tuna, salmon, bass, and flounder.

Bony fishes are suited to many different environments. Most bony fishes have streamlined bodies for easy movement through the water. They also have different kinds of fins. The fins are used for steering and balance. The body of a bony fish is covered with tough, overlapping scales. The scales cover and protect the skin.

▷ *Classify:* What are the three classes of fish?

LESSON SUMMARY

▶ Fishes are the simplest and oldest group of vertebrates.

▶ Fishes are coldblooded animals.

▶ Fishes breathe through gills.

▶ Fishes have a two-chambered heart.

▶ Lampreys and hagfish are jawless fishes.

▶ Sharks, skates, and rays are cartilaginous fishes.

▶ Bony fishes have skeletons made up of bone.

▶ Bony fishes have streamlined bodies, fins, and a body covered with scales.

CHECK *Find the sentence in the lesson that answers each question. Then, write the sentence.*

1. What is a coldblooded animal?

2. What are gills?

3. How many chambers does a fish's heart have?

4. What are the only two kinds of jawless fishes?

5. What is the main characteristic of a cartilaginous fish?

6. How do fishes use fins?

APPLY *Complete the following.*

👁 7. **Observe:** Look at the picture of the perch on page 196. How many fins does a perch have?

8. **Classify:** Places each of the following fishes into its correct class:

a. sea horse
b. eel
c. hagfish
d. sea skate

e. lamprey
f. hammerhead shark
g. trout
h. whale shark

....
Skill Builder..

▶ *Inferring* When you infer, you form a conclusion based upon facts. A flounder is shown below. Both of a flounder's eyes are on the same side of its head. A flounder also is the color of sand or mud. Both of these features are adaptations. How do you think a sandy color and eyes on one side of the head made a flounder suited to life on the ocean floor?

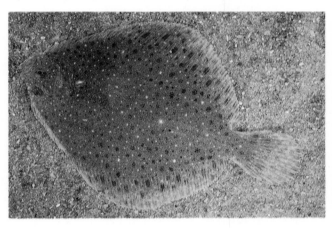

SCIENCE CONNECTION ◆○◆○◆○◆○◆○◆○◆○◆○◆○◆○◆○◆○◆○◆○○

UNUSUAL FISHES

The fishes with unusual names and traits are the sea horse and the porcupine fish. Can you predict why a fish would be named after a horse or a porcupine?

The sea horse looks like a miniature horse. It is a bony fish, with an unusually shaped skeleton. It swims holding its horselike head in an upright position. Female sea horses lay their eggs in a pouch found only on the males. The males carry the eggs in the pouches until the eggs hatch.

The porcupine fish is another unusual bony fish. It is covered with spines. The spines protect the porcupine fish from other animals. When threatened, the porcupine fish makes its spines stick out by filling itself with water. As it takes in water, it becomes larger and rounder. It puffs up like a spiny balloon. Later, when it no longer needs to protect itself, it lets out the water. It becomes smaller and its spines lie flat. Its shape then looks more like that of other fishes.

What are amphibians?

Objective ▶ Identify the characteristics of amphibians.

TechTerm

▶ **amphibian** (am-FIB-ee-un): animal that lives part of its life in water and part on land

Amphibians An **amphibian** (am-FIB-ee-un) is an animal that lives part of its life in water and part on land. In fact the word "amphibian" comes from a Greek word meaning "double life." Most young amphibians live in water. As they grow, they undergo changes. Adult amphibians live mainly on land. Scientists think that the first amphibians developed from fish that could stay out of water for long periods of time. These fish had lungs and could breathe air.

Amphibians are coldblooded. Except for toads, amphibians have smooth, moist skin. Toads have bumpy skin. Amphibians also have webbed feet. Because they live part of their life in water and part on land, they use gills, lungs, and their skin to exchange oxygen and carbon dioxide. Amphibian eggs do not have shells and are laid in water.

❯ *Predict:* As the temperature of the mud that a salamander is in drops, what will happen to its body temperature?

Classifying Amphibians Scientists have classified about 2500 different kinds of amphibians. Scientists classify amphibians into three orders based upon their body structures. The best-known amphibians are frogs and toads. They are the tailless amphibians. Salamanders belong to a different order. They are the amphibians with tails. The third order of amphibians do not have legs. They are wormlike animals. These animals live in the forests of South America.

▶ *List:* What are the three orders of amphibians?

More About Amphibians Amphibians have a closed circulatory system. They have a three-chambered heart. The heart has two atria and one ventricle. Almost all amphibians have gills during the early stages of development. As adults they have lungs. Some salamanders have gills throughout their lives. Many amphibians also breathe through their skin. Their skin must stay moist so that oxygen and carbon dioxide can be exchanged through the skin.

▶ *Infer:* If some kinds of salamanders have gills throughout their lives, in what kind of environment would they live?

LESSON SUMMARY

► Amphibians are animals that live part of their lives in water and part on land.

► Amphibians are coldblooded animals with smooth skin and webbed feet.

► Amphibians are classified into three orders based upon their body structures.

► Amphibians have three-chambered hearts, gills when young, and lungs as adults.

CHECK *Find the sentence in the lesson that answers each question. Then, write the sentence.*

1. What does the word "amphibian" mean?

2. From what animals do most scientists think amphibians developed?

3. What kind of skin do toads have?

4. What kind of amphibians are frogs and toads?

5. Why must amphibians keep their skin moist?

APPLY *Complete the following.*

6. **Contrast:** How does the body structure of a frog differ from that of a salamander?

7. **Infer:** Why do you think most amphibians have webbed feet?

8. List six characteristics of amphibians.

InfoSearch

Read the passage. Ask two questions that you cannot answer from the information in the passage.

Caecilians Caecilians (see-SIL-ee-uns) are amphibians without legs. Caecilians look like long, colorful earthworms. In fact, a caecilian may be as much as 30-cm long. Like earthworms, caecilians live in soil. Some caecilians have very small eyes. Most caecilians, however, are considered blind.

SEARCH: Use library references to find answers to your questions.

TECHNOLOGY AND SOCIETY

COMPUTER DISSECTIONS

Computers have become useful tools in science. They can be used to store data and make calculations. They can be used to add color to photographs. In fact, many of the cell photographs in this book have had color added to them by computers. Computers also can make charts, tables, and graphs to keep track of research. They also can be used to run simulations (sim-yuh-LAY-shuns). A simulation is a moving model.

In many schools, animal dissections are no longer done on preserved specimens. Instead, computers are used to run a dissection simulation. A special computer program, such as a frog dissection, is on a computer disk. You can dissect the frog by using the dissection simulation disk. On the computer screen, you use laboratory equipment to open the frog. You can explore the different organs and organ systems of the frog on the computer screen.

Many people prefer using computer dissection. They do not think animals should be killed for dissections. Other people think that dissections on preserved specimens is important to the study of life science. What do you think?

How do frogs develop?

Objective ▶ Explain metamorphosis in frogs.

TechTerms

▶ **metamorphosis** (met-uh-MOWR-fuh-sis): changes during the stages of development of an organism

▶ **tadpole:** larval stage of a frog

Metamorphosis The changes during the stages of development of an organism are called **metamorphosis** (met-uh-MOWR-fuh-sis). Amphibians and insects change during their life cycles. During its life cycle, an amphibian changes the way it looks. During early stages, amphibians can live only in water. They breathe with gills. As an adult, amphibians live on land. They use lungs and their moist skin to breathe.

▶ *Name:* What are two groups of animals that change during their life cycle?

Amphibian Eggs All amphibians lay eggs. The eggs do not have a shell. They usually are laid in water. For example, frogs and toads lay their eggs in freshwater lakes, ponds, and other bodies of water. They lay their eggs in the spring. Toads lay long strings of eggs. Frogs lay clumps of eggs. The eggs are surrounded by a jellylike substance. The eggs cling to water plants.

👁 *Observe:* Do frogs lay many eggs, a few eggs, or one egg?

Tadpoles In about 12 days, frog eggs hatch into **tadpoles.** Tadpoles are the early stage of a frog. They are the larval stage. The larval stage does not look like a frog at all. Tadpoles look like very small fish. Their bodies are streamlined. They have a thin tail. They breathe with gills.

▶ *Name:* What is the early stage of a frog called?

Tadpole to Adult Frog As a tadpole grows, its body changes. Legs begin to develop. As the tadpole begins to develop into a young frog, its tail grows shorter. Gradually the tail disappears. In most amphibians, lungs develop and the gills disappear. At this point, the young frog leaves the water. It is adapted, or suited, to a life on land.

▶ *Infer:* What adaptation does the young frog have for breathing air?

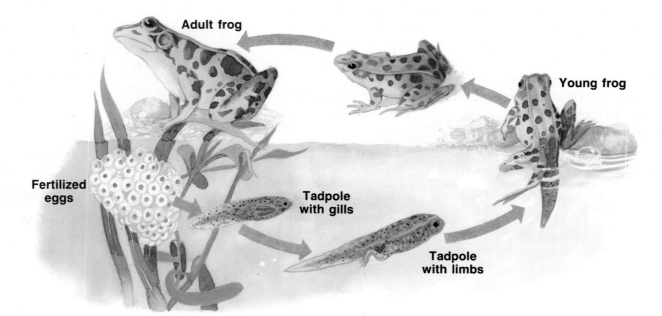

Adult frog

Young frog

Fertilized eggs

Tadpole with gills

Tadpole with limbs

LESSON SUMMARY

▶ The changes during the stages of development of an organism are called metamorphosis.

▶ Amphibians lay eggs without shells in water.

▶ A tadpole is the larval stage of a frog.

▶ As a tadpole develops into an adult frog, it develops legs, loses its tail, develops lungs, and loses its gills.

CHECK *Write true if the statement is true. If the statement is false, change the underlined term to make the statement true.*

1. During early stages in their development, amphibians breathe with <u>lungs</u>.

2. Amphibians and <u>insects</u> undergo metamorphosis.

3. Most amphibians lay their eggs <u>on land</u>.

4. The larval stage in the development of a frog is a <u>toad</u>.

5. Frogs hatch into tadpoles in about <u>12 weeks</u>.

6. The eggs of frogs are surrounded by a <u>jellylike</u> substance.

APPLY *Complete the following.*

7. **Sequence:** Place the events in the development of a frog in the correct order.
   a. Lungs develop and gills disappear. **b.** Eggs are laid in water. **c.** The tail begins to disappear. **d.** Tadpoles hatch from the eggs. **e.** Legs are almost fully developed and the tail is almost gone.

▲ 8. **Model:** Develop a flowchart that shows the stages in frog development.

▶ 9. **Infer:** Why do you think frogs lay so many eggs at one time?

10. **Hypothesize:** Why do you think amphibians lay their eggs in water?

Health & Safety Tip

A wart is a bumpy growth of the skin. People sometimes say that you can get warts by touching a toad. However, this is not true. Use library references to find out what causes warts. Also try to discover why many people associate warts with toads.

ACTIVITY

SEQUENCING TADPOLE DEVELOPMENT

You will need a pencil and a sheet of paper.

1. Study the diagram of the development of a tadpole.

2. **Sequence:** Place the drawings of tadpole development in the correct order. Write the correct order of the letters.

3. Study the diagram of bullfrog development.

Questions

1. Describe the most noticeable changes that take place in each stage of tadpole development shown in the diagram.

2. **Analyze:** What is a small bullfrog called?

3. **Analyze:** How many years does it take a bullfrog to develop into an adult?

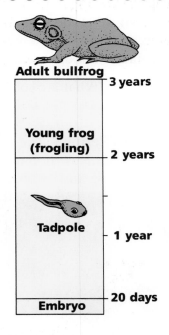

Adult bullfrog — 3 years

Young frog (frogling) — 2 years

Tadpole — 1 year

— 20 days

Embryo

What are reptiles?

Objectives ▶ Explain how reptiles are adapted to life on land. ▶ Describe the four orders of reptiles.

Life on Land Reptiles were the first true land animals. Unlike amphibians, reptiles do not need water for reproduction. Reptiles lay their eggs on land. The eggs are covered with a leathery shell. Young reptiles develop in the eggs.

Reptiles have lungs throughout their lives. They never have gills. They have hard, dry, skin covered with scales. The skin is waterproof. The waterproof skin cuts down on water loss from the body.

▐▐▐▶ *Explain:* How are reptiles adapted for life on land?

Reptiles Reptiles are coldblooded. Most have two pairs of legs with clawed feet. They have a three-chambered or four-chambered heart.

More than 100 million years ago the reptiles were the major group of animals on the earth. By about 65 million years ago, most of these reptiles including dinosaurs had died out. Today there are about 6500 different kinds of reptiles. They are classified into four orders.

▐▐▐▶ *Name:* What are some of the characteristics of reptiles?

Orders of Reptiles The smallest order of reptiles are the tuataras (too-uh-TAY-ruhs). They are called "living fossils" because they have many

features of ancient reptiles. These animals have changed very little since they first appeared about 200 million years ago.

The largest order of reptiles includes the snakes and lizards. The big difference between these two kinds of animals is that snakes do not have legs. There are about 4500 different kinds of snakes and lizards.

Turtles and tortoises make up another order of reptiles. Turtles live mostly in water. Tortoises live mostly on land. Turtles and tortoises have shells. Turtles have flat, streamlined shells. Tortoises have high, domed shells.

The fourth order of reptiles includes alligators (AL-uh-gayt-urs) and crocodiles. These animals spend much of their time in water. You can tell the difference between them by looking at their heads. Alligators have broad rounded heads. Crocodiles have more triangle-shaped heads.

▶ *Infer:* Why do you think a flat, streamlined shell is helpful to a turtle in water?

Crocodile

Alligator

LESSON SUMMARY

▶ Reptiles are land animals that lay eggs.

▶ Reptiles have lungs and scaly, waterproof skin.

▶ Reptiles are coldblooded, have a three- or four-chambered heart and most have two pairs of legs.

▶ Reptiles are classified into four orders.

▶ Tuataras are the smallest order of reptiles.

▶ Snakes and lizards make up the largest order of reptiles.

▶ Turtles and tortoises are reptiles with shells.

▶ Alligators and crocodiles make up an order of reptiles that spend much of their time in water.

CHECK *Complete the following.*

1. Where do young reptiles develop?
2. Why are tuataras called "living fossils?"
3. How do snakes differ from lizards?
4. Where do most tortoises live?
5. How could you tell an alligator from a crocodile?

APPLY *Complete the following.*

6. **Contrast:** How do snakes differ from all other reptiles?
7. Look at the dinosaurs in the feature. What characteristics of reptiles do these dinosaurs have?

InfoSearch

Read the passage. Ask two questions that you cannot answer from the information in the passage.

A Misunderstood Group Many people are afraid of snakes. They think that snakes are slimy and poisonous. Actually most snakes are not poisonous. Many snakes are helpful. They eat mice, rats, and other pests.

Poisonous snakes have long, curved hollow teeth called fangs. When a snake bites its victim, venom (VEN-um) in the fangs is injected into the wound. Venom is snake poison. Many venoms paralyze the muscles used for breathing, causing death.

SEARCH: Use library references to find answers to your questions.

⌄'⌄' LOOKING BACK IN SCIENCE ⌄'⌄'⌄'⌄'⌄'⌄'⌄'⌄'⌄'⌄'⌄'⌄'⌄'⌄'⌄'⌄'

THE AGE OF DINOSAURS

The first dinosaurs appeared on the earth more than 200 million years ago. For nearly 140 million years dinosaurs were the dominant animals on the earth. Then about 65 million years ago, the dinosaurs died out. Scientists are not sure why dinosaurs became extinct.

Dinosaurs lived during a time in the history of the earth called the Mesozoic Era. Other reptiles lived at the same time as the dinosaurs. The Mesozoic Era is sometimes called the Age of Reptiles. Because dinosaurs were the dominant animals during this time, it also is called the Age of Dinosaurs.

You may think that all dinosaurs were large animals. In fact, some were quite large, but others were as small as hens. The largest known dinosaur is *Brachiosaurus.* It was 30 meters tall. *Tyrannosaurus rex* and triceratops are two other large dinosaurs. *Compsognathus* was one of the smallest dinosaurs. It was less than 1 meter tall.

Contour feather   **Down feather**

Objectives ▶ Describe the characteristics of birds.
▶ Classify different kinds of birds.

TechTerm

▶ **warmblooded:** having a body temperature that remains about the same

Features of Birds Unlike fishes, amphibians, and reptiles, birds are **warmblooded.** The body temperature of a warmblooded animal remains about the same. It does not change when the temperature of the animal's surroundings changes. Birds maintain their body temperatures by using heat produced by the breakdown of food.

Birds are very easy to recognize. They are the only animals with feathers. Birds also have other features.

▶ Birds have two wings and two legs.

▶ Birds have lightweight bones.

▶ Birds have a beak without teeth.

▶ Birds lay hard-shelled eggs.

▶ *Compare:* What is the difference between coldblooded and warmblooded animals?

Five Groups All birds are classified in the class *Aves.* They are classified into five groups based on their beaks and feet. Figure 1 shows the five groups of birds. Look at the swimming birds. All swimming birds, such as ducks, have webbed feet for paddling through water. Hawks and other meat-eating birds have long, sharp claws to capture animals.

Look at the structure of the bill, or beak in each group. The beak varies with eating habits. Birds that feed on seeds have short strong beaks. Hunting birds, such as owls and eagles, have curved beaks for tearing meat.

▶ *Infer:* Why do you think hunting birds need strong beaks?

Body Plan of Birds Most birds have a streamlined, cigar-shaped body. The feathers that cover the bird serve several functions. They give the bird its shape. Some feathers are important in flight. Other feathers, called down feathers, help to keep in body heat.

Birds have well-developed organ systems. Their respiratory system, in particular, works very well. It has to supply the muscles with large amounts of oxygen during flight. Birds have a four-chambered heart. The two upper chambers are the atria. The two lower chambers are the ventricles.

▶ *State:* How many chambers does a bird's heart have?

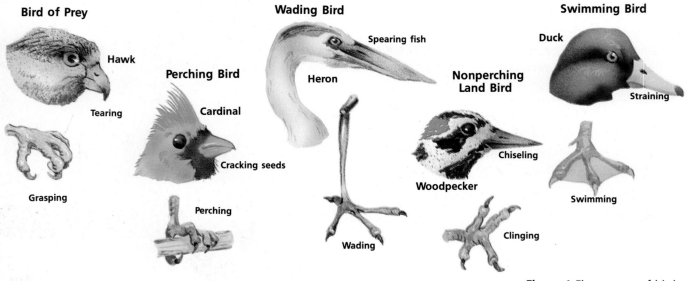

Bird of Prey — Hawk — Tearing — Grasping

Perching Bird — Cardinal — Cracking seeds — Perching

Wading Bird — Spearing fish — Heron — Wading

Nonperching Land Bird — Woodpecker — Chiseling — Clinging

Swimming Bird — Duck — Straining — Swimming

Figure 1 Five groups of birds

LESSON SUMMARY

▶ Birds are warmblooded animals.

▶ Birds are the only animals with feathers, have two wings and two legs, have lightweight bones, a beak, and lay eggs.

▶ Birds are classified into five groups based upon the shapes of their beaks and feet.

▶ Birds have a cigar-shaped body covered with feathers.

▶ Birds have well-developed organ systems.

CHECK *Write true if the statement is true. If the statement is false, change the underlined term to make the statement true.*

1. Birds are <u>coldblooded</u> animals.

2. Birds are the only animals with <u>feathers</u>.

3. Birds that eat <u>seeds</u> have strong, curved beaks for tearing.

4. All <u>perching</u> birds have webbed feet.

5. Birds have <u>down</u> feathers to help keep in body heat.

6. The upper chambers of a bird's heart are the <u>ventricles</u>.

APPLY *Complete the following.*

7. **Classify:** Use the diagram of bird classification on page 204 to classify each of the birds listed.
   a. hawk   c. cardinal   e. heron
   b. duck   d. woodpecker   f. eagle

8. **Apply:** Sea gulls live near ocean waters. Describe what kind of feet and beaks you think sea gulls have.

9. **Infer:** People, like birds, have four-chambered hearts. What do you think the upper and lower chambers of the human heart are called?

10. What characteristic can you use to classify an animal as a bird?

Health & Safety Tip............................

The meat and eggs of poultry contain a bacteria called *Salmonella. Salmonella* causes food poisoning. Use library references to find out what the symptoms of *Salmonella* poisoning are. Also, find out two ways to help reduce your risk of poisoning from this bacterium.

ACTIVITY

OBSERVING A BIRD EGG

You will need a chicken egg and a small dish.

1. Obtain a raw chicken egg.

2. **Observe:** Describe the shell of the chicken egg.

3. Crack open the egg into the dish.

4. **Observe:** Look at the inside of the shell. Find the shell membrane and the air space. Draw a model of the inside of the egg shell. Label the parts.

5. **Model:** Look at the egg in the dish. Draw a model of the inside of the egg. Label the yolk and egg white.

Questions

1. a. **Observe:** What color is the egg shell?   b. Is the shell smooth or slightly bumpy?

2. a. **Observe:** What color is the egg white?   b. The yolk?

3. What two parts make up most of the bird egg?

4. a. **Infer:** What is the function of the shell?   b. The yolk?

5. **Hypothesize:** Why do you think birds make nests for their eggs?

What are mammals?

Objective ▶ Describe the characteristics of mammals.

TechTerms

▶ **mammary** (MAM-ur-ee) **glands:** glands where milk is produced in female mammals

▶ **placenta** (pluh-SEN-tuh): structure in pregnant female placental mammals that connects the mother and developing mammal

Conquerors of the Earth More than 250 million years ago, animals called therapsids (THER-up-sidz) roamed the earth. Fossils show this animal had characteristics of reptiles and mammals. Biologists infer that all modern mammals came from the therapsids. When the dinosaurs died out, early mammals did not have to compete for food and living space. The mammals survived and reproduced. Today, mammals are one of the most successful groups of animals on the earth.

◼ *Analyze:* Why did mammals survive when the dinosaurs died out?

Characteristics of Mammals Mammals are a class of vertebrates that include humans and most of the best-known large animals. Mammals are warmblooded. Because they are warmblooded, mammals can live almost everywhere. They live on the plains, in forests, in deserts, and in swamps. They live in oceans, and a few can even fly.

Besides being warmblooded, mammals have these characteristics.

▶ Mammals have body hair.

▶ Mammals have a four-chambered heart.

▶ Most mammals have a highly-developed brain and nervous system.

▶ Female mammals nurse their young with milk. The milk is produced by **mammary** (MAM-ur-ee) **glands.**

▶ *List:* What are the characteristics of mammals?

Kinds of Mammals There are three basic mammal groups. One group is made up of egg-laying mammals. There are only two kinds of animals in this group. These are the duckbill platypus (PLAT-uh-pus) and the spiny anteater. In these animals, the young develop in an egg surrounded by a shell.

The second mammal group is the pouched mammals. In pouched mammals, the young are born at a very early stage of development. They then crawl into a pouch, or pocket, on the belly of the mother. They remain in the pouch until they are big enough to survive on their own. Kangaroos, opossums, and koala bears are pouched mammals.

The third and largest group of mammals is the placental (pluh-SEN-tul) mammals. In these animals a special structure called the **placenta** (pluh-SEN-tuh) develops in the female during pregnancy. The placenta connects the mother and developing offspring. Dogs, cats, cattle, seals, whales, bats, apes, and humans are placental mammals.

▶ *List:* What are the three kinds of mammals?

LESSON SUMMARY

▶ Scientists think that modern mammals came from the therapsids.

▶ Mammals include humans and most of the best-known large animals.

▶ Mammals are warmblooded, have body hair, a four-chambered heart, a highly-developed nervous system, and nurse their young with milk produced in mammary glands.

▶ The duckbill platypus and the spiny anteater are mammals that lay eggs.

▶ The young of opossums, kangaroos, and koalas develop in the pouch of the mother.

▶ The young of most mammals develop in an organ called a placenta within the mother's body.

CHECK *Complete the following.*

1. There are _____ basic mammal groups.

2. All mammals have _____ on their bodies.

3. Female mammals produce milk in _____ glands.

4. Mammals have a _____ heart.

5. The young of the largest group of mammals are connected to the mother by the _____ .

6. The young of spiny anteaters develop inside a _____ .

7. Kangaroos and koalas are _____ mammals.

APPLY *Complete the following.*

8. **Contrast:** How do spiny anteaters and duckbill platypuses differ from reptiles?

9. **Compare:** What characteristic is used to classify mammals into each of their groups?

Skill Builder.....................................

📁 *Classifying* When you classify, you place things into groups based upon similarities. The placental mammals are classified into orders. These orders include the flying mammals, toothless mammals, insect-eating mammals, gnawing mammals, rodentlike mammals, aquatic mammals, trunk-nosed mammals, carnivorous mammals, hoofed mammals, and primates. Use library references to find the characteristics each of these mammal orders. Then, cut out pictures of twenty different mammals from newspapers or magazines. Use your pictures to make a poster that classifies each of your mammals into its order.

LEISURE ACTIVITY

PETS

People keep many different kinds of animals as pets. Small mammals, birds, and fish are common pets. Some people have unusual pets such as alligators and crickets. On farms, horses and baby pigs and ducks are sometimes pets. Dogs and cats are especially popular pets for the home.

Pets can be interesting and fun companions. Some pets are workers as well as companions. For example, dogs are trained to help guide blind people. Horses can be both companions and transportation for their owners.

By caring for a pet, you learn about responsibility. It is important to learn to care for your pet's special needs. Does your pet need exercise? What does it eat? How is grooming important to your pet's health?

Treat your pet fairly and you will have its respect and affection. Your pet's main role is to be your friend. As a pet owner, your role is to enjoy your pet while treating it fairly and taking care of it properly.

How do animal embryos develop?

Objective ▶ Recognize how the development of mammals differs from other animals.

TechTerms

- ▶ **egg:** female sex cell
- ▶ **fertilization** (fur-tul-i-ZAY-shun): union of a male sex cell and a female sex cell
- ▶ **sperm:** male sex cell

Water Animals Most fish and other animals that live in water lay their **eggs** in the water. An egg is a female sex cell. The male swims over the eggs and deposits **sperm** on them. Sperm are male sex cells. The sperm joins with the eggs. Only one sperm cell joins with each egg. The union of a sperm and an egg is called **fertilization** (fur-tul-i-ZAY-shun). In these animals, the eggs are fertilized outside the body of the female. After fertilization, the embryo (EM-bree-oh) begins to develop. An embryo is a developing organism. The embryo develops inside the egg. Remember the egg is outside the body of the female. The embryo uses food stored in the egg.

▶ *Infer:* Where do the embryos of frogs develop?

Land Animals The eggs of land animals and sea mammals are fertilized inside the body of the female. In some animals, such as snakes and birds, the eggs develop in a hard shell after fertilization. The eggs are then laid. The embryos develop in the eggs outside the body of the female. The embryo uses the food stored in the egg. You know the food as the yolk.

▶ *State:* Where are the eggs of land animals fertilized?

Embryo Development in Mammals In all mammals, fertilization takes place inside the body of the female, or mother. Most mammals do not lay eggs. In these mammals, the embryo develops inside the mother's body. Most mammals give birth to living young.

▶ *State:* Where do the embryos of most mammals develop?

Embryo Nutrition The embryos of mammals get their food from the mother. Digested food and oxygen from the mother's blood pass into the bloodstream of the embryo. The embryo uses this food and oxygen for growth and development. Waste substances produced by the embryo pass into the mother's blood. These wastes are then excreted from the mother's body along with her own waste products.

▶ *State:* Where do mammal embryos get their food supply?

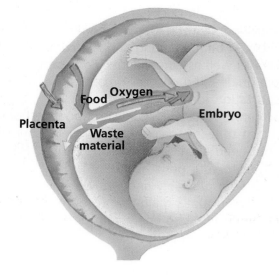

Food Oxygen

Placenta

Waste material

Embryo

LESSON SUMMARY

▶ The union of a sperm cell and an egg cell is called fertilization.

▶ The eggs of land and sea mammals are fertilized inside the body of the mother.

▶ In most mammals, the embryo develops inside the mother's body.

▶ The embryos of mammals get their food from the mother.

CHECK *Find the sentence in the lesson that answers each question. Then, write the sentence.*

1. What is an egg?
2. What is a sperm?
3. What is fertilization?
4. What is an embryo?
5. Where does an embryo that develops inside an egg get its food?
6. Where do the embryos of mammals get their food?
7. Where do the embryos of mammals develop?
8. Where do the embryos of animals such as birds and snakes develop?

APPLY *Complete the following.*

9. **Contrast:** How is the development of the embryos of most animals different from the development of a mammal embryo?

▶ 10. **Infer:** What structure in mammals is used to get food to a mammal embryo and to get rid of waste products?

Skill Builder

Researching When you do research, you gather information about a topic. The time it takes for an embryo to fully develop inside its mother's body is called the gestation (jes-TAY-shun) period. Gestation periods are different for different kinds of animals. Use library references to find out the gestation periods for the following mammals: elephant, mouse, dog, cat, human, guinea pig, horse, and cow. Present your findings in a graph.

Sperm cells

Egg cell

ACTIVITY

COMPARING GESTATION TIME AND ANIMAL SIZE

You will need a pencil and a sheet of graph paper.

1. Study the gestation times of the mammals in Table 1.
2. **Calculate:** Change all gestation times to days.
3. **Graph:** On a sheet of graph paper, make a bar graph showing the gestation times of the mammals. Show gestation time in number of days.

Questions

1. **Sequence:** Place the mammals in Table 1 in order from smallest to largest.

▲ 2. **Organize:** Use the sequence from smallest to largest, and make a table comparing relative size and gestation time in days.

3. Do larger or smaller mammals seem to need longer gestation times? Write one or two sentences to explain your answer.

Table 1	Mammal Gestation Times
Dog	9 weeks
Elephant	20-22 months
Hamster	16 days
Humans	9 months
Gorilla	8-9 months
Cow	40 weeks
Rabbit	26-30 days
Sheep	about 2 months

Challenges

STUDY HINT Before you begin the Unit Challenges, review the TechTerms and Lesson Summery for each lesson in this unit.

TechTerms..

amphibians (198)
cartilage (196)
coldblooded (196)
egg (208)
endoskeleton (194)

fertilization (208)
gills (196)
mammary glands (206)
metamorphosis (200)
notochord (194)

placenta (206)
sperm (208)
tadpole (200)
vertebrate (194)
warmblooded (204)

TechTerm Challenges..

Matching *Write the TechTerm that matches each description.*

1. union of a sperm cell and an egg cell
2. organs used for obtaining oxygen dissolved in water
3. animal with a backbone
4. organs in female mammals that produce milk
5. structure in pregnant female mammals that connects the mother and developing mammal
6. change during the development of an organism
7. larval stage of a frog

Applying Definitions *Explain the difference between the terms in each pair.*

1. sperm, egg
2. coldblooded, warmblooded
3. amphibians, reptiles
4. cartilage, bone
5. endoskeleton, exoskeleton
6. notochord, nerve cord
7. gills, lungs
8. vertebrate, invertebrate

Content Challenges..

Multiple Choice *Write the letter of the term that best completes each statement.*

1. The three features common to all chordates are a hollow nerve cord, gill slits at some time during their development, and
   a. cartilage.   **b.** a notochord.   **c.** an endoskeleton.   **d.** a placenta.

2. In most adult chordates, the notochord is replaced by
   a. gills slits.   **b.** lungs.   **c.** a backbone.   **d.** an endoskeleton.

3. Fish are classified as either jawless, cartilaginous, or
   a. bony.   **b.** tailless.   **c.** tailed.   **d.** venomous.

4. Fish, amphibians, and reptiles all
   a. are cold blooded.   **b.** are warmblooded.   **c.** have a three chambered heart.   **d.** have smooth, moist skin.

5. In the larval stage, a frog is called a(n)
   a. toad.   b. egg.   c. adult.   d. tadpole.
6. The largest order of reptiles is made up of
   a. tuataras.   b. snakes and lizards.   c. turtles and tortoises.   d. alligators and crocodiles.
7. Both birds and mammals
   a. have lightweight bones.   b. have mammary glands.   c. are warmblooded.   d. are cold-blooded.
8. Biologists believe that mammals come from a group of animals called
   a. dinosaurs.   b. therapsids.   c. carnivores.   d. marsupials.
9. The duckbill platypus and the spiny anteater are
   a. placental mammals.   b. pouched mammals.   c. egg-laying mammals.   d. oceanic mammals.
10. The developing organism that results from the union of an egg and a sperm cell is called a(n)
   a. embryo.   b. fetus.   c. baby.   d. mammal.

True/False *Write true if the statement is true. If the statement is false, change the underlined term to make the statement true.*

1. Fishes, amphibians, reptiles, birds, and mammals are classified in the phylum *porifera*.
2. All chordates have three tissue layers.
3. Fish breathe through lungs.
4. In coldblooded animals, body temperature changes as the outside temperature changes.
5. Amphibians usually lay their eggs on land.
6. Frog eggs have leathery shells.
7. Dinosaurs are an extinct group of mammals.
8. All birds belong to the class *Aves*.
9. Birds are coldblooded animals.
10. Birds have a two-chambered heart.
11. Fish and other animals that live in water lay eggs that are fertilized outside the female's body.
12. The embryo of an egg-laying land animal gets nutrition from the egg's yolk.

Understanding the Features...

Reading Critically *Use the feature reading selections to answer the following. Page numbers for the features follow each question in parentheses.*

1. What are the two educational requirements for a career as an animal technician? (195)
2. What is the purpose of the spines on the porcupine fish? (198)
3. **Infer:** What might be some disadvantages of using computer dissection instead of dissection of preserved animals? (199)
4. During which era did the dinosaurs reign? (203)
5. Name three responsibilities of a pet owner. (207)

Concept Challenges..

Critical Thinking *Answer each of the following in complete sentences.*

1. How do coldblooded animals differ from warmblooded animals? Are humans coldblooded or warmblooded?
2. What are the major characteristics of an amphibian?
3. What are three differences between amphibians and reptiles?
4. **Infer:** Why do you think birds that feed on seeds have short strong beaks?
5. In what two ways are bird and mammals alike?

Interpreting a Diagram *Use the diagram to complete the following.*

1. What is the part of the fish labeled C?
2. What is the part of the fish labeled E?
3. **Infer:** What does the part labeled E protect?
4. **Analyze:** Which parts help the fish swim through the water?
5. What is the part of the fish labelled D?
6. What does the fish use the part labeled D for?

Finding Out More..

1. **Research:** Scientists have come up with several different theories to explain why the dinosaurs became extinct. Using library reference books, research as many of these theories as you can. State which of the theories seems most logical to you and why.
2. Dianne Seale is a biologist who did research to find out how nutrients in a pond affect the development of a tadpole. Visit the library and find out about Dianne Seale's work. Report your findings to the class in the form of a scientific article.
3. **Classify:** With the assistance of your teacher or a field guide, find out the names of ten kinds of birds that live in your area. Then classify each bird you have identified into one of the five groups discussed in this unit.
4. **Classify:** Using old books or magazines, cut out five pictures of animals from each class of vertebrates: fishes, amphibians, reptiles, birds, and mammals. (You will have a total of 25 pictures.) Use your pictures to make a poster that indicates the class in which each animal belongs. Include the name of each class above each group of pictures.
5. Many animals today are currently in danger of becoming extinct. Use an encyclopedia to find out what some of these endangered species are. Also find out why they are becoming extinct and what you can do to prevent their total disappearance from the earth.

NUTRITION AND DIGESTION

CONTENTS

STUDY HINT As you read each lesson in Unit 11, write the topic sentence for each paragraph in the lesson on a sheet of paper. After you complete each lesson, compare your list of topic sentences to the Lesson Summary.

11-1 What are nutrients?

Objective ▶ Identify the nutrients used by the body.

TechTerms

- **carbohydrate** (kar-buh-HY-drayt): nutrient that supplies energy
- **nutrient** (NOO-tree-unt): chemical substances in food needed by the body for growth, energy, and life processes
- **protein** (PRO-teen): nutrient needed to build and repair cells

Nutrients All foods contain **nutrients** (NOO-tree-unts). Nutrients are chemical substances in food that your body needs for growth and energy. Nutrients also are needed to carry out life processes. There are five kinds of nutrients. They are **carbohydrates** (kar-buh-HY-drayts), fats, **proteins** (PRO-teenz), vitamins, and minerals. Some foods are rich in one nutrient. Most foods contain more than one nutrient.

▶ **Define:** What are nutrients?

Carbohydrates Carbohydrates are nutrients that supply your body with energy. They are your main source of energy. Starches and sugars are carbohydrates.

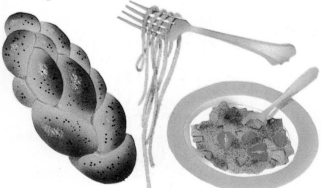

There are two kinds of carbohydrates, simple and complex. Sugars are simple carbohydrates. They are used quickly by your body. Sugars give your body short, quick bursts of energy. Starches are complex carbohydrates. They give your body energy over long periods of time. Breads and pasta contain starch.

▶ **Name:** What are the two kinds of carbohydrates?

Fats Fats are the energy-storage nutrients. The stored energy in fats can be used if energy from carbohydrates is used up. Fat that is stored in the body is used for insulation (in-suh-LAY-shun). It keeps your body warm. Fat also is used to protect your body organs.

Fats can be either solids or liquids. Solid fats come mostly from animals. Liquid fats are called oils. They usually come from vegetables, such as corn. Many oils are used in cooking.

▶ **List:** Identify ways the body uses fat.

Proteins Your body needs proteins for growth and repair. Proteins also are used to build tissues, such as muscles. You get some proteins from foods you eat. Other proteins are made in your body. When carbohydrates and fats have been used up, proteins also can be used for energy. Milk, fish, and cheese are good sources of protein. Peas, peanuts, and beans also are good sources of protein.

▶ **Explain:** Why are proteins needed by the body?

LESSON SUMMARY

▶ Nutrients are chemical substances in food that are needed by the body for growth, energy, and life processes.

▶ Carbohydrates are nutrients that supply the body's main source of energy.

▶ The two kinds of carbohydrates are simple and complex.

▶ Fats are the energy-storage nutrients.

▶ Fats can be either liquids or solids.

▶ Proteins are used to build and repair cells.

CHECK *Write true if the statement is true. If the statement is false, change the underlined term to make the statement true.*

1. Your body needs <u>nutrients</u> for growth, energy, and to perform life processes.

2. Simple carbohydrates give your body energy over a <u>long</u> period of time.

3. The energy-storage nutrients are <u>carbohydrates</u>.

4. Your body uses <u>fat</u> to build and repair cells.

5. Starches and grains are good sources of <u>complex</u> carbohydrates.

APPLY *Answer the following.*

6. **Hypothesize:** Why do long distance runners eat foods rich in starch the evening before a race?

7. **Infer:** Why should you eat more complex carbohydrates than simple carbohydrates?

InfoSearch

Read the passage. Ask two questions that you cannot answer from the information in the passage.

Water Although it is not a nutrient, your body needs a constant supply of water. More than two-thirds of your body is made up of water. Water is needed to carry out your life processes. All of the chemical changes that take place in your body require water. Water also helps control your body temperature. You should try to drink at least six glasses of water each day.

SEARCH: Use library references to find answers to your questions.

ACTIVITY

TESTING FOR FATS

You will need a brown paper bag, scissors, and 5 small samples of different foods.

1. Cut the bag open so that you have a flat surface of brown paper.

2. Rub a piece of food on the bag. Write the name of the food near the spot.

3. Repeat Step 2 for each of your foods.

4. Let the spots dry. If the spots disappear, there is no fat in the food. If the spots remain, there is fat in the food.

5. Copy Table 1. Record your observations in the table by checking the correct box.

Table 1			
FOOD	PREDICT	FAT	NO FAT
1.			
2.			
3.			
4.			
5.			

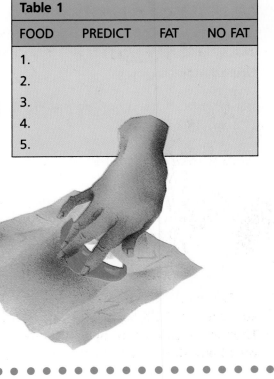

Questions

1. How do you test foods for the presence of fats?

2. **Observe:** Which foods left a spot on the paper?

3. **Summarize:** What can you conclude about the foods you tested?

4. Make a list of the foods you found contain fat.

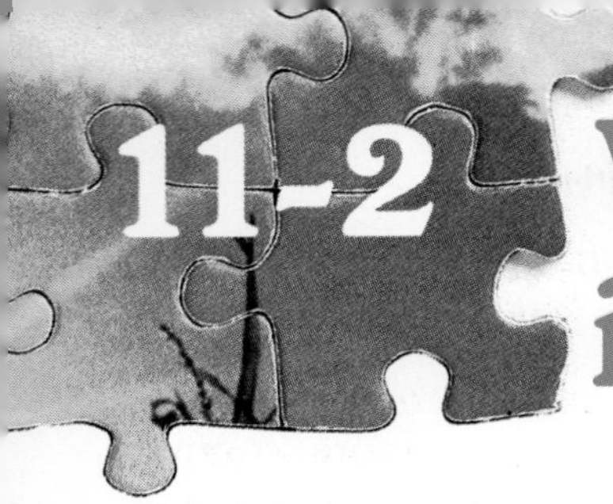

11-2 Why are proteins important?

Objective ▸ List five ways the body uses proteins.

TechTerms

- **amino acid:** building block of proteins
- **enzyme** (EN-zym): protein that controls chemical activities
- **molecule** (MAHL-uh-kyool): smallest part of a substance that has all the properties of that substance

Protein Molecules Proteins are giant **molecules** (MAHL-uh-kyoolz). A molecule is the smallest part of a substance that has all the properties of that substance. In fact, proteins are the largest molecules in living things. All proteins contain atoms of carbon, hydrogen, oxygen, and nitrogen.

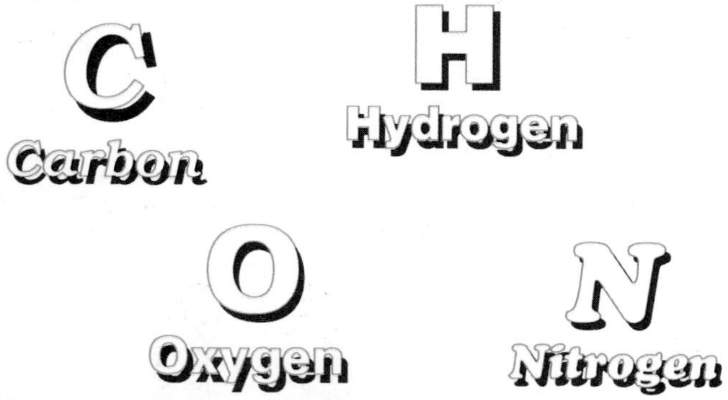

Figure 1 Atoms in proteins

▸ ***Name:*** What atoms do protein molecules contain?

Importance of Proteins Proteins are the building blocks of living material. Your body uses proteins in several ways. One important use of proteins is to build new cells. Another use of protein is to repair cells that are damaged. Proteins also are used to make **enzymes** (EN-zymz). Enzymes are substances that control many of the chemical activities in the body. Proteins also can be used as a source of energy for body cells.

▸ ***Explain:*** How are proteins used by the body?

Protein Building Blocks The building blocks of proteins are **amino acids.** Proteins are formed when many smaller amino acid molecules join together. There are about 20 different amino acids. Twelve of these amino acids are made in the body. The other eight amino acids must be taken into the body. The foods you eat contain different amino acids. The amino acids are put together in many different ways to form thousands of different proteins. This is like making thousands of words from the 26 letters of the alphabet.

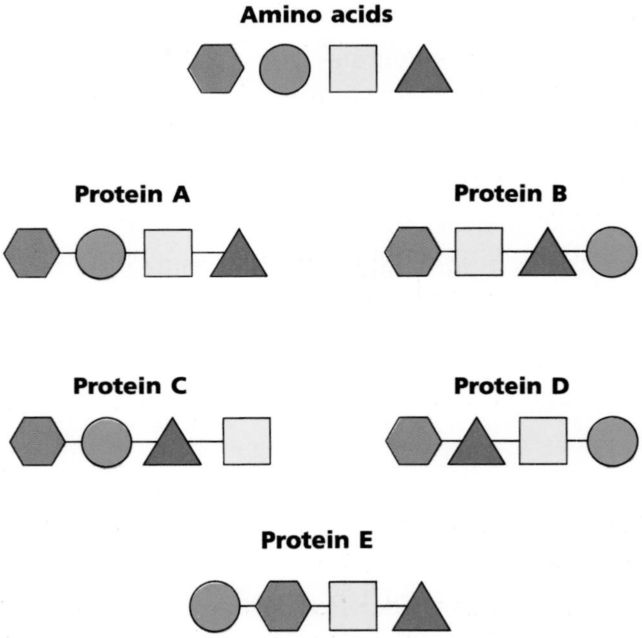

Figure 2 Protein building

▸ ***Identify:*** What are the building blocks of proteins?

Making Proteins The body uses proteins in foods to make the special proteins it needs. When proteins in food are digested, the amino acids are separated from one another. Cells in your body make their own proteins by putting these amino acids together again in their own special way.

▸ ***Infer:*** Why do you think the body can make so many proteins?

LESSON SUMMARY

▶ Proteins are the largest molecules in living things.

▶ Proteins are used to build new cells, repair damaged cells, make enzymes, control chemical activities, and as a source of energy.

▶ Amino acids are the building blocks of proteins.

▶ Cells in your body make their own proteins by combining amino acids.

CHECK *Complete the following.*

1. Substances that control chemical activities in cells are called _____ .

2. The building blocks of proteins are _____ .

3. There are about _____ different amino acids.

4. Proteins are the _____ molecules in living things.

5. All proteins contain atoms of carbon, oxygen, hydrogen, and _____ .

APPLY *Complete the following.*

6. **Analyze:** Why is a diet low in protein not healthful?

7. **Infer:** Plant proteins are incomplete proteins because they do not contain all eight of the amino acids needed by the body. Why do you think it is important for people who do not eat animal proteins to eat a variety of plant proteins?

Use Figure 2 on page 216 to complete the following.

8. **Model:** Draw two proteins that could be formed by the amino acids shown. Do not copy any of the proteins shown.

Skill Builder...

Researching Vegetarians (vej-uh-TER-ee-unz) are people who do not eat animal products. Therefore, they do not eat any complete proteins. Vegetarians must combine different incomplete vegetable proteins in order to make complete proteins. Some vegetarians include small amounts of milk and eggs in their diet to get complete proteins. Use library resources to find out the signs and symptoms of a low protein diet.

ACTIVITY

MODELING PROTEIN MOLECULES

You will need 1 sheet of red construction paper, 1 sheet of blue construction paper, and a scissors.

1. Cut various shapes from the red and blue construction paper. **Caution: Be careful when using sharp objects.**

2. Place the shapes side by side to form a chain modeling a protein molecule.

3. Repeat Step 2 making as many protein models as you are able. Do not repeat the same pattern twice.

Questions

1. **Model:** What do the pieces of paper represent?

2. **a.** How many amino acid molecules did you use?   **b.** How many different protein models were you able to construct?

3. **a.** How many amino acids are there in your body?   **b.** Why can your body make thousands of proteins from these amino acids?

4. What happens to the number of proteins as the number of amino acids increases?

11-3 Why are vitamins important?

TechTerms

► **deficiency** (di-FISH-un-see) **disease:** disease caused by the lack of a certain nutrient

► **vitamin** (VYT-uh-min): nutrient found naturally in many foods

Vitamins Your body needs small amounts of **vitamins** (VYT-uh-minz) so it can function properly. Vitamins are nutrients that are found naturally in many foods. You get most of the vitamins you need from food. However, two vitamins—D and K—are made in your body.

▐▐▐▶ *Define:* What are vitamins?

Importance of Vitamins Vitamins are important for proper growth. They also are important for keeping bones and teeth strong and healthy. Your muscles and nerves are kept healthy by vitamins. Vitamins also help control many of the chemical activities that take place in your body.

Most vitamins work with other vitamins or nutrients. Some vitamins help change carbohydrates and fats to energy. Unlike carbohydrates and fats, vitamins do not give off energy.

▐▐▐▶ *Explain:* How are vitamins used in your body?

Deficiency Disease Your body tissues need small amounts of vitamins every day. If your diet does not include enough of a certain vitamin, you can become sick. This kind of sickness is called a **deficiency** (di-FISH-un-see) **disease.** Rickets is a deficiency disease that causes soft bones and teeth. Rickets is caused by a lack of vitamin D. Look at Table 1 to see other deficiency diseases.

▐▐▐▶ *Define:* What is a deficiency disease?

Table 1 Vitamins and Deficiency Diseases			
VITAMIN	USE IN BODY	SOURCES	DEFICIENCY DISEASE
A	healthy skin, eyes, ability to see well at night, healthy bones and teeth	orange and dark green vegetables, eggs, fruit, liver milk	night blindness
B_1 thiamin	healthy nerves, skin, and eyes; helps body get energy from carbohydrates	liver, pork, whole grain foods	beriberi
B_2 riboflavin	healthy nerves, skin and eyes; helps body get energy from carbohydrates, fats, and proteins	eggs, green vegetables, milk	skin disorders
B_3 niacin	works with B vitamins to get energy from nutrients in cells	beans, chicken, eggs, tuna	pellagra
C	healthy bones, teeth, and blood vessels	citrus fruits, dark green vegetables	scurvy
D	healthy bones, teeth; helps body use calcium	eggs, milk, made by skin in sunlight	rickets
E	healthy blood and muscles	leafy vegetables, vegetable oil	none known
K	normal blood clotting	green vegetables, tomatoes	poor blood clotting

LESSON SUMMARY

▶ Vitamins are nutrients found naturally in foods.

▶ Vitamins are important for growth and for proper body functions.

▶ Most vitamins work with other vitamins or nutrients.

▶ A deficiency disease is caused by a lack of a certain nutrient.

CHECK *Complete the following.*

1. Nutrients that are found naturally in foods are _____ .

2. A _____ is caused by a diet that is missing a certain nutrient.

3. An example of a deficiency disease that causes soft bones is _____ .

APPLY *Use Table 1 on page 218 to complete the following.*

◀ 4. **Analyze:** You have cut yourself and note that your blood is slow to clot. What vitamin are you lacking in your diet? Explain.

5. **Interpret:** What vitamins help the body get energy from carbohydrates?

◀ 6. **Analyze:** If you have scurvy, what vitamin are you missing in your diet?

7. **Interpret:** What is another name for vitamin B_2?

8. **Interpret:** What is the deficiency disease for vitamin A?

Skill Builder..

▲ *Organizing* Using library references, look up the deficiency diseases listed in Table 1. Find out the signs and symptoms of each deficiency disease. Organize the information in a table.

Health & Safety Tip..........................

The skins and peels of many fruits and vegetables are rich in vitamins and minerals. Fruits and vegetables often are sprayed with pesticides (PES-tuh-sydz). Pesticides are chemical substances used to kill pests such as insects. Pesticides can be harmful to people. Use library references to find out how to clean the skins and peels of fruits and vegetables so they are safe to eat.

▼▼▼ LOOKING BACK IN SCIENCE ▼▼▼▼▼▼▼▼▼▼▼▼▼▼▼▼▼▼▼▼▼▼

SCURVY

Until the 1700s, many sailors died at sea from a mysterious disease. The sailors would suffer from bleeding gums and loose teeth. Their skin would bruise easily. The disease the sailors suffered from is called scurvy (SKUR-vee). A few hundred years ago, the cause of scurvy was unknown. However, many sailors suffering from scurvy got better after their ships reached port. Therefore, people thought that scurvy was caused by "sea air."

In the 1700s, British doctors noticed that sailors ate mostly salted beef and biscuits. The doctors learned that the juice from citrus fruits, such as oranges, lemons, and limes, could prevent and cure scurvy. English ships started carrying a supply of lemons and limes. Citrus juice was given to the sailors daily. How did citrus juices prevent and cure scurvy? Scientists discovered that scurvy is a deficiency disease caused by a lack of vitamin C in the diet. Citrus fruits are an excellent source of vitamin C.

11-4 Why are minerals important?

Objective ▶ Explain why minerals are important.

TechTerm

▶ **mineral:** nutrient needed by the body to develop properly

Minerals Your body needs **minerals** as well as vitamins in order to grow. Minerals are nutrients needed by the body to develop properly. You need small amounts of some minerals. For example, you need small amounts of iron, iodine, sulfur, and zinc. You need large amounts of other minerals. Calcium, phosphorus, and sodium are minerals that you need larger amounts of.

▶ **Define:** What are minerals?

Important Uses of Minerals Each mineral has a different job. For example, iron is needed to form red blood cells. Calcium and phosphorus (FAHS-fur-us) are needed to build strong teeth and bones. Bones are made up of a lot of calcium. It makes them hard. If calcium is removed from bones, they become soft and rubbery. Sodium is needed for healthy muscles and nerves. Chlorine is needed to make a chemical used in digestion. Iodine controls body growth and the oxidation of food.

▶ **Explain:** What happens when calcium is removed from bones?

Deficiency Diseases A deficiency disease can be caused if certain minerals are missing from the diet. For example, if you take in too little iron, a deficiency disease called anemia can result. Anemia is sometimes called iron-poor blood. If you are deficient in iodine, a deficiency disease called goiter can result. Table 1 lists some important minerals and signs of deficiency.

▶ **Identify:** What deficiency disease is caused by too little iodine?

Table 1	Minerals and Their Uses		
MINERAL	USES	SOURCES	SIGNS OF DEFICIENCY
calcium	builds bones and teeth	milk and milk products, canned fish, green leafy vegetables	soft bones poor teeth
phosphorus	builds bones and teeth	red meat, fish, eggs, milk products, poultry	none known
iron	builds red blood cells	red meat, whole grains, liver, egg yolk, nuts, green leafy vegetables	anemia (paleness, weakness, tiredness) brittle fingernails
sodium	helps keep muscles and nerves healthy	table salt, found naturally in many foods	none known
iodine	used to make a chemical that controls oxidation	seafood, iodized salt	goiter
potassium	helps keep muscles and nerves healthy	bananas, oranges, meat, bran	loss of water from cells, heart problems, high-blood pressure
Sulfur	formation of body cells	beans, peanuts, wheat germ, beef	none known
Magnesium	strong bones and muscles nerve action	nuts, whole grains, green leafy vegetables	none known
Zinc	formation of enzymes	milk, eggs, seafood, milk, whole grains	none known

LESSON SUMMARY

▶ Minerals are nutrients needed by the body to develop properly.

▶ Each mineral has a different job.

▶ Deficiency diseases can be caused when certain minerals are missing from your diet.

CHECK *Complete the following.*

1. Anemia can result from too little _____ .

2. The body needs large amounts of calcium, phosphorus, and _____ .

3. Sodium is needed for healthy muscles and _____ .

4. Bones are hard because they contain _____ .

APPLY *Use Table 1 on page 220 to answer the following.*

5. **Interpret:** What foods contain iodine?

6. **Infer:** You are feeling tired and weak. What mineral might you be deficient in?

7. **Analyze:** Which minerals help build strong teeth and bones?

8. **Interpret:** What foods are good sources of potassium?

9. **Interpret:** Why is sulfur important to your diet?

Skill Builder

▲ *Organizing* Enriched foods are foods that have had vitamins and minerals added to them. Milk, bread, orange juice, and pasta are foods that often are enriched. Find ten labels from foods that have been enriched. What vitamin or mineral was added to each of these foods? Organize your findings into a table.

CAREER IN LIFE SCIENCE

NUTRITIONIST

Do you have an interest in food and nutrition? If so, you may want to become a nutritionist. Nutritionists are concerned with the nutrients in food and the relationship between diet and health.

Nutritionists may work in hospitals or in private practice. They also may work for companies that make food products. In hospitals, nutritionists study patients' nutritional needs and design diets suited to those needs. Nutritionists also may teach nutrition classes to pregnant women, heart disease patients, and people with weight problems. In industry, nutritionists help companies to research, develop, and test new food products.

If you are interested in becoming a nutritionist, you should take science and math courses in high school. Nutritionists also need a college degree with a major in nutrition. For more information, write to the American Dietetic Association, 620 North Michigan Ave., Chicago, Illinois 60611.

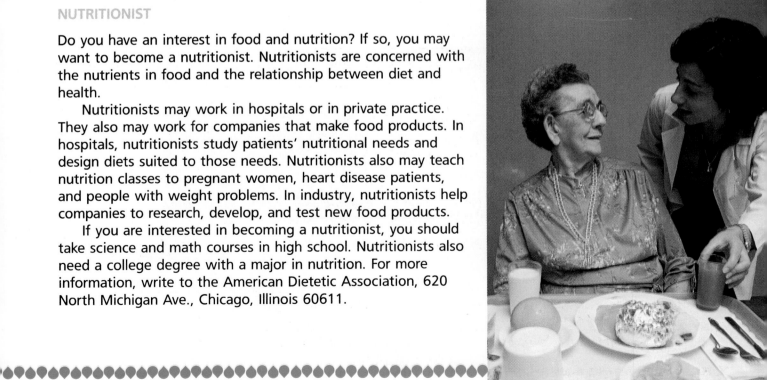

What is a balanced diet?

Objectives ► Recognize the parts of a balanced diet. ► Name the four basic food groups.

Four Food Groups All foods have been classified into four groups. These groups are called the four basic food groups. The four food groups are the dairy group, the meat group, the vegetable-fruit group, and the bread-cereal group.

Each food group is a good source of different nutrients. For example, the dairy group is rich in vitamins A and D, as well as calcium. Table 1 shows the four food groups and the nutrients that each group provides.

▶ *Identify:* Name the four basic food groups.

A Balanced Diet Diets that contain the right amounts of nutrients are called balanced diets. A balanced diet is important for good health. You can plan a balanced diet by including foods from the four basic food groups. This diet will contain all the nutrients you need for good health.

▶ *State:* What does a balanced diet consist of?

Importance of a Balanced Diet Have you ever heard the expression, "You are what you eat"? In a way this is true. Many people do not eat balanced diets. They eat large amounts of fats and proteins. By planning your diet, you can eat the most healthful foods. If you improve your eating habits, you can improve your health. You may even increase your life span.

▶ *Explain:* Why is a balanced diet important?

Planning a Diet You can use the basic food groups as a guide to good eating. Your daily diet should contain the right amount of foods from each group. People who are growing or very active, usually need more servings from the meat group and the vegetable-fruit group.

▶ *Explain:* Why should you plan your diet?

Malnutrition When your body is not properly nourished, it becomes weak. This disorder is called malnutrition (mal-noo-TRISH-un). Malnutrition means bad nutrition. Many people think malnutrition occurs only from a lack of food. However, anyone can be malnourished. If you eat too many foods from one group and not enough foods from another group, you can become malnourished.

▶ *State:* What happens if you eat too much from one food group and not enough from another food group?

Figure 1 Bread-cereal group

Figure 3 Dairy Group

Figure 2 Fruit-vegetable group

Figure 4 Meat Group

LESSON SUMMARY

▶ All foods are placed into four basic food groups.

▶ Each food group is a good source of different nutrients.

▶ A balanced diet contains all the nutrients you need for good health.

▶ A balanced diet is important because it keeps you healthy.

▶ Your diet should be planned to include the proper balance of nutrients.

▶ Malnutrition occurs when you eat too much food from one food group and not enough food from another food group.

CHECK *Complete the following.*

1. All foods are placed into _____ basic food groups.

2. Diets that contain the right nutrients are called _____ diets.

3. Balanced diets contain the proper amounts of _____ .

4. If you do not eat dairy products, you may suffer from _____ .

APPLY *Complete the following.*

5. **Identify:** Name two foods from each of the four food groups.

6. **Analyze:** Which food group is the best source of proteins?

7. **Hypothesize:** Can an overweight person be malnourished?

InfoSearch

Read the passage. Ask two questions that you cannot answer from the information in the passage.

Eating Disorders Bulimia (byoo-LIM-ee-uh) and anorexia (an-uh-REK-see-uh) are harmful eating disorders. The number of people who have these eating disorders has increased in the last few years. Bulimia and anorexia threaten good health and even life. Bulimics are of normal weight. They eat large amounts of food at one time. Then they either throw up the food, or use laxatives. Anorexics are usually underweight. They refuse to eat food. There are many self-help groups that can help bulimics and anorexics.

SEARCH: Use library references to find answers to your questions.

ACTIVITY

PLANNING A BALANCED DIET

You will need paper and a pencil.

1. Divide a piece of paper into four columns. Label the columns Milk Group, Vegetable-Fruit Group, Meat Group, and Bread-Cereal Group.

2. Classify each of the foods listed in the Data Table into the proper food group. Write each food in the correct column.

3. Use the information in the Data Table to plan a balanced breakfast. Be sure to include one serving from the milk or meat group, the bread-cereal group, and the fruit-vegetable group. Your meal should provide about 400–500 calories.

4. Find the total number of calories for your meal.

Questions

1. Which foods listed in the Data Table belong to the bread-cereal group?

2. Which foods belong to the milk group?

3. What foods did you select for your balanced breakfast?

4. What is the total number of calories for your breakfast?

Data Table	
FOOD	CALORIES
Apple	80
Banana	100
Bran flakes (1 cup)	110
Egg	80
Ham slice	250
Margarine (1 Tbs.)	100
Milk (2% fat)	120
Orange juice (6 oz.)	110
Slice of bread	70
Strawberry jam (1 tsp.)	18
Whole milk (8 oz.)	160

How do living things get energy?

Objective ▶ Explain how organisms "burn" food to get energy.

TechTerms

▶ **Calorie** (KAL-uh-ree): unit used to measure energy from foods

▶ **oxidation** (ahk-suh-DAY-shun): slow burning of foods in your body

Energy Your body needs energy to stay alive. Everything you do requires energy. You need energy to walk, to run, and even to sleep. Your heart needs energy to pump blood through the body. Right now, your body is growing. You need energy to grow.

||||▶ *State:* Why does your body need energy?

Oxidation You get the energy you need from food. As food gets broken down in the body, it produces energy. The slow burning of foods in your body is called **oxidation** (ahk-suh-DAY-shun). Oxidation produces the energy that your body needs. Living cells get most of their energy by the oxidation of sugar. Your body cells use energy to carry on their life activities. Some energy also is given off as heat.

||||▶ *Define:* What is oxidation?

Waste Products of Oxidation During oxidation, carbon dioxide and water are given off as waste products. To show that your body gives off water, breathe on a mirror. What do you see? Water collects on the mirror. The water is a waste product of oxidation going on in your body.

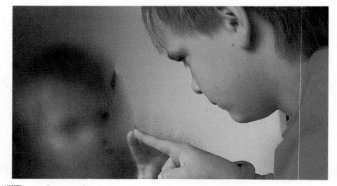

||||▶ *Identify:* What are the waste products of oxidation?

Measuring Food Energy Different foods contain different amounts of energy. The amount of energy food gives off is measured in **Calories** (KAL-uh-reez). A Calorie is a unit used to measure food energy. Fat gives off the most energy. Each gram of fat gives off 9 Calories of energy. Proteins and carbohydrates each give off the same amount of energy. Each gram of protein or carbohydrate gives off 4 Calories of energy.

||||▶ *Define:* What is a Calorie?

LESSON SUMMARY

▶ Your body needs energy to stay alive.

▶ Oxidation of foods produces the energy that your body needs.

▶ Carbon dioxide and water are waste products of oxidation.

▶ The amount of energy food gives off is measured in Calories.

CHECK *Find the sentence in the lesson that answers each question. Then, write the sentence.*

1. What process produces the energy your body needs?

2. What are the waste products of oxidation?

3. Do all foods give off the same amount of energy?

4. What unit is used to measure food energy?

5. Where do you get the energy you need?

6. What nutrient supplies the most energy?

APPLY *Complete the following.*

7. **Calculate:** If food contains 5 grams of fat, how many Calories of fat are in the food?

8. **Infer:** Why do foods with high amounts of fat have more Calories than foods that are high in carbohydrates?

9. **Hypothesize:** Boys ages 12 to 15 need about 2800 Calories each day. Girls the same age need about 2400 Calories each day. What do you think would happen if you ate more Calories than what you needed?

Skill Builder.................................

Calculating Calculate how many Calories are in the following foods:

A: 8 g protein, 11 g carbohydrates, 1 g fat
B: 4 g protein, 10 g carbohydrates, 8 g fat
C: 0 g protein, 1 g carbohydrates, 5 g fat
D: 14 g protein, 4 g carbohydrates, 3 g fat
E: 5 g protein, 7 g carbohydrates, 6 g fat

Skill Builder.................................

Comparing Keep a daily log of the foods you eat for one week. Also keep track of any physical activity you do. Use library references to calculate the number of Calories used up in activity. Also, calculate the number of Calories in the foods you ate. Compare the two totals. According to your records, did you take in more Calories, or burn up more Calories?

SCIENCE CONNECTION

CHEMOSYNTHESIS

Deep in the ocean, no sunlight can reach the ocean floor. Photosynthesis is not possible. However, bacteria that live on the bottom of the ocean floor have another way of making their own food. These bacteria live near vents, or openings in the ocean floor. The vents are formed by underwater volcanoes. Sulfur compounds shoot out of the vents. The bacteria change the compounds into food. The process of making food from chemicals is called chemosynthesis.

The temperature near the ocean vents is so great that most living things could not survive. However, tube worms and some clams and crabs live near the vents. They feed on the deep-sea bacteria. Because the bacteria are able to use chemicals to produce food, a chain of unusual life forms flourishes.

What is the digestive system?

Objective ▶ Identify the organs in the human digestive system.

TechTerms

- ▶ **digestion** (dy-JES-chun): process by which foods are changed into forms the body can use
- ▶ **epiglottis** (ep-uh-GLAT-is): flap of tissue that prevents food from entering the windpipe
- ▶ **esophagus** (i-SAF-uh-gus): tube that connects the mouth to the stomach
- ▶ **peristalsis** (per-uh-STAWL-sis): wavelike movement that moves food through the digestive tract

Digestion The foods that you eat are not in a form that your body can use. The foods you eat must be changed so that they can be used by the body. The process by which foods are changed into usable forms is called **digestion** (dy-JES-chun). During digestion, larger pieces of food are broken down into smaller pieces. Complex chemicals in food also are changed into simpler ones.

▶ *Define:* What is digestion?

The Digestive Tract The foods that you eat move through a coiled tube inside the body. This tube is called the digestive tract. It is about 10 meters long. Some parts are narrow. Other parts are wide. The digestive tract and other digestive organs make up your digestive system. Some digestive organs are the liver, pancreas (PAN-kree-us), and gall bladder. These organs are not part of the digestive tract. However, they do help with digestion.

▶ *Identify:* What organs help with digestion, but are not part of the digestive tract?

Parts of the Digestive System Figure 1 shows the digestive system. The mouth is the beginning of the digestive system. Once food is swallowed, it moves into a long tube called the **esophagus**

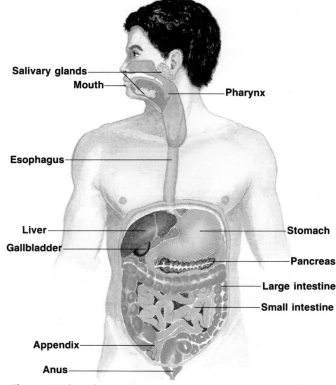

Salivary glands
Mouth
Pharynx
Esophagus
Liver
Gallbladder
Stomach
Pancreas
Large intestine
Small intestine
Appendix
Anus

Figure 1 Digestive system

(i-SAF-uh-gus). The esophagus connects the mouth to the stomach. As you swallow, a thin flap of tissue keeps food from entering the windpipe. This flap of tissue is called the **epiglottis** (ep-uh-GLAT-is). From the stomach, food moves into another narrow tube called the small intestine. The small intestine leads to the large intestine. The end of the large intestine is called the rectum.

▶ *Name:* What are the organs of the digestive system?

Peristalsis Once food leaves the mouth, it enters the esophagus. The walls of the esophagus secrete mucus. The mucus helps the food move through the esophagus easily. Once food leaves the esophagus, it continues through the digestive system. Food is moved through the digestive system by wavelike movements of muscle. The wavelike movement that moves food through the digestive tract is called **peristalsis** (per-uh-STAWL-sis).

▶ *Define:* What is peristalsis?

LESSON SUMMARY

▶ Digestion is the process by which foods are changed so that they can be used by the body.

▶ The digestive tract and other digestive organs make up your digestive system.

▶ The parts of the digestive system are the mouth, esophagus, small and large intestines, and rectum.

▶ Food is moved through the digestive system by mucus and peristalsis.

CHECK *Write true if the statement is true. IF the statement is false, changed the underlined term to make the statement true.*

1. The process by which foods are changed into usable forms is called <u>peristalsis</u>.

2. The end of the large intestine is called the <u>pancreas</u>.

3. Food is moved through the digestive tract by a wavelike movement called <u>digestion</u>.

Answer the following.

4. What connects the mouth to the stomach?

5. What prevents food from entering the windpipe?

APPLY *Complete the following.*

6. **Sequence:** Develop a flowchart to show the parts of the digestive tract through which food passes, in order, beginning with the mouth.

7. **Infer:** Why do you think food must be broken down into smaller pieces?

Health & Safety Tip.........................

Sometimes a piece of food enters the windpipe. When this happens, choking occurs. To help prevent choking, you should chew your food carefully, eat slowly, and avoid talking or running with food in your mouth. Use library references to find out what to do if someone you are eating with begins to choke.

Figure 2 Universal sign of choking

SCIENCE CONNECTION ◆○◆○◆○◆○◆○◆○◆○◆○◆○◆○◆○◆○◆○◆○◆○◆

VESTIGIAL ORGANS

Did you know that at the end of your spine is a small tailbone? The tailbone has no function in humans. It is a vestigial (ves-TIJ-ee-ul) organ. A vestigial organ is a small part of the body that seems to have no use.

Many animals have vestigial organs. For example, some snakes have small hip bones that have no function. Some scientists think that vestigial organs did have a function at one time. Scientists think the organs had a useful purpose in the ancestors of the animals that now have the vestigial organs. Snakes with vestigial hip bones probably evolved from animals with hips.

Humans have more than 100 vestigial structures, including the muscles used to move the ears, the third eyelid, and the appendix. You are probably most familiar with the appendix. It is a small fingerlike sac attached to the large intestine. It has no function. However, if food gets trapped in the appendix, it may become infected. An infected appendix must be removed surgically.

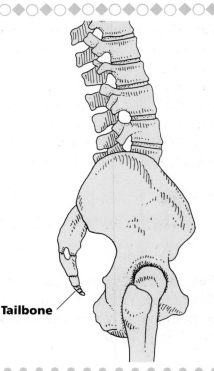

Tailbone

What is digestion?

Objectives ▶ Compare mechanical and chemical digestion. ▶ Explain the function of enzymes.

TechTerms

- ▶ **chemical digestion:** process by which large food molecules are broken down into smaller food molecules
- ▶ **mechanical digestion:** process by which large pieces of food are cut and crushed into smaller pieces
- ▶ **saliva:** liquid in the mouth that helps in digestion

Digestion Digestion is the process of changing foods into usable forms. Food is changed two ways—physically and chemically. Changes in shape and size are examples of physical changes. A chemical change results in new substances. For example, a log burning is a chemical change. The wood is changed into ash and soot.

▶ **Identify:** What are two ways food is changed?

Mechanical Digestion You use your teeth to cut and crush foods. There are four different kinds of teeth. The four kinds of teeth and their jobs are shown in Figure 1. When food is crushed by the teeth, it is broken into small pieces. Your tongue moves the food around. This crushing of food is a physical change. The physical change of food is called **mechanical digestion.** Mechanical digestion breaks large pieces of food into smaller pieces.

▶ **State:** How is food broken down mechanically in the mouth?

Chemical Digestion The process by which large food molecules are broken down into smaller molecules is called **chemical digestion.** Chemical digestion begins in the mouth with **saliva.** Saliva is a liquid found in the mouth. It mixes with foods and makes them soft and moist.

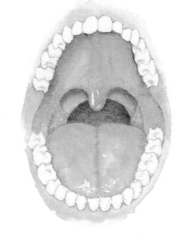

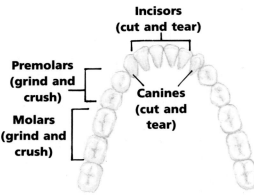

Incisors
(cut and tear)

Premolars
(grind and crush)

Canines
(cut and tear)

Molars
(grind and crush)

Saliva contains enzymes. Enzymes are chemicals that control chemical reactions in the body. Enzymes help digest food. They break down complex chemicals in food into simpler chemicals. Each enzyme can break down only one specific kind of food molecule. For example, the enzyme in saliva helps digest starch. It changes starch into sugar.

▶ **Define:** What is an enzyme?

Process of Digestion Enzymes act only on the outside surface of food particles. The mechanical breakdown of food provides a larger surface area for the enzymes to work on. The digestion of food takes a number of steps. With each step, the food is broken down into smaller chemicals. The body can use only the smallest, simplest chemicals.

▶ **Infer:** Why is the mechanical breakdown of food important?

LESSON SUMMARY

▶ Food is changed both physically and chemically.

▶ Large pieces of food are cut and crushed into smaller pieces of food during mechanical digestion.

▶ Large food molecules are broken down into smaller food molecules during chemical digestion.

▶ Enzymes are chemicals that control chemical reactions in the body.

▶ The body can only use the smallest, simplest chemicals in food.

CHECK *Write true if the statement is true. If the statement is false, change the underlined term to make the statement true.*

1. Shape and size are <u>chemical</u> properties.

2. The crushing of food by the teeth is a <u>mechanical</u> change.

3. The liquid found in the mouth is called <u>enzymes</u>.

4. Enzymes in saliva change <u>fat</u> into sugars.

5. The breakdown of large food molecules into small food molecules is <u>mechanical</u> digestion.

APPLY *Answer the following.*

▶ 6. **Infer:** Why do you think it is important to chew your food into small pieces?

7. What nutrient begins to be digested in the mouth?

InfoSearch

Read the passage. Ask two questions that you cannot answer from the information in the passage.

Tooth Decay Tooth decay is a problem of the digestive system. Tooth decay causes cavities. A cavity is a hole in the tooth. Saliva, food, and bacteria in the mouth mix to form a film called plaque (PLAK). Plaque breaks down the hard, outer covering of the tooth. If plaque is allowed to build up, it spreads to the soft parts of the tooth. Brushing your teeth and flossing daily can help prevent the buildup of plaque.

SEARCH: Use library references to find answers to your questions.

CAREER IN LIFE SCIENCE

DENTAL HYGIENIST

A trip to the dentist often includes a session with a dental hygienist. Dental hygienists clean and polish patients' teeth and remove plaque from under the gums. They also look for signs of tooth decay or gum disease and report these problems to the dentist. Dental hygienists also may take X-rays or assist a dentist with procedures such as pulling teeth. They also teach people the best way to brush their teeth and to remove plaque on a daily basis.

To be a dental hygienist, a person needs to complete a two-year course at a college or professional school. Some students take additional training to earn a bachelor's or master's degree. Dental hygienists need to be skillful in working with their hands. They also need to be skillful in working with worried patients.

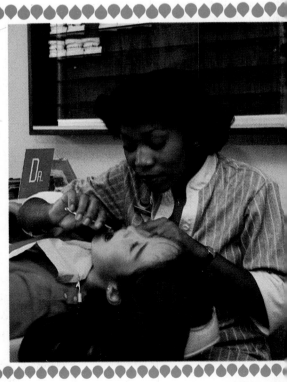

What happens to food in the stomach?

Objective ▶ Describe what happens to food once it enters the stomach.

TechTerms

- ▶ **chyme** (KYM): thick liquid form of food
- ▶ **gastric juice:** juice produced in the stomach that contains mucus, pepsin, and hydrochloric acid
- ▶ **pepsin** (PEP-sin): enzyme that digests proteins

The Stomach Once food leaves the esophagus, it enters the stomach. The stomach is a J-shaped, baglike organ that stores food. The stomach also breaks down food. In fact, mechanical digestion takes place in the stomach as well as in the mouth. The walls of the stomach are made up of layers of strong muscles. These muscles tighten and squeeze the food. The food is broken into small pieces.

�IIIII▶ *Describe:* How does the stomach aid in mechanical digestion?

Gastric Juice The small pieces of food in the stomach are mixed with stomach juice. This juice makes the food soft. The juice the stomach produces is called **gastric juice.** Gastric juice contains mucus, **pepsin** (PEP-sin), and hydrochloric (hy-druh-KLAWR-ik) acid.

▐IIII▶ *Identify:* What is gastric juice made up of?

Chemical Digestion in the Stomach One of the enzymes in gastric juice is pepsin. Pepsin is an enzyme. It begins the digestion of proteins. Hydrochloric acid is a very strong acid. It is needed to make the stomach acidic. Pepsin can work only in an acidic environment. Hydrochloric acid kills bacteria in the stomach. It also helps to break down food. The mucus in gastric juice protects the stomach lining from the hydrochloric acid and pepsin. Figure 1 shows how chemical digestion takes place.

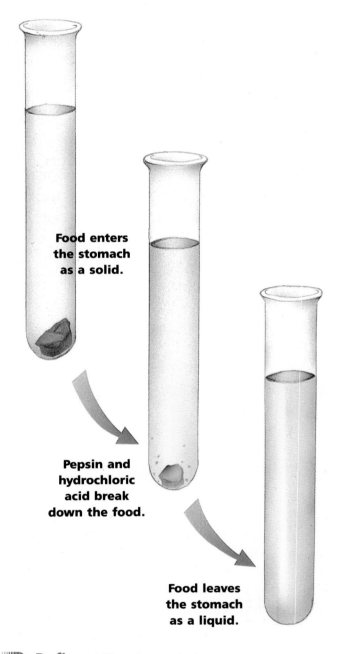

Food enters the stomach as a solid.

Pepsin and hydrochloric acid break down the food.

Food leaves the stomach as a liquid.

▐IIII▶ *Define:* What is pepsin?

Chyme Once food has been crushed by the stomach and mixed with gastric juice, it is ready to leave the stomach. Food that leaves the stomach is in the form of a thick liquid. This liquid is called **chyme** (KYM). Chyme is released slowly from the stomach into the small intestine.

▐IIII▶ *Define:* What is chyme?

LESSON SUMMARY

► The stomach is a baglike organ that stores food.

► Gastric juice contains mucus, pepsin, and hydrochloric acid and is produced by the stomach.

► Hydrochloric acid makes the stomach acidic so that pepsin can digest proteins. Mucus protects the stomach from the hydrochloric acid and pepsin.

► Chyme is a thick liquid form of food that leaves the stomach.

CHECK *Find the sentence in the lesson that answers each question. Then, write the sentence.*

1. What does mucus in the gastric juice do?
2. How is food mechanically digested in the stomach?
3. What enzyme digests proteins?
4. In what form is food that leaves the stomach?
5. What does the stomach look like?

APPLY *Answer the following.*

6. Why is the stomach an important part of the digestive system?

7. **Infer:** What would happen if the stomach did not produce hydrochloric acid?
8. **Analyze:** How can you tell that digestion has taken place in test tube 3?
9. **Sequence:** Describe the change that food undergoes from the time it enters the mouth to the time it enters the small intestine.

Skill Builder

Researching An ulcer occurs when the stomach acids digest the lining of the stomach. The hole that occurs in the stomach lining is called an ulcer. Ulcers also can occur in the small intestine. Use library references to find out the causes of ulcers, the signs and symptoms of ulcers, and the treatment of ulcers. Present your findings in a table.

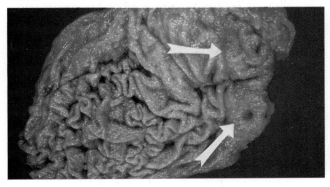

LOOKING BACK IN SCIENCE

WILLIAM BEAUMONT (1785-1853)

William Beaumont was an American army doctor in the early 1800s. On June 6, 1822, an 18-year-old named Alexis St. Martin was accidentally shot in the stomach. Dr. Beaumont saved the young man's life and eventually the wound healed. However, the wound never completely closed. For the rest of his life, St. Martin had a two and a half inch opening in his left side.

William Beaumont discovered that he could view the workings of the stomach through the opening. For the next nine years, with St. Martin's cooperation, Beaumont studied the stomach. At the time, most information about human digestion was obtained by examining the remains of the deceased. The use of X-rays had not been discovered yet. Therefore, the ability to view a functioning body system was extraordinary. Beaumont published his findings in 1833, providing other doctors with valuable information about human digestion. Much of what people know today about the functions of the stomach is based on the observations of Dr. Beaumont.

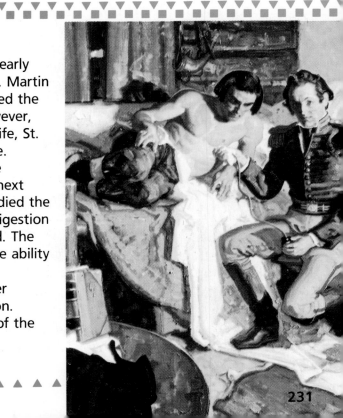

Objective
▶ Describe what happens to foods in the small intestine.

TechTerms

- **bile:** green liquid that breaks down fats and oils
- **emulsification** (i-mul-suh-fi-KAY-shun): process of breaking down large droplets of fat into small droplets of fat
- **lipase** (LY-pays): enzyme that digests fats and oils

The Small Intestine The small intestine is a long coiled tube. It is about 6.5 m long and 2.5 cm wide. Like the stomach, the walls of the small intestine are muscular. Food moves through the small intestine by peristalsis. Most of the chemical digestion of food takes place in the small intestine.

▶ **Explain:** What happens in the small intestine?

Digestion in the Small Intestine Digestive juices in the small intestine contain many enzymes. These enzymes complete digestion. One of these enzymes is **lipase** (LY-pays). Lipase digests fats and oils. Unlike starch and proteins, fats are not digested in the mouth or stomach. Fats are digested only in the small intestine. Table 1 reviews the digestion of the three main nutrients. Notice that all digestion is completed in the small intestine.

Table 1 Summary of Digestion		
NUTRIENT	DIGESTION BEGINS	DIGESTION COMPLETED
proteins	stomach	small intestine
carbohydrates	mouth	small intestine
fats	small intestine	small intestine

▶ **Define:** What is lipase?

The Pancreas The pancreas is a small organ that lies below the stomach. When food first enters the small intestine, the pancreas releases digestive juices. These digestive juices travel to the small intestine through a small tube. Pancreatic digestive juices contain enzymes. These enzymes change starches, proteins, and fats into simpler forms.

▶ **Identify:** How does the pancreas aid in chemical digestion?

The Liver The liver is the largest organ in the human body. One job of the liver is to produce **bile.** Bile is a green liquid that breaks down large droplets of fat into smaller droplets of fat. The breaking down of large fat droplets into smaller fat droplets is called **emulsification** (i-mul-suh-fi-KAY-shun).

Bile does not move directly from the liver to the small intestine. Bile is stored in a small sac under the liver. This sac is called the gall bladder. Bile moves from the gall bladder into the small intestine through a small tube.

▶ **Define:** What is emulsification?

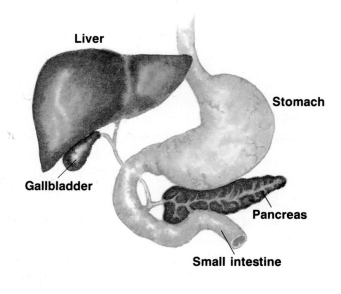

Figure 1 Organs that aid in the digestion of fat

LESSON SUMMARY

▶ Most of the chemical digestion of food takes place in the small intestine.

▶ Lipase is an enzyme that digests fats and oils.

▶ The pancreas releases digestive juices into the small intestine to break up starch, protein, and fat.

▶ The liver produces bile which digests fats and oils.

▶ Bile is stored in the gall bladder.

CHECK *Write true if the statement is true. If the statement is false, change the underlined term to make the statement true.*

1. Most <u>mechanical</u> digestion takes place in the small intestine.

2. Lipase is an enzyme that digests <u>fats</u>.

3. The gall bladder stores <u>digestive juices</u>.

4. The <u>pancreas</u> produces bile.

5. Bile is responsible for the <u>emulsification</u> of fat.

6. Food moves through the small intestine by <u>peristalsis</u>.

APPLY *Use table 1 on page 232 to answer the following.*

7. Where does the digestion of carbohydrates begin?

8. Where does the digestion of protein begin?

9. What is the only nutrient that begins its digestion in the small intestine?

InfoSearch

Read the passage. Ask two questions that you cannot answer from the information in the passage.

Digestive System Problems There are many different problems of the digestive system. One problem of the digestive system is indigestion. Indigestion is caused by eating too much, too little, or too fast. Another problem of the digestive system is heartburn. Heartburn occurs when stomach acids go up into the esophagus. This causes a burning sensation in the esophagus. The esophagus is located behind the heart, so the burning feels like it is coming from the heart.

SEARCH: Use library references to find answers to your questions.

ACTIVITY

INVESTIGATING FAT DIGESTION

You will need 2 test tubes with stoppers, cooking oil, baking soda, a medicine dropper, a measuring spoon, and water.

1. Fill each test tube halfway with water.

2. Using the medicine dropper, put 4 drops of cooking oil in each test tube.

3. Use the measuring spoon to add ¼ teaspoon of baking soda to one of the test tubes.

4. Stopper each test tube and shake the test tubes well.

Questions

1. **Observe:** What happened in the test tube containing only oil and water?

2. What happened in the test tube containing oil, water, and baking soda?

3. Which test tube contains fats that are not broken down?

4. Which test tube contains fats that are partly broken down?

5. **Compare:** How is the action of the baking soda similar to that of enzymes?

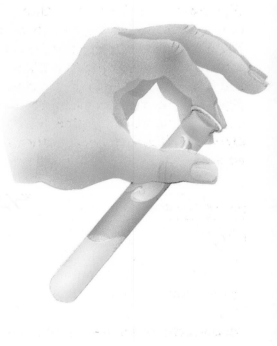

How is food absorbed by the body?

▶ Describe absorption. ▶ Describe what happens to food after it leaves the small intestine.

TechTerms

▶ **absorption** (ab-SAWRP-shun): movement of food from the digestive system to the blood

▶ **colon:** large intestine

▶ **villi:** fingerlike projection on the lining of the small intestine

Absorption in the Small intestine After food has been changed in usable forms, it is ready to be absorbed (ub-ZOWRBD), or taken into the bloodstream. **Absorption** (ab-SAWRP-shun) is the movement of food from the digestive system to the blood. Absorption of fats, proteins, carbohydrates, vitamins, and water takes place through the walls of the small intestine. Once inside the blood, digested food is carried to all of your body cells.

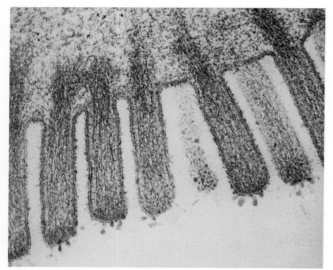

The inner lining of the small intestine is folded. The folds have millions of tiny fingerlike projections called **villi** (VIL-y). Villus is the singular form of villi. The folds and villi make the surface area of the small intestine bigger. Digested food passes through the absorptive layer of the villus and into the blood vessels.

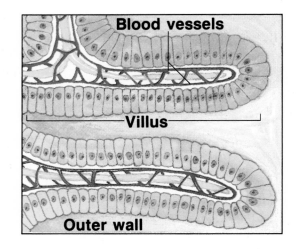

▶ **Explain:** What happens to food in the small intestine?

The Large Intestine The large intestine is the last part of the digestive system. It also is called the **colon.** Some undigested food, minerals, and water are not absorbed in the small intestine. They form a watery mixture. This mixture enters the large intestine. Water and minerals are absorbed into the blood in the large intestine. The remaining solid wastes are stored temporarily in the rectum. Then they are eliminated from the body.

▶ **Describe:** What happens to food in the large intestine?

The Appendix There is a small, thin sac located where the small intestine and large intestine meet. This sac is called the appendix (uh-PEN-diks). Scientists do not know what the appendix does in humans.

Sometimes food gets trapped in the appendix. The appendix becomes infected with bacteria. This infection of the appendix is called appendicitis (uh-PEN-duh-SY-tis). An infected appendix must be removed.

▶ **State:** What is the job of the appendix?

LESSON SUMMARY

▶ Absorption is the movement of food from the digestion system to the blood.

▶ Villi on the walls of the small intestine make the surface area of the small intestine bigger.

▶ The large intestine, or colon, is the large part of the digestive system where the absorption of minerals and water takes place.

▶ The appendix is a small sac located where the small intestine and large intestine meet, that has no known purpose.

▶ An infected appendix is called appendicitis.

CHECK *Answer the following.*

1. **Explain:** How does food get to all the body cells?

2. What are villi?

3. What happens to undigested food?

4. Where is the appendix located?

Complete the following.

5. The process by which food leaves the bloodstream and enters the body is called _____ .

6. Solid wastes are stored in the _____ .

APPLY *Answer the following.*

7. **Infer:** Why must food be absorbed?

8. **Hypothesize:** How do folds increase the surface area of the small intestine?

9. **Infer:** Why do you think it is important to have an infected appendix removed?

Ideas in Action

IDEA: Whole wheat and other whole grains are more healthful than processed grains, such as white flour.

ACTION: List five ways that you can substitute whole grains for processed grains in your diet.

Health & Safety Tip

A healthful diet includes fiber. Fiber is the part of a plant that cannot be digested by humans. There are different kinds of fiber. Some kinds of fiber are effective in decreasing fat absorption. Other kinds of fiber help lower cholesterol levels. Scientific studies have shown that diets high in fiber also offer some protection against colon cancer. Use library references to write a report about foods that are high in fiber. Include ways to substitute these foods into your everyday diet.

ACTIVITY

CALCULATING SURFACE AREA

You will need string, scissors, tape, and a ruler.

1. Cut a piece of string to extend across the top of your desk. Tape the ends of the string to opposite sides of the desk.

2. Tape the end of another piece of string to one side of the desk. Form a series of "S" curves across the desk with the string similar to line AB. Tape the other end of the string to the desk.

3. Use a ruler to measure the distance between the taped ends of each piece of string.

4. Place the curved string in a straight line along the floor. Measure the distance between the string's ends.

Questions

1. What was the distance between the taped ends of each piece of string?

2. What was the distance between the ends of the curved string when it was straightened on the floor?

3. Which piece of string had more surface area?

4. **Analyze:** How do folds in the inner lining of the small intestine aid in digestion?

A

C

B

D

STUDY HINT Before you begin the Unit Challenges, review the TechTerms and Lesson Summary for each lesson in this unit.

TechTerms

absorption (234)	digestion (226)	molecule (216)
amino acid (216)	emulsification (232)	nutrient (214)
bile (232)	enzyme (216)	oxidation (224)
Calorie (224)	epiglottis (226)	pepsin (230)
carbohydrates (214)	esophagus (226)	peristalsis (226)
chemical digestion (228)	gastric juice (230)	protein (214)
chyme (230)	lipase (232)	saliva (228)
colon (234)	mechanical digestion (228)	villi (234)
deficiency disease (218)	mineral (220)	vitamin (218)

TechTerm Challenges

Matching *Write the TechTerm that matches each description.*

1. fingerlike projections on the lining of the small intestine
2. enzyme that digests fats and oils
3. enzyme that begins the digestion of protein
4. breaking down of large food molecules into small food molecules
5. long tube that connects the mouth to the stomach
6. process by which foods are changed into forms the body can use
7. slow burning of foods in the body
8. nutrient found naturally in food
9. smallest part of a substance that has all the properties of that substance
10. chemical substances in food needed by the body for growth, energy, and life processes
11. protein that controls chemical activities
12. disease caused by the lack of a certain nutrient
13. nutrient needed by the body to develop properly
14. amount of energy given off by food
15. thin flap of tissue that keeps food from entering the windpipe
16. wavelike motion that moves food through the digestive tract
17. thick liquid form of food
18. breakdown of fat
19. large intestine

Identifying Word Relationships *Explain how the words in each pair are related. Write your answers in complete sentences.*

1. amino acid, protein
2. saliva, carbohydrates
3. bile, fat
4. gastric juice, stomach

Content Challenges

Multiple Choice *Write the letter of the term that best completes each statement.*

1. The wavelike movement that moves food through the digestive system is called
   a. chyme.   **b.** mechanical digestion.   **c.** peristalsis.   **d.** chemical digestion.
2. The largest organ inside the body is the
   a. stomach.   **b.** gallbladder.   **c.** liver.   **d.** pancreas.
3. The largest amount of food energy comes from
   a. fats.   **b.** carbohydrates.   **c.** vitamins.   **d.** proteins.
4. Anemia is caused by a diet that lacks
   a. zinc.   **b.** protein.   **c.** iron.   **d.** vitamin A.
5. Starches are
   a. proteins.   **b.** simple carbohydrates.   **c.** fats.   **d.** complex carbohydrates.
6. Mechanical digestion begins in the
   a. esophagus.   **b.** pharynx.   **c.** mouth.   **d.** stomach.
7. Undigested food from the small intestine moves into the
   a. pancreas.   **b.** appendix.   **c.** stomach.   **d.** large intestine.
8. Saliva begins the chemical digestion of
   a. protein.   **b.** starches.   **c.** fats.   **d.** nutrients.
9. Amino acids make up
   a. proteins.   **b.** fats.   **c.** water.   **d.** carbohydrates.
10. More than two-thirds of the human body is made up of
    a. protein.   **b.** fat.   **c.** water.   **d.** carbohydrates.

Completion *Write the term that best completes each statement.*

1. Chemical reactions that take place in the body are controlled by _____ .
2. Gastric juice contains pepsin, _____, and mucus.
3. Bile is produced in the _____ .
4. The _____ releases digestive juices into the small intestine.
5. The _____ stores and breaks down food.
6. When food is swallowed, it enters the _____ .
7. Liquid fats usually are called _____ .
8. Nutrients made up of amino acids are _____ .
9. A weakened condition that results from a lack of a certain nutrient is called a _____ .

Understanding the Features

Reading Critically *Use the feature reading selections to answer the following. Page numbers for the features follow each question in parentheses.*

1. **Identify:** What are some of the signs and symptoms of scurvy? (219)
2. **Name:** Where are three places where a nutritionist may work? (221)
3. **Define:** What is chemosynthesis? (225)
4. **Name:** What are four vestigial organs in humans? (227)
5. **Infer:** Why do dental hygienists need good hand control and coordination? (229)
6. What is William Beaumont most known for? (231)
7. **Infer:** Why should a person studying to be a nutritionist take science and math classes? (221)

Concept Challenges

Critical Thinking *Answer each of the following in complete sentences.*

1. **Explain:** Is the action of bile on fat part of mechanical digestion or chemical digestion? Explain your answer.
2. **Infer:** What role does the large surface area of the small intestine play in absorption?
3. Why is absorption important?
4. **Hypothesize:** Why do you think low-carbohydrate 'liquid diets' are unhealthy?
5. What would you eat before an athletic competition, a bowl of spaghetti or a steak? Explain your answer.

Interpreting a Diagram *Use the diagram of the human digestive system to answer the following.*

1. What is the function of the part labeled B?
2. What letter indicates a vestigial organ?
3. What letter indicates where gastric juice is produced?
4. What is the function of the part labeled H?
5. What letter indicates where bile is produced?
6. What does letter F represent?

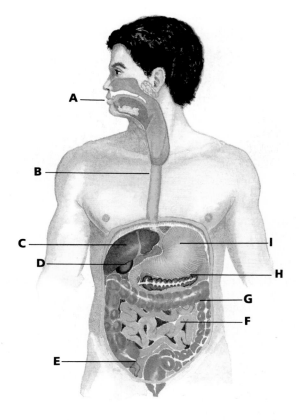

Finding Out More

1. Most nutritionists believe that fiber plays an important role in the prevention of cancer. Use library references to find out the role of fiber in the diet and how it helps to prevent cancer.
2. Two problems of the digestive system are diarrhea and constipation. Use library references to find out what causes each of the these two problems. Find out what can be done to treat each of these digestive system problems.
3. Studies have shown that people who eat a lot of fish have less heart disease than people who eat a lot of red meat. Use library references to find out about Omega 3, an oil found in fish. Find out the benefits of Omega 3, as well as information on fish oil capsules found in many health food stores.
4. Sodium is an important part of the diet. However, most people consume too much salt. An excess of salt can cause many health problems. Develop a chart that lists the effects of too much salt on the different body systems.
5. Some people drink "sports drinks" after they have exercised. Visit your local food store and find a "sports drink." Write down the ingredients of the drink, as well as the claims the drink makes. Using this book as a reference, decide whether the claims that are made are true. Present your findings to the class.

SUPPORT AND MOVEMENT

CONTENTS

STUDY HINT After you read each lesson in Unit 12, write a brief summary on a sheet of paper explaining how the information in each lesson applies to your everyday life.

12-1 What is the skeletal system?

Objective ▶ Describe the functions of the skeletal system and its parts.

TechTerms

- ▶ **cartilage** (KART-ul-idj): tough, flexible connective tissue
- ▶ **endoskeleton** (en-duh-SKEL-uh-tun): internal skeleton
- ▶ **exoskeleton** (ek-soh-SKEL-uh-tun): external skeleton

The Skeletal System Have you ever seen a house being built? If you have, you have probably seen the wooden framework that makes up a house. The framework of a house is important. It gives a house its shape. The framework also supports a house. You also have a frame that supports your body. This frame is your skeleton. Most of your skeletal system is made up of bone. Bone is a very hard tissue.

▐▌▶ *Compare:* How is a skeleton similar to the frame of a house?

Kinds of Skeletons Some living things do not have a skeleton. Their bodies are entirely soft. Other organisms, such as lobsters and insects, have a skeleton outside their bodies. An external skeleton is called an **exoskeleton** (ek-soh-SKEL-uh-tun). An exoskeleton is tough and hard. It protects the animal. Human beings and many other animals have a skeleton inside their bodies. An internal skeleton is called an **endoskeleton** (en-duh-SKEL-uh-tun).

▶ *Infer:* List three organisms with an exoskeleton and three with an endoskeleton.

Cartilage Feel your knee. The bones of your knee are very hard. Some parts of your skeleton are not as hard. Move the tip of your nose. Bend one of your ears with your hand. These two parts of your body are not made of bone. They are made

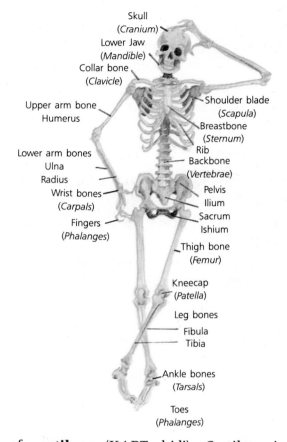

- Skull (*Cranium*)
- Lower Jaw (*Mandible*)
- Collar bone (*Clavicle*)
- Upper arm bone Humerus
- Lower arm bones Ulna Radius
- Wrist bones (*Carpals*)
- Fingers (*Phalanges*)
- Shoulder blade (*Scapula*)
- Breastbone (*Sternum*)
- Rib
- Backbone (*Vertebrae*)
- Pelvis Ilium Sacrum Ishium
- Thigh bone (*Femur*)
- Kneecap (*Patella*)
- Leg bones Fibula Tibia
- Ankle bones (*Tarsals*)
- Toes (*Phalanges*)

up of **cartilage** (KART-ul-idj). Cartilage is a tough, but flexible connective tissue. When you were a baby, most of your skeleton was made up of cartilage. However, during the second and third months of development, bone slowly replaced the cartilage in your skeleton.

▐▌▶ *Define:* What is cartilage?

Jobs of the Skeleton Besides giving support and shape to your body, your skeletal system has many important jobs. One of the these jobs is to move the body. The bones work together with muscles to move the body. Another job of the skeleton is to protect organs. For example, the ribs protect the heart and lungs. Some bones have a very special job. Blood cells are made inside some bones. Bones also store minerals that the body needs.

▶ *Infer:* What do you think the skull protects?

240

LESSON SUMMARY

▶ The skeleton supports the body and gives it its shape.

▶ Most organisms have either an exoskeleton or an endoskeleton.

▶ Cartilage is a tough, flexible connective tissue that makes up parts of the skeletal system.

▶ The skeletal system has many important jobs and functions, including protecting organs and making red blood cells.

CHECK *Complete the following.*

1. A very hard tissue that makes up the skeletal system is _____ tissue.

2. As a developing baby, most of your skeleton was made up of _____ .

3. The skeletal system works with _____ to move the body.

Answer the following.

4. What kind of skeleton do humans have?

APPLY *Complete the following.*

5. **Classify:** Which animal, a grasshopper or a fish, has an endoskeleton? Which has an exoskeleton?

Use the diagram of the skeletal system on page 240 to answer the following.

6. **Analyze:** How many bones make up the lower arm?

7. **Analyze:** What is another name for the breastbone?

....
Skill Builder.................................

Relating Roots and Word Parts Many words that begin with "oss" and "oste" refer in some way to bones. "Os" is the Latin word for bone. "Osteon" is the Greek word for bone. Use a dictionary to find five words that begin with "oss" and "oste" and refer to bone. Write definitions for each of the five words. Circle the part of the word that relates to bone.

TECHNOLOGY AND SOCIETY

ARTHROSCOPIC SURGERY

Can you imagine having surgery done through a small tool inserted into your body? Due to advances in both medicine and technology, operations are now being performed in this way. This kind of operation is called arthroscopic (ahr-thruh-SKAHP-ik) surgery. Most often arthroscopic surgery is done on the knee. A small cut is made on the body. A small instrument, called an arthroscope, is put into the cut. A special light also is inserted into the cut. The arthroscope, working with the light, projects an image on the screen. By looking at the screen, doctors can see what they are doing.

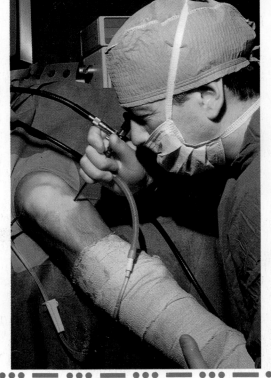

One of the advantages of arthroscopic surgery is that instead of being put to sleep, the person stays awake during surgery. Doctors have discovered that when people are awake during surgery, they recover faster. People who have arthroscopic surgery recover faster than people who have regular surgery. Because the cut is so small, less tissue is damaged. The body does not need to repair itself as much. The other advantage of arthroscopic surgery is only a very small scar is left.

What are bones?

Objective ▶ Describe the parts of a bone.

TechTerms

▶ **compact bone:** hardest part of bones

▶ **periosteum** (per-i-AS-tee-um): thin membrane that covers a bone

▶ **spongy bone:** bone cells that make up the soft and spongy ends of bones

Bones The human skeleton is made up of about 206 bones. Bones come in all shapes and sizes. Some bones are very small. There are three small bones in your ear that help you hear. They are the smallest bones in your body. Other bones are quite large. The bone in your thigh is the largest bone in the body. Some bones are tubelike. Others are flat. Although bones are different in size and shape, they all have a similar structure.

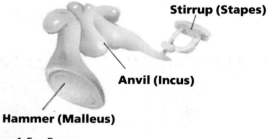

Stirrup (Stapes)

Anvil (Incus)

Hammer (Malleus)

Figure 1 Ear Bones

▶ *Infer:* Name some places in your body where there are small bones.

Structure of Bones Bones are made up of both living and nonliving material. Each bone is cov-

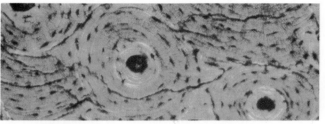

Figure 2 Bone Cells

ered by the **periosteum** (per-i-AS-tee-um). It is a thin membrane. The periosteum has many blood vessels in it. These blood vessels bring food and oxygen to the living bone cells.

The hardest part of a bone is called **compact bone.** Compact bone is made up of living bone cells, protein fibers, and nonliving minerals. The mineral calcium makes compact bone hard. Calcium in your diet helps keep your bones hard and strong. Dairy products are rich in calcium.

Bones are not entirely hard, however. The ends of bones are soft and spongy. The soft part of bones is called **spongy bone.** Spongy bone looks like a sponge. It has many holes in it. Spongy bone is lightweight and gives the bone its strength.

▶ *Describe:* What is the function of the periosteum?

Marrow The spaces in spongy bone are filled with bone marrow. Bone marrow is a soft tissue. It is red or yellow in color. Spongy bone contains red bone marrow. New blood cells are made in the red bone marrow. Long bones contain yellow marrow. Yellow marrow is mostly fat.

▶ *Name:* In which kind of bone marrow are red blood cells made?

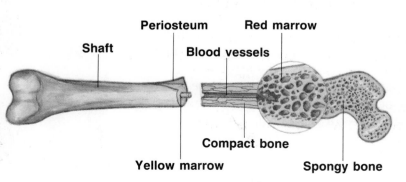

Shaft

Periosteum

Blood vessels

Red marrow

Compact bone

Yellow marrow

Spongy bone

Figure 3 Structure of a Bone

- ▶ Bones are different in size and shape, but they have similar structures.
- ▶ Bones have both living and nonliving materials in them.
- ▶ The hardest part of a bone is called compact bone.
- ▶ The soft and spongy ends of bones are called spongy bone.
- ▶ The spaces in spongy bone are filled with red or yellow bone marrow.

CHECK *Write true if the statement is true. If the statement is false, change the underlined term to make the statement true.*

1. New <u>bone</u> cells are made in spongy bone.
2. The mineral <u>calcium</u> keeps bones strong and hard.
3. Blood vessels in <u>spongy bone</u> supply the bones with food and oxygen.
4. Most bones have <u>the same</u> sizes and shapes.
5. Bone cells and protein fibers make up <u>compact bone</u>.

APPLY *Complete the following.*

➤ 6. **Infer:** Is the periosteum living or nonliving? Explain your answer.

➤ 7. **Infer:** What would happen if you did not eat foods containing calcium?

InfoSearch

Read the passage. Ask two questions about the topic that you cannot answer from the information in the passage.

Vitamin D Sunshine and the right foods are important for proper bone growth. Bones are made up of the minerals calcium and phosphorus. To make sure your bones are healthy, you should eat foods containing these minerals each day. Vitamin D is also needed for healthy bones. Vitamin D is found in many foods. Your body can also produce vitamin D, by being exposed to sunlight. This is why vitamin D is also called the sunshine vitamin.

SEARCH: Use library references to find answers to your questions.

TECHNOLOGY AND SOCIETY

BONE MARROW TRANSPLANTS

Bone marrow transplants are used to treat many different bone disorders. Certain kinds of anemia and leukemia may be treated with bone marrow transplants. A bone marrow transplant is needed when bone marrow produces abnormal blood cells, instead of normal cells.

A healthy donor is needed for a bone marrow transplant. Radiation treatments are given to the person with the abnormal bone marrow cells. The radiation kills the abnormal marrow cells in the body. Healthy bone marrow is then taken from the donor's hip. The healthy marrow is inserted into the patient's bones.

The success of a bone marrow transplant depends on whether the healthy bone marrow produces more healthy bone marrow. Another factor in the success of a bone marrow transplant is if the patient's body accepts or rejects the new bone marrow. Rejection of the new bone marrow is one of the major drawbacks of bone marrow transplants. Another drawback is infection.

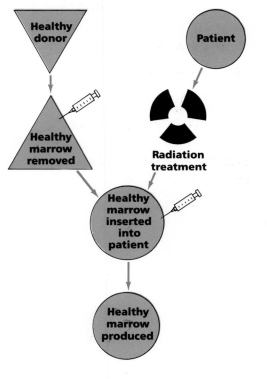

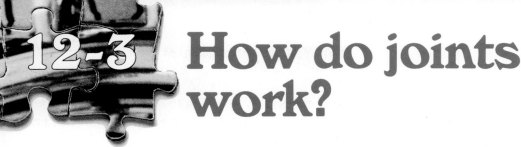

12-3 How do joints work?

Objective ▶ Identify the motions and locations of the four kinds of movable joints.

TechTerms

▶ **joint:** place where two or more bones meet

▶ **ligaments** (LIG-uh-ments): tissue that connects bone to bone

Joints The place where two or more bones meet is called a **joint.** Some bones are connected to other bones at the joint. However, most bones are connected at joints by **ligaments** (LIG-uh-ments). A ligament is a tough band of tissue that connects one bone to another bone.

▶ **Define:** What is a joint?

Kinds of Joints There are three main kinds of joints in the body. They are fixed joints, partly-movable joints, and movable joints. Fixed joints do not allow any movement. The joints in your skull are fixed. Partly movable joints allow a little bit of movement. The joints between your ribs and breastbone move a little bit. However, most of the joints in the body are movable. Your arms and legs have several movable joints.

▶ **State:** What are the three kinds of joints in the body?

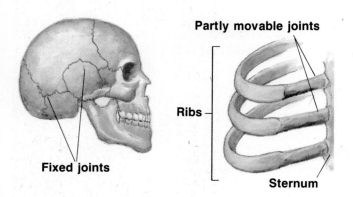

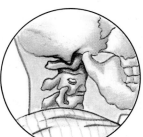
Pivotal joint

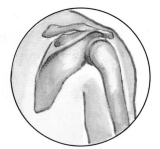
Ball-and-socket joint

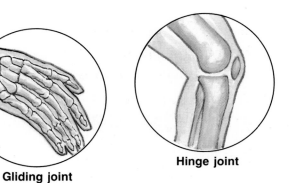
Gliding joint

Hinge joint

▶ **Ball-and-socket Joints** Ball-and-socket joints allow bones to move in all directions. The joint between your upper arm and shoulder is a ball-and-socket joint. Your arm can move up and down, side to side, front to back, and around in a circle.

▶ **Gliding Joints** Gliding joints allow some movement in all directions. In a gliding joint, the bones slide along each other. Your wrist has gliding joints.

▶ **Hinge Joints** Hinge joints allow bones to move forward and backward in only one direction. This movement is similar to a door opening and closing. Hinge joints are located in your elbows and knees.

▶ **Pivotal Joints** Pivotal joints allow bones to move side to side and up and down. The joint between your skull and neck is a pivotal joint.

Movable Joints There are four kinds of movable joints. These joints are ball-and-socket joints, gliding joints, hinge joints, and pivotal joints.

▶ **List:** List the four kinds of movable joints and give examples of where they are found in the body.

LESSON SUMMARY

▸ A joint is where two bones meet.

▸ The three kinds of joints are fixed joints, partly-movable joints, and movable joints.

▸ The four kinds of movable joints are ball-and-socket joints, gliding joints, hinge joints, and pivotal joints.

CHECK *Complete the following.*

1. Joints can be either fixed, partially movable, or _____.

2. The joints between your breastbone and ribs are ____.

3. A _____ is a band of tissue that connects one bone to another bone.

4. The joint in your elbow is a _____ joint.

APPLY *Complete the following.*

5. **Hypothesize:** What would happen if the joint between your arm and shoulder was not a ball-and-socket joint?

6. **Infer:** Why do you think the joints between your rib cage and sternum are partly-movable?

InfoSearch

Read the passage. Ask two questions about the topic that you cannot answer from the information in the passage.

Arthritis Do you know anyone that has arthritis? Arthritis affects people who are both young and old. Arthritis is a term that describes many different joint problems. The most common form of arthritis occurs when the cartilage between the bones is replaced with bone deposits. A person with arthritis has swollen joints. Movement in these joints is limited and can be very painful.

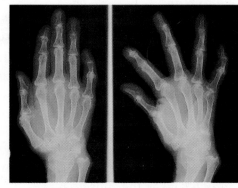

SEARCH: *Use library references to find answers to your questions.*

ACTIVITY

OBSERVING JOINT MOVEMENTS

You will need a paper and pencil.

1. Move your ankle in as many different ways as possible. Write down all the movements your ankle can make.

2. Move your fingers in as many different ways as possible. **CAUTION:** Do not force movements at the joint. Write down all the movements your fingers can make.

3. Move your head in as many different ways as possible. Write down all the movements your head can make.

4. Move your leg and hip joint in as many different ways as possible. Write down all the movements your leg can make at the hip.

Questions

1. **a.** What kind of joint was used in each step? **b.** How can you tell?

2. **Observe:** Look at the X-ray of the hand. **a.** How many joints can you count? **b.** What kind of joints are shown?

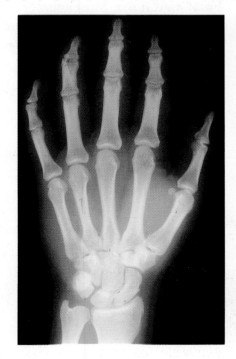

What is the muscular system?

Objective ▶ Describe how muscles work.

TechTerms

▶ **extensor** (ik-STEN-sur): muscle that straightens a joint

▶ **flexor** (FLEK-sur): muscle that bends a joint

▶ **tendons** (TEN-duns): tissue that connects muscle to bone

Muscles More than 600 muscles make up the muscular system. Muscles are tissues that can shorten along their length. Without muscles, the bones of the skeletal system could not move the body. Muscles are attached to bones by **tendons** (TEN-duns). A tendon is a strong elastic band of tissue. Tendons make movement possible. When a muscle contracts, or shortens, it pulls on the tendon, which makes the muscle move.

▶ *Contrast:* What is the difference between a tendon and a ligament?

Muscle Actions Muscles only move bones when they contract. For this reason, muscles can only pull bones. They cannot push bones. For example, there are muscles that bend, or flex your knee joint. These muscles are called **flexors** (FLEK-surs). There are other muscles that straighten, or extend your knee joint. These muscles are **extensors** (ik-STEN-surs).

▶ *Define:* What does a flexor muscle do?

Teamwork In order to move your body, most muscles must work in teams of two. The muscles that bend and straighten the arm are good examples of flexors and extensors working together. These muscles are called the biceps (BY-seps) and triceps (TRY-seps). The biceps are the flexors. They bend the arm at the elbow. The triceps are the extensors. They straighten the arm at the elbow. As you bend your arm, the biceps contract, and the triceps relax. As you straighten your arm, the opposite takes place. The biceps relax, while the triceps contract and pull the arm straight.

▶ *State:* Which muscle in the arm is the flexor? The extensor?

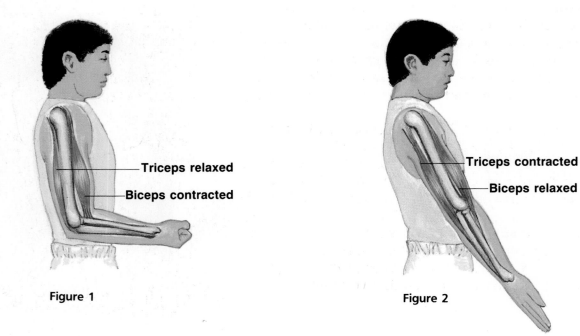

Triceps relaxed
Biceps contracted

Figure 1

Triceps contracted
Biceps relaxed

Figure 2

LESSON SUMMARY

▶ Muscles are tissues that can shorten along their length.

▶ Muscles can only pull bones.

▶ Muscles usually work in pairs to move the body.

CHECK *Find the sentence or sentences that answers the question. Then write the sentence.*

1. What is a tendon?

2. Are muscles relaxed or contracted when they move bones?

Answer the following.

3. What is the main function of the muscular system?

4. What do extensor muscles do?

5. Why must muscles work together?

APPLY *Use Figures 1 and 2 one page 246 to answer the following.*

6. a. In Figure 1, which muscle is contracted?   b. Which muscle is relaxed?

7. a. In Figure 2, which muscle is contracted?   b. Which muscle is relaxed?

Skill Builder.....................................

Researching A good way to make muscles strong is to exercise. Design your own individual exercise program. Keep in mind the following important parts of a good workout: warm up, exercise time, and cool down. Other factors to include are mode, duration, intensity, and frequency. Use library references to find out what mode, duration, intensity, and frequency mean in regard to exercise. Then in a report, write up an individualized exercise program for yourself.

Skill Builder.....................................

Researching Find out what each of the following muscle and skeletal problems are: sprains, strains, charley horses, muscle cramps, and tendonitis. Circle the word or words in each definition that is part of the muscular system. Draw a square around the word or words that are part of the skeletal system. What can cause each of these problems.

◆○◆ SCIENCE CONNECTION ◆○◆○◆○◆○◆○◆○◆○◆○◆○◆○◆○◆○◆○◆○◆○◆
ANABOLIC STEROIDS

You have probably read about athletes being disqualified from sporting events for taking steroids (STIR-oyds). These athletes have taken anabolic (an-uh-BAHL-ik) steroids. Anabolic steroids are synthetic, or made in a lab. Steroids are male hormones. Hormones are chemicals that regulate body functions. Steroids were developed to help people with muscle diseases. However, athletes began taking steroids to increase their muscle strength and muscle mass.

Humans are born with a fixed number of muscle fibers. The size of muscle fibers can increase, but the number will never change. People take steroids because they believe that steroids will increase the number of muscle fibers. The only way to increase muscle mass and strength is through weight-training exercise.

Steroids are considered illegal unless taken under a doctor's care. There are many harmful side effects of steroids. Some of the side effects include high blood pressure, liver damage, kidney damage, and damage to reproductive organs.

WARNING

Never take STEROIDS unless they are prescribed by a doctor

12-5 What are the kinds of muscles?

Objective ► Name the three kinds of muscle and identify where they are located in the body.

TechTerms

- ► **cardiac** (KAHR-dee-ak) **muscle:** type of muscle found only in the heart
- ► **smooth muscle:** muscle that causes movements that you cannot control
- ► **striated** (STRY-ayt-ed) **muscle:** muscle attached to the skeleton, making movement possible

Striated Muscle There are three different kinds of muscle tissue in the human body. Each kind of muscle tissue has a different job. One kind of muscle tissue is **striated** (STRY-ayt-ed) **muscle.** Striated means striped. If you look at Figure 1, you can see how striated muscle gets its name.

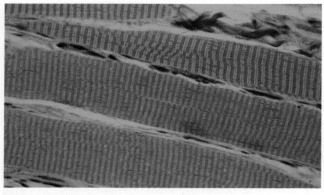

Figure 1 Striated muscle

Striated muscle also is called skeletal muscle because it is attached directly to the skeleton. Striated muscles are what makes the body move. Move your foot. Open and close your fingers. These are movements you can control. Striated muscles are usually voluntary (VAHL-un-ter-ee) muscles. They are voluntary muscles because you can control their movements.

▐▐▶ *Name:* Give three places on your body where striated muscle is found.

Smooth muscle Muscle tissue that is found in the walls of blood vessels, the stomach, and other internal organs is called **smooth muscle.** Smooth muscle causes movements that you cannot control. For this reason, smooth muscle is sometimes called involuntary muscle. For example, after you eat, you cannot stop the muscles lining your stomach from digesting the food.

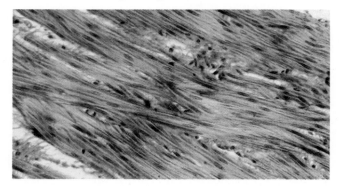

Figure 2 Smooth muscle

▶ *Infer:* Why do you think smooth muscle is often called involuntary muscle?

Cardiac Muscle The third kind of muscle is called **cardiac** (KAHR-dee-ak) **muscle.** The word 'cardiac' means 'heart.' Cardiac muscle is found only in the heart. Cardiac muscle is very strong. The heart must be strong to pump blood throughout the body. Cardiac muscle is striated. Unlike most striated muscle, cardiac muscle is involuntary. You have no control over your heart beating.

Figure 3 Cardiac muscle

▐▐▶ *Identify:* Where is the only place cardiac muscle is found?

LESSON SUMMARY

▶ There are three different kinds of muscle tissue in the body, each with a different job.

▶ Striated muscles are muscles that you control for movement.

▶ Smooth muscle causes movements that you cannot control.

▶ Cardiac muscle is involuntary muscle found only in the heart.

CHECK *Write true if the statement is true. If the statement is false, change the underlined term to make the statement true.*

1. The word "cardiac" means "heart."

2. Smooth muscles are voluntary muscles.

3. Blinking your eyes is caused by involuntary muscles.

4. Striated muscle is usually voluntary.

5. Smooth muscle causes the heart to beat.

APPLY *Complete the following.*

6. Which kind of muscle tissue is responsible for walking?

7. **Infer:** What kind of muscle tissue is responsible for breathing?

8. **Observe:** Look at Figures 2 and 3 on page 248. What is similar in striated and cardiac muscle?

Skill Builder

📁 ***Classifying*** When you classify, you group things together based upon similarities. Using reference materials, classify the following muscles as being voluntary or involuntary:

a. muscles in the eye that move your eye back and forth

b. cardiac muscle

c. muscles in your stomach that you use to do sit-ups

d. muscles in the stomach that digest your food

e. muscles in your blood vessels

f. muscles in your jaw

g. muscles in your legs

CAREER IN LIFE SCIENCE

EXERCISE SPECIALIST

Do you like to exercise? Do you enjoy working with people? If you do, you may enjoy a career as an exercise specialist. Exercise specialists work for corporations, health clubs, and even hospitals. Exercise specialists perform many different jobs. Exercise specialists may teach classes or give lectures. They may teach people how to use exercise equipment. Exercise specialists also help people start exercising. They set up individualized exercise programs. Hospitals often use exercise specialists. They develop exercise programs for patients who have had heart attacks or who are overweight.

If a career as an exercise specialist interests you, you should take science and health classes. A four year college degree is needed for an exercise specialist. Many community colleges, and organizations, such as the YMCA, offer certification programs in exercise. For more information, write to: The American Alliance for Health, Physical Education, Recreation and Dance, 1900 Association Drive, Reston, VA 22091.

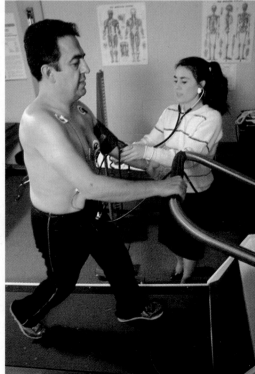

12-6 What is skin?

Objective ► Identify the layers and some of the structures of skin.

TechTerms

- **dermis:** living, inner layer of skin
- **epidermis** (ep-uh-DUR-mis): dead, outer layer of skin
- **pores:** tiny openings on the skin's surface

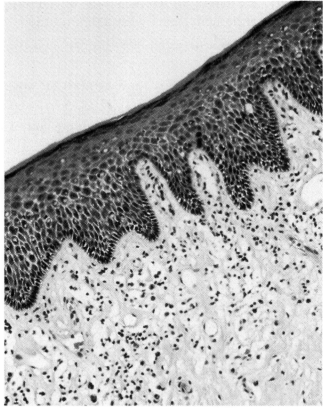

The Largest Organ The skin is the largest organ of the human body. It covers the entire outside of your body. In addition, skin also is inside your body. Skin covers most of your organs. The main function of skin is to support and protect the body. The skin is made up of two layers. There is an outer layer and the inner layer.

👁 *Observe:* Look at the photo. Is the outer layer or the inner layer of skin thicker?

Epidermis The outer layer of skin is called the **epidermis** (ep-uh-DUR-mis). The epidermis covers and protects the body. It is made up of dead skin cells. These dead skin cells are replaced constantly by the living skin cells beneath them. Each time you scrape your knee, or even wash your hands, thousands of dead skin cells are wiped away.

☛ *Infer:* Why do you think skin cells are constantly being replaced?

Dermis The inner layer of the skin is called the **dermis.** The dermis is the living layer of skin. The dermis is much thicker than the epidermis.

The dermis has many different structures in it. One of these structures is sweat glands. Sweat glands remove liquid wastes from the body in the form of sweat, or perspiration (pur-spuh-RAY-shun). Perspiration leaves the body through tiny openings in the skin called **pores.**

Hair follicles (FAHL-ih-kuls) also are in the dermis. Each hair on your body grows from a hair follicle. Oil glands are located near the hair follicles. Oil glands produce oil. The oil waterproofs the skin and keeps it soft. Nerve endings and many tiny blood vessels also are located in the dermis.

▥▶ *Name:* What are three structures found in the dermis?

Sense Receptors in the Skin The nerve endings in the skin receive many messages. These nerve endings are called sense receptors (ri-SEP-turs). When you think of the skin, you probably think of the sense of touch. The skin also is sensitive to heat, cold, pain and pressure. There are special nerve endings for each of these. The skin does not have the same number of each kind of receptor. For example, touch receptors are far apart on the back. They are close together at the tip of your nose.

▥▶ *Name:* To what five things is the skin sensitive?

250

LESSON SUMMARY

▶ The main function of skin is to support and protect the body.

▶ The outer layer of skin, the epidermis, covers and protects the body.

▶ The dermis is the living inner layer of the skin.

▶ Sweat glands, located in the dermis, remove liquid wastes from the body through pores.

▶ Hair follicles, oil glands, nerve endings, and blood vessels, are all found in the dermis.

▶ The skin has sense receptors for touch, heat, cold, pain, and pressure.

CHECK Answer the following.

1. Where is skin found?

2. Is the outer layer of skin alive or dead?

3. What is the function of oil glands?

4. What are pores?

5. What are the nerve endings in the skin called?

Explain the difference between the words in each pair.

6. dermis; epidermis

7. oil gland; sweat gland

APPLY Complete the following.

☛ 8. **Infer:** Why are organs covered with skin?

📋 9. **Hypothesize:** Why do skin cells reproduce so rapidly?

10. If you cut yourself and it bleeds, which layer of skin have you cut into? How do you know?

☛ 11. **Infer:** Do you think there are more touch receptors on the tip of your tongue or your wrist? Explain.

Health & Safety Tip.........................

Getting a "healthy" tan can be damaging to your skin. You should always wear a sunscreen to protect your skin from the ultraviolet rays of the sun. Long-term abuse of the sun can cause skin cancer. Using library references, make a chart that lists signs and symptoms of skin cancer, as well as ways to prevent skin cancer.

⁌⁗ ACTIVITY ⁗⁗⁗⁗⁗⁗⁗⁗⁗⁗⁗⁗⁗⁗⁗

TOUCH RECEPTORS

You will need a pencil and a paper clip.

1. Work with a partner.

2. Straighten the paper clip. Bend the paper clip until its ends are about 2 cm apart.

3. Have your partner close their eyes. Touch the ends of the paper clip to your partner's arm.

4. Ask your partner how many points touched their arm.

5. Repeat this three times touching different parts of your partner's arm, hand, and fingers. Record responses.

6. Repeat Steps 3-5 with your partner recording your responses.

Questions

☛ 1. **Infer:** Were your responses always correct? Explain your answer.

2. **a.** Which part of your arm, hand, or fingers was most sensitive? ☛**b. Infer:** Why do you think this part of your body is so sensitive?

STUDY HINT Before you begin the Unit Challenges, review the TechTerms and Lesson Summary for each lesson in this unit.

TechTerms..

cardiac muscle (248)
cartilage (240)
compact bone (242)
dermis (250)
endoskeleton (240)
epidermis (250)

exoskeleton (240)
extensor (246)
flexor (246)
joint (244)
ligament (244)
periosteum (242)

pores (250)
smooth muscle (248)
spongy bone (242)
striated muscle (248)
tendons (246)

TechTerm Challenges..

Matching *Write the TechTerm that matches each description.*

1. outer covering of a bone
2. striated, involuntary muscle found in the heart
3. place where two or more bones meet
4. connective tissue that connects bone to bone
5. bone made up of bone cells, protein fibers, and nonliving minerals
6. connective tissue that connects muscle to bone
7. strong, lightweight bone with many holes in it
8. tiny openings on the skin's surface
9. involuntary muscle
10. dead, outer layer of skin

Identifying Word Relationships *Explain how the terms in each pair are related. Write your answers in complete sentences.*

1. extensor, flexor
2. bone, cartilage
3. striated muscle, smooth muscle
4. dermis, epidermis
5. endoskeleton, exoskeleton

Content Challenges..

Multiple Choice *Write the letter of the term that best completes each statement.*

1. The human skeletal system is made up of 206
   a. bones.   **b.** muscles.   **c.** tendons.   **d.** ligaments.

2. Blood cells are made in the
   a. periosteum.   **b.** compact bone.   **c.** spongy bone.   **d.** red bone marrow.

3. The kind of muscle found in the walls of the blood vessels is
   a. cardiac muscle.   **b.** skeletal muscle.   **c.** smooth muscle.   **d.** striated muscle.

4. Joints that allow movement in only one direction are
   a. hinge joints.   **b.** ball-and-socket joints.   **c.** fixed joints.   **d.** gliding joints.

5. The joints that are found in the shoulder are
   a. gliding joints.   b. pivotal joints.   c. hinge joints.   d. ball-and-socket joints.
6. As an embryo develops, the cartilage in the skeleton is replaced by
   a. bone.   b. tendon.   c. ligaments.   d. muscle.
7. The connective tissue that connects a muscle to a bone is a
   a. ligament.   b. tendon.   c. skeletal muscle.   d. bone marrow.
8. The main job of the skin is to
   a. make blood cells.   b. move muscles.   c. support and protect the body.   d. store minerals.
9. The living part of the skin is the
   a. epidermis.   b. pore.   c. sweat gland.   d. dermis.
10. Skeletal muscle is
    a. found in the heart.   b. found in the stomach.   c. involuntary.   d. voluntary.
11. Sweat leaves the body through
    a. hair follicles.   b. sweat glands.   c. pores.   d. oil glands.
12. Skin is sensitive to all of the following EXCEPT
    a. touch.   b. temperature.   c. pain.   d. atmosphere.
13. Spongy bone is strong and
    a. flexible.   b. lightweight.   c. fat.   d. heavy.

True/False *Write true if the statement is true. If the statement is false, change the underlined term to make the statement true.*

1. The hardest part of bone is compact bone.
2. Spongy bone is filled with cartilage.
3. Fixed joints are found in the neck.
4. You have over 600 muscles.
5. The tip of your nose is bone.
6. Muscles only can push bones.
7. Your skin is made up of two layers.
8. Human beings have an exoskeleton.
9. Muscles that bend a joint are called extensors.
10. Smooth muscle is voluntary muscle.
11. Striated muscle also is called skeletal muscle.
12. The epidermis is the inner layer of skin.
13. Touch receptors on the skin of the back are far apart.

Understanding the Features..

Reading Critically *Use the feature reading selections to answer the following. Page numbers for the features follow each question in parentheses.*

1. **Infer:** Why do people recover quicker from arthroscopic surgery than from regular surgery? (241)
2. **Name:** What are two problems that affect the success of bone marrow transplants? (243)
3. **Hypothesize:** What benefits do athletes think they can get by taking steroids (247)
4. **Infer:** Why should you take science classes if you are interested in becoming an exercise specialist? (249)

Concept Challenges...

Critical Thinking *Answer each of the following in complete sentences.*

1. **Explain:** How do muscles work to move bones?

2. **Infer:** Discs of cartilage are found between the vertebrae. What function do you think these discs serve?

3. **Infer:** What happens to the cartilage between your ribs and breastbone when you breathe?

4. **Hypothesize:** What might happen to you if your skin was not sensitive to pain or pressure?

5. **Locate:** Name three places on the human skeleton that are cartilage rather than bone.

Interpreting a Diagram *Use the diagram of the human skeletal system to answer the following.*

1. **Observe:** What three bones make up the pelvis?
2. How many bones are there in the lower leg?
3. What is the scientific name for the skull?
4. **Observe:** Where is your patella located?
5. **Analyze:** What bones are important for holding a pen or pencil?
6. What is the common name for clavicle?

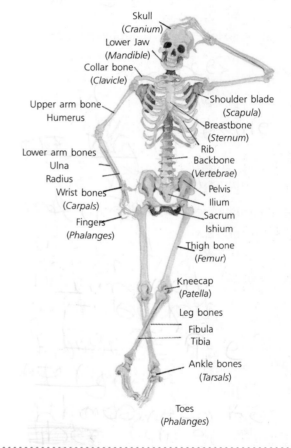

Skull (*Cranium*)
Lower Jaw (*Mandible*)
Collar bone (*Clavicle*)
Upper arm bone Humerus
Lower arm bones Ulna Radius
Wrist bones (*Carpals*)
Fingers (*Phalanges*)
Shoulder blade (*Scapula*)
Breastbone (*Sternum*)
Rib
Backbone (*Vertebrae*)
Pelvis
Ilium
Sacrum
Ishium
Thigh bone (*Femur*)
Kneecap (*Patella*)
Leg bones
Fibula
Tibia
Ankle bones (*Tarsals*)
Toes (*Phalanges*)

Finding Out More...

1. Use a first-aid manual to prepare a table on sports injuries. List whether bones, muscles, or both are involved, symptoms, and the first aid treatment for each injury. Include: sprains, strains, dislocations, tendonitis, muscle cramps, and fractures.

2. Scoliosis is an abnormal curvature of the spine. Use library references to find out how people get scoliosis and the different methods of treating people with scoliosis.

3. Muscular dystrophy is a muscular system disease. Using library references, find out the causes of muscular dystrophy, its effect on the body, and treatments that are being tested.

4. Acne is a combination of skin diseases. Prepare a chart that lists some of the causes of acne as well as the ways of controlling acne. Present your chart to the class.

5. Using library references, develop a chart that lists five careers that involve muscles, bones, and skin. Write the name of the career, the job description, and the educational requirements of each career.

TRANSPORT

CONTENTS

STUDY HINT After you read each lesson in Unit 13, write a brief summary on a sheet of paper explaining how the information in each lesson applies to your everyday life.

What are the parts of the heart?

Objectives ▶ Describe the heart. ▶ Explain how blood moves through the heart.

TechTerms

▶ **atrium** (AY-tree-um): upper chamber of the heart

▶ **septum:** thick tissue wall that separates the left and right sides of the heart

▶ **valve:** thin flap of tissue that acts like a one-way door

▶ **ventricle** (VEN-tri-kul): lower chamber of the heart

A Muscular Organ The heart is a muscular organ. Its job is to pump blood. The heart is divided into four parts, or chambers. There are upper and lower chambers. Each upper chamber of the heart is called an **atrium** (AY-tree-um). The plural of atrium is atria. The atria receive blood. The lower chambers are called **ventricles** (VEN-tri-kuls). The ventricles pump blood out of the heart.

▶ *Identify:* How many chambers does the heart have?

Blood Flow in the Heart Look at Figure 1. You can see that the heart is divided into two sides—a left side and a right side. A thick tissue wall separates the two sides of the heart. This tissue wall is called the **septum.**

Blood flows into the atria of the heart. When the atria are full of blood, they contract. This motion pumps the blood into the ventricles. Once the ventricles are full of blood, they contract. This motion pushes the blood out of the heart.

▶ *Sequence:* List the flow of blood from when it enters the heart to when it leaves the heart.

Heart Valves Inside the heart, there are four **valves.** A valve is a thin flap of tissue. It acts like a one-way door. The valves keep the blood moving

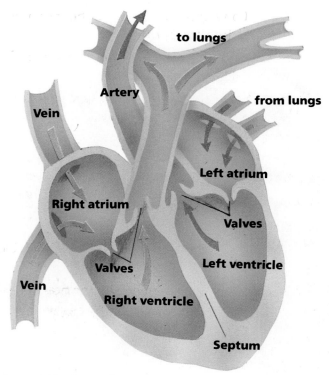

Figure 1 The human heart

in only one direction. Blood can flow only from the atria to the ventricles. Blood cannot flow backwards. Between the atria and ventricles there are valves. If the blood tries to go backwards, the valve shuts. There also are valves between the ventricles and the blood vessels. As blood leaves the ventricles, it goes through the valves.

▶ *Identify:* What keeps blood from flowing backwards in the heart?

Heartbeat Your heartbeat is the rhythm of your heart pumping blood. A stethoscope (STETH-uh-skohp) is an instrument doctors use to listen to heartbeat. If you were to listen to your heartbeat, you would hear a lub-dub sound. The lub-dub sound is made by your valves opening and closing. When the valves between the atria and ventricles snap shut, they make a "lub" sound. When the valves between the ventricles and blood vessels snap shut, they make a "dub" sound.

▶ *Define:* What is a stethoscope?

LESSON SUMMARY

▶ The heart is divided into atria, or upper chambers, and ventricles, or lower chambers.

▶ The septum is a thick tissue wall that separates the left and right sides of the heart.

▶ Blood flows from the atria to the ventricles then out into the body.

▶ Heart valves prevent blood from flowing backwards.

▶ Heartbeat is the rhythm of your heart pumping blood.

CHECK *Complete the following.*

1. The _____ divides the heart into left and right sides.

2. The upper chambers of the heart are called _____ .

3. There are _____ chambers in the heart.

4. The _____ pump blood into the body.

Answer the following.

5. Where are valves found?

6. What causes the sound of the heartbeat?

APPLY *Answer the following.*

▶ 7. **Infer:** What do you think would happen if a valve was damaged?

8. **Hypothesize:** What might cause your heartbeat to increase?

Ideas in Action...........................

IDEA: Cholesterol is a fatlike substance found in the body. Too much cholesterol can clog your arteries. A buildup of cholesterol may be the cause of high blood pressure, heart attack, or stroke.

ACTION: Look at different food products that contain nutrition labels. Find the section of the label that lists the amount of cholesterol in the product. Which foods have a high level of cholesterol? Which foods have a low level of cholesterol? Present your findings in a table.

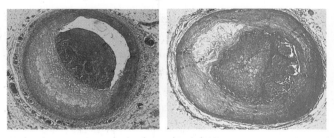

ACTIVITY

COMPARING ANIMAL HEART RATES

You will need a pencil and paper.

1. Study Table 1.

2. **Sequence:** List the animals according to their heart rates from fastest to slowest.

3. **Sequence:** List the animals according to their relative sizes from smallest to largest.

Questions

1. **a. Analyze:** Which animal has the slowest heart rate? **b.** Is the animal large or small?

2. **a. Analyze:** Which animal has the slowest heart rate? **b.** Is the animal large or small?

3. **Infer:** Do you think there is a relationship between animal size and heart rate? Explain.

4. **a. Hypothesize:** Do you think heart rates varies among humans according to their body sizes? **b.** How could you find out?

Table 1 Heartbeat Rates	
ANIMAL	HEART RATE/MINUTE
elephant	25
mouse	1000
human	72
cat	120
cow	65
bird	570

What are blood vessels?

Objective ▶ Name and describe the three kinds of blood vessels.

TechTerms

- ▶ **aorta** (ay-OWR-tuh): largest artery in the body
- ▶ **arteries** (ART-ur-ees): blood vessels that carry blood away from the heart
- ▶ **capillaries** (KAP-uh-ler-ees): tiny blood vessels that connect arteries to veins
- ▶ **veins** (VANES): blood vessels that carry blood to the heart

Blood Vessels Blood moves in the body through a closed system of tubes. These tubes are called blood vessels. The human body has three kinds of blood vessels. The **arteries** (ART-ur-ees) are blood vessels that carry blood away from the heart. Blood vessels that carry blood back to the heart are **veins** (VANES). Veins and arteries are connected by **capillaries** (KAP-uh-ler-ees). Capillaries are tiny blood vessels.

▶ *Name:* What are the three kinds of blood vessels?

Arteries Arteries have thick muscular walls. As the heart beats, it pumps blood through the arteries at high pressure. The arteries must be strong to be able to handle this pressure. The strong walls prevent the arteries from bursting. The largest artery in the body is the **aorta** (ay-OWR-tuh).

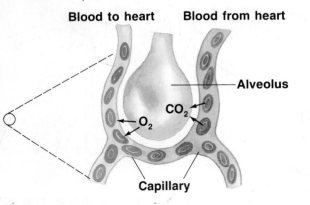

Blood to heart   **Blood from heart**

Alveolus

CO_2

O_2

Capillary

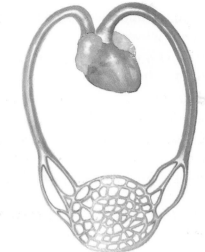

▶ *List:* What are some characteristics of arteries?

Pulse As your heart beats, it pushes blood through the arteries in spurts. With each spurt of blood, a beat can be felt. The beat you feel is your pulse. You can feel a pulse wherever an artery is close to the skin's surface. Your pulse rate and heartbeat rate are the same.

▶ *Identify:* Can you feel your pulse in a vein, artery, or capillary?

Veins Veins have thinner walls than do arteries. Blood pumps through the veins at less pressure than it does in arteries. Blood does not flow as easily through veins as it does through arteries. The contraction of muscles keeps the blood flowing. Some veins also have valves that keep the blood from flowing backwards.

▶ *Explain:* Why are the walls of veins thinner than the walls of arteries?

Capillaries Capillaries have walls that are only one cell thick. Blood cells travel through capillaries in single file. The capillaries are where every substance in the blood is exchanged with body cells. For example, carbon dioxide and waste products move from body cells into the blood through capillaries. Food and oxygen in the blood move through the capillaries into the body cells.

▶ *Describe:* What happens in capillaries?

LESSON SUMMARY

▶ The three kinds of blood vessels are arteries, veins, and capillaries.

▶ Arteries have thick muscular walls and are strong and elastic.

▶ A pulse is felt in an artery each time the heart beats.

▶ Veins have thin walls and have valves to keep blood flowing towards the heart.

▶ Capillaries are where the exchange of oxygen, carbon dioxide, food, and wastes take place between the blood and body cells.

CHECK *Write true if the statement is true. If the statement is false, change the underlined term to make the statement true.*

1. Arteries, veins, and <u>valves</u> are the three kinds of blood vessels.

2. Blood is pumped through the arteries at a <u>low</u> pressure.

3. You can feel your pulse only in an <u>artery</u>.

4. Blood is prevented from flowing backward in veins by <u>capillaries</u>.

5. Because blood is pumped through them at low pressure <u>arteries</u> have thin walls.

6. The aorta is the largest <u>vein</u> in the body.

APPLY *Answer the following.*

7. **Compare:** What are the differences between arteries and veins?

8. **Infer:** Why do you think you cannot feel a pulse in a vein?

InfoSearch

Read the passage. Ask two questions that you cannot answer from the information in the passage.

High Blood Pressure Blood pressure is a measure of the force of blood on the arteries. An average blood pressure is 120/80. The top number, 120, is the pressure on the arteries when the ventricles are contracting. The bottom number, 80, is the pressure on the arteries when the ventricles are at rest. High blood pressure causes the heart to overwork. Some people inherit high blood pressure. Other people get it from stress. Still other people get high blood pressure from their diets. Over time, high blood pressure causes the arteries to weaken.

SEARCH: Use library references to find answers to your questions.

ACTIVITY

MEASURING PULSE RATE

You will need a clock or watch with a second hand.

1. Sit quietly for two minutes. Place your middle and index finger over the inside of your wrist. Find your pulse.

2. **Measure:** Take your pulse for 30 seconds. Multiply this number by 2. Record your answer.

3. Stand up for two minutes. Take your pulse for 30 seconds. Multiply this number by 2. Record your answer

4. Jog in place for two minutes. Take your pulse for 30 seconds. Multiply this number by 2. Record your answer.

5. Rest for two minutes. Take your pulse for 30 seconds. Multiple this number by 2. Record your answer.

Questions

1. **Analyze:** How did your pulse change when you stood up?

2. **Analyze:** How did your pulse change when you stopped jogging?

3. **Analyze:** What affect did exercise have on your heart rate?

What is blood?

Objective ▶ Describe the different parts of blood.

TechTerms

- ▶ **hemoglobin** (HEE-moh-gloh-bin): iron compound in red blood cells
- ▶ **plasma** (PLAZ-muh): liquid part of blood
- ▶ **platelets** (PLAYT-lits): tiny colorless pieces of cells

Blood Blood is a liquid tissue. You have about 5 liters of blood in your body. Blood makes up about 9 percent of your body weight.

Blood is a mixture. It has a liquid part and a solid part. Scientists use a centrifuge (SEN-truh-fyooj) to separate blood into its two parts. The blood is spun around in the centrifuge. The solid part of the blood is forced to the bottom of a test tube. The liquid part of blood remains on top.

▶ *Define:* What is a centrifuge?

Plasma The liquid part of blood is called **plasma** (PLAZ-muh). Plasma is a straw-colored liquid. It is made up mostly of water. Digested nutrients, dissolved vitamins, and minerals are found in plasma. Carbon dioxide also is dissolved in plasma.

▶ *List:* What are some things found in plasma?

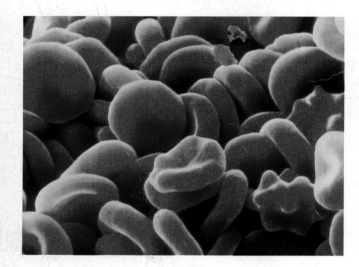

Red Blood Cells The red blood cell is different from any other cell in the body. A red blood cell has no nucleus. The job of red blood cells is to carry oxygen. Red blood cells contain **hemoglobin** (HEE-moh-gloh-bin). Hemoglobin is an iron compound found in blood. Hemoglobin gives red blood cells their color.

▶ *Define:* What is hemoglobin?

White Blood Cells White blood cells are larger than red blood cells. White blood cells have a nucleus, but they do not contain hemoglobin. White blood cells destroy germs that are harmful to the body. By destroying germs, white blood cells help fight disease. There are many more red blood cells than white blood cells. For every white blood cell there are about 1000 red blood cells.

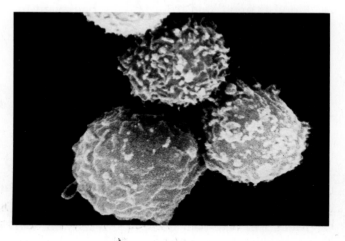

▶ *Describe:* What does a white blood cell do?

Platelets Have you ever cut yourself? What happens to the wound? Soon after you cut yourself, a clot forms. Clotting is controlled by **platelets** (PLAYT-lits). Platelets are tiny, colorless pieces of cells. When tissues are injured, chemicals are given off by the blood. These chemicals form tiny, sticky threads. The threads stick together and form a clot. The clot prevents the body from losing blood.

▶ *Name:* What controls blood clotting?

LESSON SUMMARY

▶ Blood is a liquid tissue.

▶ Scientists use a centrifuge to separate blood into its liquid and solid parts.

▶ Plasma is the liquid part of blood.

▶ Red blood cells transport oxygen and give blood its color.

▶ White blood cells destroy germs and help fight disease.

▶ Platelets form clots, which prevent the body from losing blood.

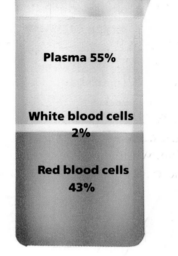

Plasma 55%

White blood cells 2%

Red blood cells 43%

CHECK *Complete the following.*

1. Red blood cells do not have a _____ .
2. Your body contains about _____ liters of blood.
3. The liquid part of blood is called _____ .
4. Blood gets its red color from _____ .
5. Platelets help the body to form _____ .

APPLY *Answer the following.*

▶ 6. **Infer:** When a person is sick, the number of white blood cells increases. Why do you think this happens?

▶ 7. **Predict:** What would happen is there were no platelets in the blood?

Health & Safety Tip.........................

The four major blood types are A, B, AB, and O. You should know what your blood type is in case of an emergency. If you do not know your blood type, ask your doctor. Record your blood type on a piece of identification and always it with you.

◆▷■◀ PEOPLE IN SCIENCE ▷■◀◀◆▷■◀◀◆▷■◀◀◆▷■◀◀◆▷■◀◀◆▷■◀◀◆▷■◀◀◆▷■◀◀◆▷■◀◀◆▷■◀◀◆▷■◀

CHARLES DREW (1904–1950)

Charles Drew was an American doctor, medical teacher, and hospital administrator. He also was the first director of the Red Cross. Drew was born in Washington, DC. In 1933, he graduated from McGill University Medical School. Drew is best known for his research on blood plasma, and setting up blood banks to collect blood plasma. He did most of his research in New York City at Columbia University between 1938 and 1940. From his research, Drew concluded that plasma could be used in blood transfusions, instead of whole blood. Using plasma in blood transfusions had two advantages. Whole blood stays fresh only about one week. Plasma can last for a longer period of time. Plasma also can be used in a transfusion for any blood type.

Drew's research findings were timely. During World War II, Drew set up blood banks in the United States to collect plasma. The blood plasma was then sent to the armed forces. The blood plasma that was collected saved many lives.

What is circulation?

Objective ▶ Describe the circulatory system and its functions.

TechTerms

▶ **circulation** (sur-kyuh-LAY-shun): movement of blood through the body

▶ **closed circulatory** (SUR-kyuh-luh-towr-ee) **system:** system in which blood is carried through the body in tubes

Circulation Your body has a transport system. It is your circulatory (SUR-kyuh-luh-towr-ee) system. The circulatory system transports, or moves, blood throughout the body. The movement of blood through the body is called **circulation** (sur-kyuh-LAY-shun).

▶ *Define:* What is circulation?

The Circulatory System Your circulatory system is made up of your heart, blood vessels, and blood. The blood vessels form a **closed circulatory system.** In a closed circulatory system, the blood moves through the blood vessels. The arteries are connected to veins by capillaries. The arteries, veins, and capillaries form a large network of tubes that form a continuous closed system.

▶ *List:* What makes up your circulatory system?

Jobs of the Circulatory System The main job of the circulatory system is transport. However, the circulatory system has many other jobs as well. Some of these jobs are as follows.

▶ **Transport of Food and Oxygen** The circulatory system transports food and oxygen to the cells of your body. Hemoglobin in red blood cells carries the oxygen.

▶ **Transport of Wastes** The circulatory system carries away wastes produced by the cells. One important waste is carbon dioxide.

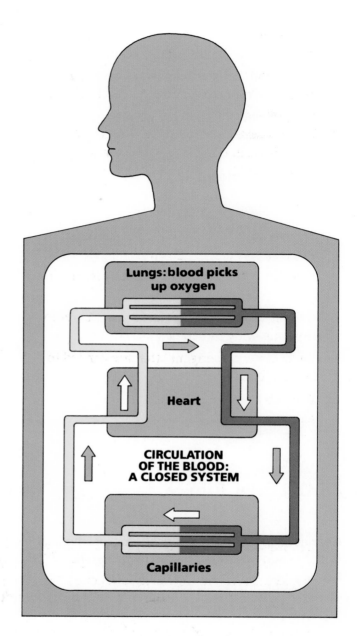

Lungs: blood picks up oxygen

Heart

CIRCULATION OF THE BLOOD: A CLOSED SYSTEM

Capillaries

▶ **Protection** Another job of the circulatory system is protection. White blood cells in your blood fight disease and harmful chemicals.

▶ **Transport of Hormones** The circulatory system carries chemicals called hormones (HAWR-mohns). Hormones carry "messages" from one part of your body to another part of the body.

▶ **Regulation** The circulatory system helps regulate your body temperature.

▶ *List:* What are some jobs of the circulatory system?

LESSON SUMMARY

▶ Circulation is the movement of blood through the body.

▶ In a closed circulatory system, blood travels within a system of blood vessels.

▶ The circulatory system transports food and oxygen, and carries away wastes such as carbon dioxide.

▶ The circulatory system protects your body from disease, carries hormones, and regulates body temperature.

CHECK *Complete the following.*

1. The circulatory system _____ blood and oxygen to all parts of the body.

2. Arteries, veins, and capillaries form a _____ circulatory system.

3. Carbon dioxide is a _____ that is removed from the body by blood.

4. Chemical "messengers" that are carried in the blood are called _____ .

5. The job of _____ blood cells is to fight disease.

APPLY *Answer the following.*

▶ 6. **Infer:** Why do you think blood is sometimes called "the river of life"?

....
InfoSearch..

Read the passage. Ask two questions that you cannot answer from the information in the passage.

Sickle-Cell Anemia Anemia (uh-NEE-mee-uh) is a disease of the blood. There are many different kinds of anemia. One kind is called sickle-cell anemia. The red blood cells are sickle-shaped. The sickle shape of the red blood cells may cause them to clog blood vessels. Their shape also causes them to break apart easily. Sickle-cells do not have as much hemoglobin as normal red blood cells. Therefore, they cannot carry enough oxygen. Sickle-cell anemia is a disease that you are born with. Sickle-cell anemia is most common among African Americans.

SEARCH: Use library references to find answers to your questions.

◄◇►◄ PEOPLE IN SCIENCE ►◄◇►◄◇►◄◇►◄◇►◄◇►◄◇►◄◇►◄◇►◄◇►◄◇►◄◇►◄

WILLIAM HARVEY (1578–1657)

William Harvey was an English physician. He was born on April 1, 1578 in Kent, England. Harvey attended a 2 1/2 year medical training course at the University of Padua, Italy. The university was considered the best medical school in Europe during the 16th century.

In the beginning of the 17th century, Harvey studied circulation in humans. Harvey also studied the circulation of blood in living animals. He thought that he could apply what he learned about circulation in snakes to circulation in humans. From his research, Harvey concluded that blood moved in a circular path through the human body. He also hypothesized that the heart acted like a pump, and helped to move blood through the body. Harvey used his knowledge about the structure of the heart and its valves to form his conclusions. Harvey's conclusions were based on careful observations and experiments.

What happens to blood as it circulates?

Objective ▶ Describe what happens to blood as it circulates.

TechTerm

▶ **pulmonary** (PUL-muh-ner-ee) **artery:** artery that carries blood from the heart to the lungs

Exchange of Substances The flow of blood throughout the body is quite simple. Blood is pumped from the left ventricle into the aorta. The aorta is the largest artery in the body. Blood that enters the aorta carries food and oxygen to the body cells.

Once the aorta leaves the heart, it branches into many smaller arteries. These arteries divide again and again until they form capillaries in all the body tissues. Substances are exchanged through the walls of the capillaries. Food and oxygen pass out of the blood in the capillaries and into the body cells. At the same time, carbon dioxide and other wastes pass from the body cells into the blood in the capillaries.

▶ *Identify:* Where are materials exchanged between the blood and the body cells?

Return to the Heart Once the exchange of substances has taken place, the blood must be returned to the heart. The capillaries in the body tissues join to form small veins. These, in turn, join to form larger veins. Blood containing carbon dioxide and other wastes is carried in the veins to the right atrium of the heart. Before it can be sent out to the body tissues again, the blood must get a fresh supply of oxygen. It also must give up its carbon dioxide. To do this the blood must be sent to the lungs.

▶ *Name:* What part of the heart does blood containing wastes return to?

Heart and Lung Circulation Once the blood is received in the right atrium, it passes into the

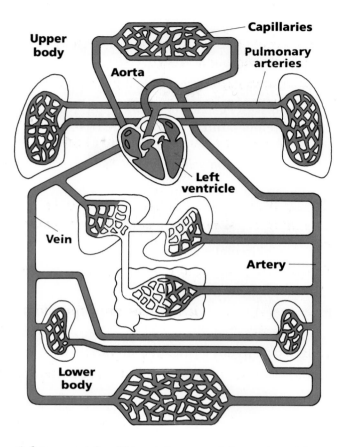

right ventricle. The right ventricle pumps blood into the **pulmonary** (PUL-muh-ner-ee) **artery.** The pulmonary artery is the artery that carries blood from the heart to the lungs. The pulmonary artery has two branches. One branch goes to each lung. In the lungs, the pulmonary arteries divide many times, until they form capillaries. As blood passes through these lung capillaries, it picks up oxygen and gives up carbon dioxide. The carbon dioxide is then exhaled from the body.

Once the blood has picked up a fresh supply of oxygen, it is ready to be circulated through the body again. The capillaries in the lungs join together to form veins. The pulmonary veins carry the blood from the lungs to the left atrium of the heart. The left atrium pumps blood into the left ventricle. The left ventricle pumps the blood throughout the body.

▶ *Identify:* Where does the blood pick up oxygen and get rid of carbon dioxide?

LESSON SUMMARY

▶ Blood containing food and oxygen is pumped through the aorta to the body cells.

▶ As the blood goes through the capillaries, the exchange of food, oxygen, and carbon dioxide takes place.

▶ Blood containing carbon dioxide is returned to the heart through veins.

▶ In the lung capillaries, blood gives off carbon dioxide and picks up oxygen.

▶ Blood containing oxygen is returned to the heart to be circulated to the body again.

CHECK *Write true if the statement is true. If the statement is false, change the underlined word to make the statement true.*

1. Blood is pumped through the aorta to the <u>lungs</u>.

2. Arteries divide many times until they form <u>capillaries</u>.

3. Blood picks up oxygen in the <u>heart</u>.

4. Blood is returned to the heart through <u>arteries</u>.

Complete the following.

5. Materials are exchanged between the blood and the body cells through the walls of _____ .

6. Blood is carried to the lungs by the _____ .

7. Blood leaving the right atrium passes into the _____ .

APPLY *Complete the following.*

8. **Sequence:** Develop a flow chart that illustrates the flow of blood through the body and lungs.

9. What changes take place in the blood as it circulates?

Skill Builder

Building Vocabulary You can sometimes infer where an artery carries blood to, just by knowing its name. Some of the major arteries of the body include: carotid artery, femoral artery, bronchial artery, brachial artery, renal artery and coronary artery. Use library references to find out where each of these arteries carries blood in the body.

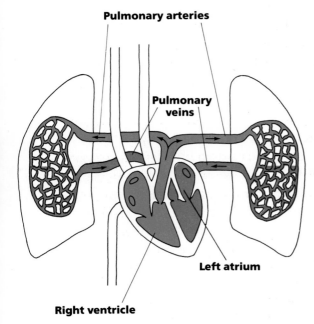

Pulmonary arteries

Pulmonary veins

Left atrium

Right ventricle

LEISURE ACTIVITY

AEROBIC EXERCISE

Aerobic (er-OH-bik) exercise is any activity that increases your heart rate and makes you breathe faster. There are many different kinds of aerobic exercise. Jogging, swimming, dancing, and walking briskly all are forms of aerobic exercise.

Aerobic exercise has many benefits. If you do aerobic exercise regularly, you will feel better. Other benefits include better physical shape, and a stronger heart. As your heart gets stronger, it pumps more blood with each beat. The heart does not have to beat as often because more blood is being pumped with each beat. This helps to keep the heart from overworking.

Aerobic exercise also can lower the risk of heart disease. Doctors believe that regular exercise lowers the amount of fatty materials in the blood. Removing fatty material from the blood helps keep them from building up on the walls of the blood vessels.

To get the most benefits from aerobic exercise, you should try to exercise at least three times a week. You should try to exercise for 30 minutes at a time. The important thing to remember is to pick an activity you enjoy.

STUDY HINT Before you begin the Unit Challenges, review the TechTerms and Lesson Summary for each lesson in this unit.

TechTerms..

aorta (258)
arteries (258)
atrium (256)
capillaries (258)
circulation (262)

closed circulatory system (262)
hemoglobin (260)
plasma (260)
platelets (260)
pulmonary artery (264)

septum (256)
valve (256)
veins (258)
ventricle (256)

TechTerm Challenges..

Matching *Write the TechTerm that best matches each description.*

1. cell parts that control clotting
2. iron compound in red blood cells that carries oxygen
3. liquid part of blood
4. movement of blood through the body
5. circulation of blood within blood vessels
6. thick wall of tissue that separates the left and right sides of the heart
7. flap of tissue that prevents blood from flowing backwards

Fill in *Write the TechTerm that completes each statement.*

1. When it reaches the lungs, the _____ divides into two branches.
2. Blood is pumped out of the heart by the _____ .
3. Blood vessels with thick muscular walls are _____ .
4. The upper chambers of the heart are the _____ .
5. The blood vessels that carry blood back to the heart are the _____ .
6. The largest artery in the body is the _____ .
7. The exchange of food, oxygen and wastes takes place through the _____ .

Content Challenges..

Multiple Choice *Write the letter of the term that best completes each statement.*

1. The heart is divided into four
   a. valves. **b.** chambers. **c.** atria. **d.** ventricles.

2. The instrument doctors use to listen to your heartbeat is a
   a. stethoscope. **b.** telescope. **c.** thermometer. **d.** centrifuge.

3. When blood is pushed through the arteries, the resulting beat felt at the skin's surface is your
   a. blood pressure. **b.** heartbeat. **c.** pulse. **d.** contraction.

4. The blood vessels through which blood flows at high pressure are
   a. arteries. **b.** veins. **c.** capillaries. **d.** valves.

5. The tiniest blood vessels are
   a. arteries. **b.** veins. **c.** capillaries. **d.** valves.

6. Disease-causing germs within the body are destroyed by
   a. red blood cells. **b.** hemoglobin. **c.** platelets. **d.** white blood cells.

LESSON SUMMARY

▶ The respiratory system is made up of the lungs, tubes, and passageways through which air moves in the body.

▶ Air enters the body through the nose and mouth, and passes into the windpipe and bronchi.

▶ The lungs are the main respiratory organs.

CHECK *Complete the following.*

1. The organ system that helps you breathe is the _____ .

2. Microscopic air sacs in the lungs are called _____ .

3. The windpipe branches into two tubes called _____ .

4. The main organs of the respiratory system are the _____ .

5. Air enters the body through your mouth and _____ .

APPLY *Answer the following.*

6. **Sequence:** Place the following words in order to show how air moves in the body.

a. windpipe b. nose c. bronchi d. throat e. lungs f. alveoli

7. **Compare:** The windpipe, bronchi, and branches of the bronchi are sometimes called the bronchial tree. How is the arrangement of these structures similar to the arrangement of a tree? (Hint: Turn your textbook upside-down and carefully study the diagram of the respiratory system.)

InfoSearch

Read the passage. Ask two questions that you cannot answer from the information in the passage.

Asthma Many people suffer from asthma (AZ-muh). Asthma usually is caused by something that irritates the smallest tubes in the lungs. The muscles in these tubes cause the tubes to get narrower. Air cannot pass easily into and out of the alveoli. Breathing becomes very difficult. Drugs often are used to help relax the muscles in the air tubes.

SEARCH: Use library references to find answers to your questions.

CAREER IN LIFE SCIENCE

RESPIRATORY THERAPIST

Few sensations are more frightening than not being able to get enough air into the lungs. A respiratory therapist works with doctors and other health care workers to help patients with breathing problems. Breathing problems can arise from diseases such as asthma or pneumonia. Accident victims or people recovering from surgery also may need help in breathing.

A respiratory therapist helps patients in different ways. A therapist may teach patients exercises that will help them to breathe more easily. A respiratory therapist also sets up and checks the various devices that are used to help treat patients with breathing disorders.

A respiratory therapist must be certified as a registered respiratory therapist, or RRT. Certification requires completion of a one- or two-year training program and then passing an examination. A respiratory therapist also should be able to work with others.

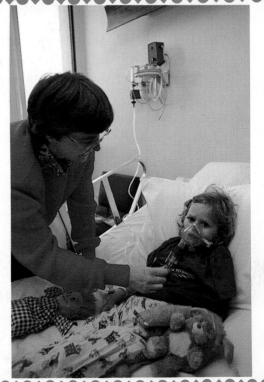

What are breathing and respiration?

Objectives ▶ Compare breathing and respiration. ▶ Explain the process of breathing.

TechTerms

- ▶ **diaphragm** (DY-uh-fram): sheet of muscle below the lungs
- ▶ **exhale:** to breathe out
- ▶ **inhale:** to breathe in
- ▶ **respiration** (res-puh-RAY-shun): the process of carrying oxygen to cells, getting rid of carbon dioxide, and releasing energy

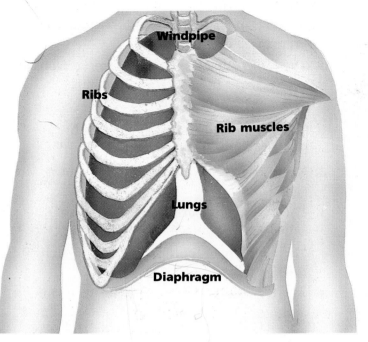

Comparing Breathing and Respiration Breathing is the process by which air is taken into the body. It is a mechanical (muh-KAN-i-kul) process. When you breathe, oxygen is not carried to your cells. Breathing does not release energy for your body to use. Carrying oxygen to your cells, getting rid of carbon dioxide, and releasing energy is called **respiration** (res-puh-RAY-shun).

Respiration is a chemical process. It has three parts.

- ▶ **External Respiration** During external respiration, oxygen and carbon dioxide are exchanged between the lungs and the blood.
- ▶ **Internal Respiration** During internal respiration, oxygen and carbon dioxide are exchanged between the blood and the cells of the body.
- ▶ **Cellular Respiration** Cellular (SEL-yoo-lur) respiration is the chemical process by which energy is released by cells. The cells change food into energy. Carbon dioxide and water are given off as waste products.

Classify: Is breathing a chemical or mechanical process?

The Diaphragm Below the lungs, there is a sheet of muscle. It is the **diaphragm** (DY-uh-fram). The diaphragm helps you breathe. It works with the ribs and rib muscles. Many organs work together to help you breathe.

▶ **List:** What parts of the body work together to help you breathe?

Breathing In When you **inhale,** or breathe in, your ribs move up and out. The diaphragm moves downward, away from the lungs. The space inside the chest becomes larger. Because of this, there is less air pressure in the lungs than outside the body. The outside air pressure causes air to rush into the lungs. The lungs fill with air and expand.

▶ **Define:** What does the word ''inhale'' mean?

Breathing Out When you **exhale,** or breathe out, the ribs move down and in. The diaphragm moves upward, toward the lungs. The space inside the chest becomes smaller. Because of this, the air pressure in the lungs is greater than the air pressure outside the body. Air moves out of the lungs. The lungs contract and take up less space in the chest.

▶ **Describe:** What happens to the size of the space in the chest when you exhale?

LESSON SUMMARY

► Breathing is the process by which air enters and leaves the body. Respiration is the process of carrying oxygen to cells, getting rid of carbon dioxide, and releasing energy.

► The diaphragm, ribs, and rib muscles work together to help you breathe.

► When you inhale, the lower air pressure in the lungs causes air to rush into the lungs.

► When you exhale, the air pressure in the lungs is higher than the air pressure outside the body, causing air to move out of the lungs.

CHECK *Find the sentence in the lesson that answers each question. Then, write the sentence.*

1. How does the diaphragm move when you exhale?

2. What is breathing?

3. Does breathing release energy?

4. What are the waste products of cellular respiration?

5. What happens to the space inside your chest when you inhale?

6. What does "inhale" mean?

7. What does "exhale" mean?

APPLY *Use the model to answer the questions.*

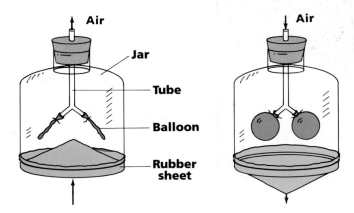

Air — Jar — Tube — Balloon — Rubber sheet

8. **a.** Which model shows inhaling? **b.** Which model shows exhaling?

9. What do the balloons represent?

10. What does the rubber sheet represent?

11. What does the jar represent?

Ideas in Action

IDEA: Breathing is a mechanical process. A mechanical process is one that involves only physical changes.

ACTION: List three examples of things that you do that involve only mechanical processes.

ACTIVITY

EXERCISE AND BREATHING RATE

You will need a watch or clock with a second hand.

1. **Measure:** Breathe in and out normally. Have a partner time the number of breaths you take in one minute. Record the number of breaths.

2. Jog in place for 20 seconds. Then stop. Have a partner time the number of breaths you take in one minute. Record this number.

3. Jog in place for 40 seconds. Then stop. Have a partner time the number of breaths you take in one minute. Record.

4. Change places with your partner and repeat the activity.

Questions

1. **a.** What was your breathing rate at rest? **b.** After 20 seconds of jogging? **c.** After 40 seconds of jogging?

2. **Compare:** How did your breathing rates compare to your partner's breathing rates?

3. **Analyze:** What effect does exercise have on breathing rate?

What happens to air before it reaches the lungs?

Objective ► Explain how air is cleaned, warmed, and moistened as it moves through the respiratory system.

TechTerms

► **cilia** (SIL-ee-uh): microscopic hairs
► **mucus** (MYOO-kus): sticky liquid

Filtering Air You normally breathe through your nose. The air that you breathe in contains dirt and dust particles. These particles may be harmful to the lungs. Inside your nose, there are many hairs. These hairs filter out and trap many dust and dirt particles.

► *Infer:* Why do you think that it is better to breathe through your nose than through your mouth?

Mucus The cells inside the nose and windpipe form a sticky liquid called **mucus** (MYOO-kus). Mucus covers the inside of the nose and windpipe. Dust, dirt, bacteria, and other harmful particles stick to the mucus. Mucus stops many particles from reaching the lungs. Mucus also keeps the tissues of the respiratory system from drying out.

► *Explain:* What are the two jobs of mucus?

Cilia Many single-celled living things have **cilia** (SIL-ee-uh). Cilia are microscopic hairs on cells. Your windpipe is lined with millions of cilia. The cilia move back and forth, pushing mucus toward the nose. Trapped particles are pushed into the nose with the mucus. Some mucus with its trapped particles is swallowed.

Sometimes mucus can irritate your nose. When this happens, you respond by sneezing. A sneeze is a burst of air. Sneezing blows harmful particles out of the nose.

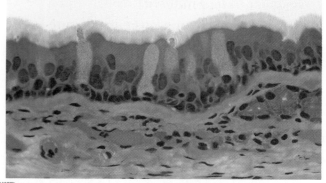

► *Name:* Where are cilia located in the respiratory system?

Warm, Moist Air Sometimes the air you inhale is cold and dry. When the air enters your body, it is warmed by heat from the body. Remember you are warmblooded. Your body works to keep your body temperature at about 37 °C, except when you have a fever. Air that enters the lungs has been warmed in your nose and throat. The body also

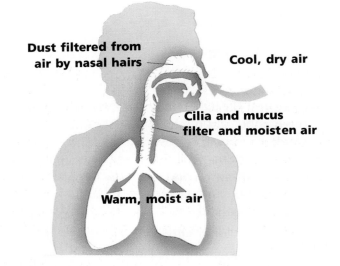

Dust filtered from air by nasal hairs

Cool, dry air

Cilia and mucus filter and moisten air

Warm, moist air

adds water vapor to the air you inhale. The air is made moist as it moves through the nose and windpipe. Air reaching the lungs is warm and moist. Warm, moist air prevents damage to the lungs.

► *Infer:* What happens to dry, warm air before it reaches the lungs?

LESSON SUMMARY

▶ Hairs in the nose filter dust particles from the air.

▶ Mucus in the nose and windpipe helps trap harmful particles contained in air.

▶ Cilia in the windpipe push mucus and trapped particles toward the nose.

▶ A sneeze is a burst of air that blows harmful particles out of the nose.

▶ Air that enters the body is warmed and moistened before it reaches the lungs.

CHECK *Complete the following.*

1. A burst of air that blows harmful particles out of the nose is a _____ .

2. The windpipe is lined with tiny hairs called _____ .

3. Air that enters the lungs is warm and _____ .

4. The sticky liquid that traps harmful particles in the respiratory tubes is _____ .

5. You normally breathe through your _____ .

APPLY *Complete the following.*

6. How is the air you breathe in changed before it reaches the lungs?

7. Which of the following have cilia?
   a. the human windpipe b. *Paramecium*
   c. sponges d. *Amoeba*

8. **Diagram:** Develop a flowchart that traces the pathway of air from the nose to the lungs. List each organ the air passes through. Beneath each organ, identify how air passing through the organ is changed.

Health & Safety Tip

Smoking is harmful to the lungs. Smoking can cause the cilia in the windpipe to stop moving. When this happens, dust, dirt, and germs are not removed. These substances are now able to get into the lungs. They may cause disease. Some diseases of the lungs can cause death. The choice you make about smoking can affect your health for years to come. Cigarette packages have warnings about the dangers of smoking. Make a list of the different warnings written on cigarette packs or in cigarette advertisements.

SCIENCE CONNECTION

RESPIRATORY DISEASES

Air taken into the body contains the oxygen needed for life. Air also contains substances that can be harmful to the body. Sometimes these substances cause respiratory diseases.

In the past, coal miners often became weak and short of breath after working in dusty mines for many years. The miners sometimes became so ill that they were unable to work. They were suffering from a disease known as "black lung." Tiny particles of coal dust caused the spongy, flexible tissue of the lungs to become stiff and thick. Today, working in mines is much safer. Miners wear masks over their mouths to keep out the coal dust.

Dangers to the lungs also are found outside the workplace. One such danger is from asbestos, a material widely used for insulation in buildings. People can inhale fibers from asbestos. The fibers cause lung disease. Today most asbestos is being removed from buildings to eliminate the risk of getting lung disease from asbestos.

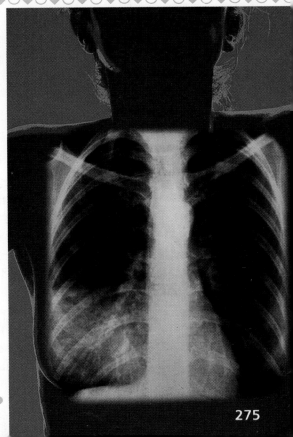

How does oxygen get into the blood?

Objectives ▶ Explain gas exchange in the lungs and between the blood and body cells. ▶ Compare the gas makeup of inhaled air and exhaled air.

Oxygen in the Air Air is a mixture of gases. It is made up mostly of nitrogen and oxygen. Your body cells need oxygen to carry out respiration. Oxygen enters the lungs in the air you inhale. One of the jobs of the lungs is to take in oxygen from the air.

▰ *Analyze:* What percentage of air is made up of oxygen?

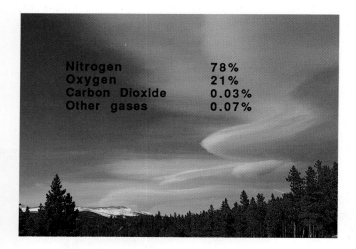

Nitrogen	78%
Oxygen	21%
Carbon Dioxide	0.03%
Other gases	0.07%

Alveoli The alveoli are the place where gases are exchanged in the lungs. Alveoli look like a bunch of grapes. The alveoli have very thin walls. They are surrounded by many capillaries (KAP-uh-ler-ees). Capillaries are very tiny blood vessels. Red blood cells move through the capillaries in single file.

In the lungs, oxygen and carbon dioxide are exchanged between the alveoli and the blood. Oxygen molecules pass through the walls of the alveoli into the capillaries. The oxygen molecules join red blood cells. The red blood cells carry oxygen to the cells of the body. Carbon dioxide molecules pass from the red blood cells through

the capillary walls into the alveoli. The carbon dioxide is removed from your body when you exhale.

▐▶ *Describe:* What happens to oxygen molecules in the lungs?

Gas Exchange in Cells Oxygen and carbon dioxide are exchanged between the body cells and red blood cells. Oxygen moves from the red blood cells into the body cells. Carbon dioxide moves from the body cells into the capillaries. The carbon dioxide is carried back to the lungs.

▐▶ *Describe:* What happens to carbon dioxide molecules in the body cells?

Air In and Air Out The gas makeup of the air you breathe in is different from the gas makeup of the air you breathe out. Inhaled air contains more oxygen than does exhaled air. Exhaled air contains more carbon dioxide than does inhaled air. Some oxygen from the air you breathe in is removed by the lungs. Your body cells add carbon dioxide to the air you breathe out. Water also is added to the air you exhale.

▰ *Analyze:* Why is there more carbon dioxide in exhaled air than inhaled air?

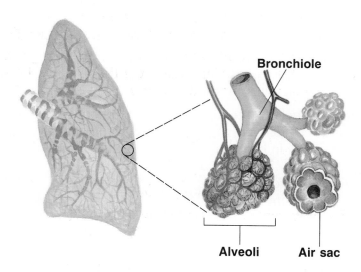

Bronchiole

Alveoli   Air sac

LESSON SUMMARY

▶ Air is a mixture of gases including nitrogen, oxygen, and carbon dioxide.

▶ The alveoli are the places where gases are exchanged in the lungs.

▶ In the lungs, oxygen and carbon dioxide are exchanged between the alveoli and the blood.

▶ Oxygen and carbon dioxide are exchanged between body cells and red blood cells at the capillaries.

▶ The air that you inhale has more oxygen and less carbon dioxide in it than the air that you exhale.

CHECK *Find the sentence in the lesson that answers each question. Then, write the sentence.*

1. What is air?
2. Where are gases exchanged in the lungs?
3. What are capillaries?
4. How is carbon dioxide removed from the body?
5. Does inhaled air contain more oxygen or more carbon dioxide than exhaled air?

6. Does exhaled air contain more oxygen or more carbon dioxide than inhaled air?

APPLY *Complete the following.*

7. How much of the air is made up of nitrogen?

▶ 8. **Infer:** Which percentages would most likely be inhaled air?
   a. 78% nitrogen; 21% oxygen
   b. 78% nitrogen; 17% oxygen

▶ 9. **Infer:** Which percentage would most likely be air you exhaled? How do you know?
   a. 21% oxygen; 0.03% carbon dioxide
   b. 17% oxygen; 4% carbon dioxide

····································
Skill Builder···································

Graphing A graph is a good way to organize information. Use the percentages of the gases contained in air shown on page 276 to make a graph of this data. Use either a bar graph or a pie graph.

⋱ ACTIVITY ⋱••••••••••••••••••••••••••••••••••••

ANALYZING EXHALED AIR

You will need a flat piece of glass, such as a window pane, 25 mL of limewater, a drinking glass, and a drinking straw.

1. Breathe out onto the flat piece of glass.
2. Add the 25 mL of limewater to the drinking glass. Limewater is used to test for carbon dioxide. If carbon dioxide is present, the limewater becomes cloudy.
3. Put the straw into the limewater. Blow through the straw into the glass of limewater. **Caution: Do not inhale while the straw is in the limewater.**

Questions

👁 1. a. **Observe:** What forms on the pane of glass? b. Where did it come from?

▶ 2. **Infer:** What happens to the limewater when you bubble exhaled air into it? Explain your answer.

▶ 3. **Infer:** What does this tell you about the air you exhale?

What is excretion?

Objective ▶ Understand how waste products are formed and removed by the body.

TechTerm

▶ **excretion** (ik-SKREE-shun): process of removing waste products from the body

Forming Waste Products Foods are combined with oxygen in your cells. Heat and other kinds of energy are produced. The energy is used by the body to carry out its life processes. When the foods are used to produce energy, waste products are formed.

Many waste products are formed by the cells of your body. Carbon dioxide and water are waste products of cellular respiration. Other waste products made by the cells are salts and nitrogen compounds. Heat also is a waste product.

▶ *List:* What are some waste products formed by your body?

Excretion The many waste products formed by your body must be removed from your body. If these waste products build up in your body, they can be harmful to you. The process of removing waste products from the body is called **excretion** (ik-SKREE-shun).

▶ *Define:* What is excretion?

The Excretory System Removing waste products from the body is the job of the excretory system. It is made up of many different organs. The lungs are part of the excretory system. You know that the lungs get rid of carbon dioxide and water. The kidneys also are organs of the excretory system. They get rid of liquid wastes. The largest organ of the excretory system is your skin. It gets rid of liquid wastes and helps you get rid of extra heat.

▶ *List:* What are the three main organs of excretion?

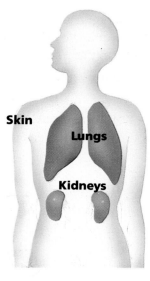

Solid Wastes Some parts of the foods that you eat cannot be digested. They cannot pass through the villi. Undigested foods form wastes. These wastes move along in the small intestine. They enter the large intestine. Water is removed from the wastes in the large intestine. As water is removed, the wastes become solid. The solid wastes move along in the large intestine and into the rectum. From there, the solid wastes are excreted through the anus (AY-nus).

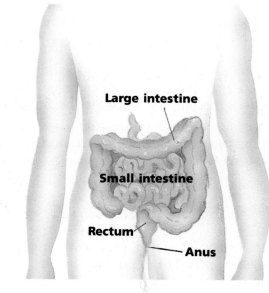

▶ *Describe:* What is the job of the large intestine?

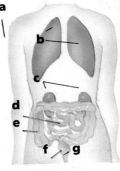

LESSON SUMMARY

▶ Waste products are formed as the body uses food to produce energy.

▶ Carbon dioxide, water, salts, nitrogen compounds, and heat are waste products formed by the body.

▶ Waste products are removed from the body in a process called excretion.

▶ The lungs, kidneys, and skin are the main organs of the excretory system.

▶ Solid wastes leave the body through the anus.

CHECK *Find the sentence in the lesson that answers each question. Then, write the sentence.*

1. What are the waste products of cellular respiration?

2. What is excretion?

3. What does the excretory system do?

4. What does skin do?

5. What is the job of the large intestine?

6. How do solid wastes leave the body?

APPLY *Use the letters in the diagram to answer the following questions.*

7. Which organs in the diagram excrete liquid wastes?

8. Which organ in the diagram is used to rid the body of excess heat?

9. Which organs in the diagram are used to eliminate solid wastes?

10. To what two organ systems do the lungs belong?

Ideas in Action...............................

IDEA: Many of the things you use ever day are thrown away as wastes.

ACTION: List five examples of products you dispose of as wastes. What happens to these products after you dispose of them?

TECHNOLOGY AND SOCIETY

USING SOUND TO BREAK APART KIDNEY STONES

Mineral compounds that are not excreted by the body can build up and form kidney stones. Kidney stones form inside the kidneys. Large kidney stones can completely block the passage of liquid wastes from the kidneys. Urine then backs up into the kidneys. Waste products can quickly destroy kidney cells.

Large kidney stones must be removed to prevent kidney damage. Medications are sometimes used to dissolve or break apart the stones. Unfortunately, these medications do not always work. Until recently, the only other way of eliminating kidney stones was to remove them surgically. However, doctors now have another way of removing kidney stones. Doctors carefully aim sound waves at the kidney stone. The sound waves cause particles in the stone to vibrate. The motion of the particles makes the stone break apart. The small parts of the kidney stone can then be passed from the body along with liquid wastes.

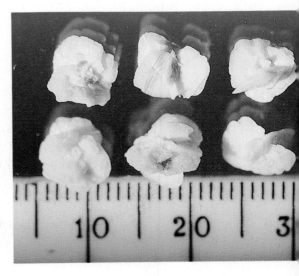

How does the skin remove wastes?

Objective ▶ Explain why the skin is an organ of excretion.

TechTerms

- **evaporation** (i-VAP-uh-ray-shun): changing of a liquid to a gas
- **perspiration** (pur-spuh-RAY-shun): waste water and salts that leave the body through the skin
- **pore:** tiny opening in the skin

Perspiration Some waste water leaves your body through your skin. This waste water is called **perspiration** (pur-spuh-RAY-shun), or sweat. Perspiration is mostly water. It also contains salts.

▶ *Define:* What is perspiration?

Sweat Glands Your skin contains many sweat glands. A sweat gland is a coiled tube surrounded by capillaries. Water molecules in the blood pass through the walls of the capillaries into the sweat gland. Each sweat gland extends to a tiny opening in the skin called a **pore.** Waste water leaves the sweat gland through the pore.

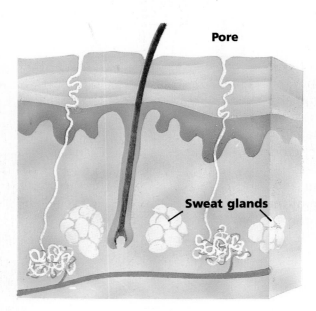

Pore

Sweat glands

▶ *Describe:* What is a sweat gland?

An Air Conditioning System Your body has a built in cooling system. When it is very hot, do you feel warm? Do you sweat after you exercise or do heavy work? Perspiration, or sweating, helps your body "cool off". The **evaporation** (i-VAP-uh-ray-shun) of perspiration from your skin cools the body. Evaporation is the changing of a liquid to a gas. As perspiration evaporates, it removes heat from your body.

▶ *Predict:* Are you likely to perspire if your body temperature drops below 37°C? Explain.

LESSON SUMMARY

▶ Perspiration is a waste excreted by the skin.

▶ Perspiration is formed by sweat glands and is excreted through pores in the skin.

▶ Evaporation of perspiration helps to cool the body.

CHECK *Complete the following.*

1. Perspiration, or sweat, is made up mostly of _____ .

2. A sweat gland is a coiled tube surrounded by _____ .

3. Perspiration leaves the body through _____ in the skin.

4. The changing of a liquid to a gas is called _____ .

5. Evaporation of perspiration helps to _____ the body.

APPLY *Answer the following.*

6. **Infer:** When you are cold, you shiver. What do you think shivering does for the body?

7. **Hypothesize:** Why do you think it is important to take in more salts when the weather is hot?

Designing an Experiment

Design an experiment to solve the problem.

PROBLEM: Does your body perspire more heavily when you exercise?

You experiment should:

1. List the materials you would need.

2. Identify safety precautions that should be followed.

3. List a step-by-step procedure.

4. Describe how you would record your data.

Skill Builder

Building Vocabulary Use a dictionary to find the meanings of the terms "deodorant" and "antiperspirant." Write the definitions in your notebook. Use the definitions to explain how a deodorant differs from an antiperspirant.

ACTIVITY

EVAPORATION AND COOLING

You will need water, isopropyl alcohol and a cotton ball.

1. Hold two fingers in front of your mouth. Blow on them. Record which finger feels cooler.

2. Dip one finger in water. Blow on the wet finger and a dry finger. Record which finger feels cooler.

3. Wet the cotton ball with isopropyl alcohol. Wet one finger. Blow on the wet finger and a dry finger. Record which finger feels cooler.

4. Wet one finger with water and one finger with isopropyl alcohol. Blow on them. Record which finger feels cooler.

Questions

1. Did the finger wet with water or the dry finger feel cooler? Explain your answer.

2. Did the finger wet with water or the finger wet with isopropyl alcohol feel cooler? Explain your answer.

3. **a.** Which step is your control? **b.** Why is it a control?

How do the kidneys work?

Objective ▶ Describe how the kidneys act as a filtering system for the blood.

TechTerms

- **urea** (yoo-REE-uh): nitrogen compound found as a waste product
- **urine** (YOOR-in): liquid waste formed in the kidneys

The Kidneys You have two kidneys. Each kidney is about 10 cm long. The kidneys are located just above your waist. One kidney is behind your liver. The other is behind the stomach. The main job of the kidneys is to filter out wastes from the blood.

▶ *Identify:* What is the main job of the kidneys?

Waste Liquid Like the skin, the kidneys help the body get rid of liquid wastes. The liquid waste formed in the kidneys is **urine** (YOOR-in). Urine is mostly water, but it also contains other materials. Urine contains many salts. Some of these salts give urine its yellow color. Urine also contains **urea** (yoo-REE-uh). Urea is a waste product formed when proteins are used by the body. It is a nitrogen compound.

▶ *Name:* What is the liquid waste formed in the kidneys called?

Kidney Tubes The inside of each kidney is made up of millions of tiny tubes. Each tube has a small cup at one end. Many coiled capillaries are found inside each cup. Blood flows through the capillaries. Water and other materials are filtered out of the blood here. These materials move through the walls of the cup and into the tube.

▶ *Sequence:* How do materials get into the kidney tubes?

Formation of Urine Water and other materials move through the kidney tube. The kidney tube is surrounded by capillaries. Some of the materials in the kidney tube are still needed by the body. These materials move out through the walls of the tube, back into the capillaries. These materials enter the blood again. Water, salts, and urea remain in the tube. This is urine. The urine forms in the kidney tube. Then it moves into the hollow, middle part of the kidney. Urine passes out of the kidneys through the ureters (yoo-REET-urz). Urine collects in the bladder. Urine passes out of the body through the urethra (yoo-REE-thruh).

▶ *Infer:* What happens to materials that pass from the kidney tube into the capillaries?

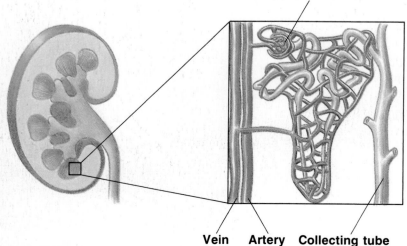

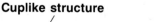
Cuplike structure

Vein   Artery   Collecting tube

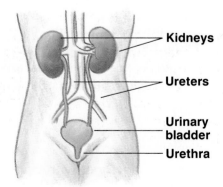
Kidneys
Ureters
Urinary bladder
Urethra

LESSON SUMMARY

▶ The main job of the kidneys is to filter wastes from the blood.

▶ Urine is made up of water, salts, and a nitrogen compound called urea.

▶ Wastes are filtered out of the blood in tubes inside the kidneys.

▶ Urine formed in the kidneys leaves the body through the urethra.

CHECK *Choose the term that makes each statement true.*

1. Urea is a (nitrogen/protein) compound.

2. You have (two/ten) kidneys.

3. The kidneys help to get rid of (solid/liquid) wastes.

4. The liquid waste formed in the kidneys is (urea/urine).

5. Urine leaves the body through the (urethra/ureters).

6. Urine leaves the kidneys through the (urethra/ureters).

APPLY *Complete the following.*

7. **Identify:** The urinary system excretes liquid wastes. What organs make up the urinary system?

▲ 8. **Model:** Develop a flowchart that traces the path of urine from the kidneys out of the body.

▶ 9. **Infer:** What element do you think is present in all proteins?

10. **Apply:** What happens to nutrients that are removed from the blood in the kidneys?

Ideas in Action

IDEA: Many things that you use every day have filters or use filters.
ACTION: List several objects that have filters. Explain what the filter does in each object.

TECHNOLOGY AND SOCIETY

DIALYSIS

A field of study that uses engineering ideas to help design machines that help or replace diseased organs is called biomedical engineering. Biomedical engineers have developed a machine that acts like a kidney. The machine is called a dialysis (dy-AL-uh-sis) machine. People who have lost their kidneys or have kidney damage are kept alive by using a dialysis machine.

A patient is connected to the machine by a tube. The tube is connected to an artery in the patient's arm. Blood from the artery flows into the tube to the dialysis machine. The machine filters the patient's blood, removing waste materials. The blood then flows out of the machine through a tube connected to a vein in the patient's arm.

Many people use a dialysis machine twice a week. However, some people need to use the machine more often. Depending on how serious kidney damage is, some patients need to use the machine every two days.

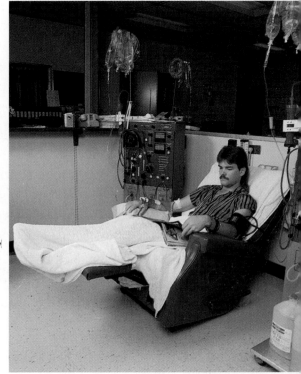

Challenges

STUDY HINT Before beginning the Unit Challenges, review the TechTerms and Lesson Summary for each lesson in this unit.

TechTerms....................

alveoli (270)
bronchi (270)
cilia (274)
diaphragm (278)
evaporation (276)

excretion (276)
exhale (272)
inhale (272)
mucus (274)
perspiration (276)

pores (276)
respiration (272)
urea (278)
urine (278)

TechTerm Challenges....................

Matching *Write the TechTerm that matches each description.*
1. microscopic air sacs in the lungs
2. breathe out
3. tubes leading into the lungs
4. liquid waste excreted by the skin
5. sticky liquid
6. liquid waste excreted by the kidneys
7. process of carrying oxygen to cells, getting rid of carbon dioxide, and releasing energy
8. sheet of muscle below the lungs

Fill-in *Write the TechTerm that best completes each statement.*
1. When you breathe in, you _____ .
2. Tiny hair-like structures called _____ filter particles from air as the air passes through the windpipe.
3. The removal of wastes from the body is called _____ .
4. The changing of a liquid to a gas is called _____ .
5. Perspiration leaves the body through openings in the skin called _____ .
6. A nitrogen waste that is found in urine is _____ .

Content Challenges....................

Multiple Choice *Write the letter of the term that best completes each statement.*
1. The waste products of cellular respiration are
   a. water and carbon dioxide.   **b.** water and oxygen.   **c.** salts.   **d.** nitrogen compounds.
2. A sweat gland is a coiled tube surrounded by
   a. alveoli.   **b.** mucus.   **c.** capillaries.   **d.** cilia.
3. Urine passes out of the body through the
   a. ureters.   **b.** urethra.   **c.** kidneys.   **d.** urea.
4. Cilia are found in the
   a. lungs.   **b.** nose.   **c.** mouth.   **d.** windpipe.
5. When you inhale, the
   a. ribs move up and out.   **b.** ribs move down and in.   **c.** diaphragm moves upward.   **d.** chest cavity becomes smaller.
6. The largest organ of the excretory system is the
   a. large intestine.   **b.** kidney.   **c.** small intestine.   **d.** skin.

7. Oxygen is carried to the cells of the body by
   a. white blood cells.   b. red blood cells.   c. alveoli.   d. cilia.

8. Air enters the body through the
   a. lungs.   b. skin.   c. nose and mouth.   d. windpipe.

9. The windpipe divides into two smaller tubes called
   a. alveoli.   b. capillaries.   c. bronchi.   d. ureters.

10. Air reaching the lungs is
    a. cold and dry.   b. warm and moist.   c. cold and moist.   d. warm and dry.

11. Oxygen and carbon dioxide are exchanged between the lungs and the blood during
    a. external respiration.   b. internal respiration.   c. cellular respiration.   d. breathing.

12. Air is 21%
    a. oxygen.   b. carbon dioxide.   c. nitrogen.   d. helium.

13. Air rushes into the lungs during
    a. inhalation.   b. exhalation.   c. cellular respiration.   d. evaporation.

14. Alveoli are found in the
    a. bronchi.   b. lungs.   c. windpipe.   d. kidneys.

15. The process of removing waste products from the body is called
    a. respiration.   b. digestion.   c. evaporation.   d. excretion.

True/False *Write true if the statement is true. If the statement is false, change the underlined term to make the statement true.*

1. Breathing is a chemical process.
2. Inhaled air contains more oxygen than exhaled air.
3. Water is removed from wastes in the small intestine.
4. Perspiration is made up mostly of salts.
5. The lungs are the main organs of the respiratory system.
6. The chemical process by which energy is released by cells is called cellular respiration.
7. Waste water leaves sweat glands through capillaries.
8. When mucus irritates your nose, you respond by sneezing.
9. Urea is a protein compound.
10. When you inhale, the diaphragm moves upward.
11. Urine collects in the bladder after passing out of the kidneys.
12. The job of the respiratory system is to take carbon dioxide into the lungs.

Understanding the Features..

Reading Critically *Use the feature reading selections to answer the following. Page numbers for the features follow each question in parentheses.*

1. Why is asbestos being removed from buildings? (275)
2. **Define:** What is biomedical engineering? (283)
3. What are three ways to treat kidney stones? (279)
4. **List:** What are two diseases that can cause breathing problems? (271)

Concept Challenges..

Critical Thinking *Answer each of the following in complete sentences.*

1. **Sequence:** Through what tubes and passageways does air pass from the outside of the body to the lungs?
2. How do mucus and cilia help you fight infection?
3. **Contrast:** What is the difference between breathing and respiration?
4. **Relate:** How do the respiratory system and circulatory system work together?
5. **Compare:** How is the skin like an air-conditioning system?

Interpreting a Diagram *Use the diagram to answer each of the following questions.*

1. Once air leaves the throat, where does it enter next?
2. What are the tubes that enter the lungs called?
3. **Locate:** Where are two places that air enters the body?
4. How does the nose help in purifying air?
5. Where are cilia found in the respiratory system?

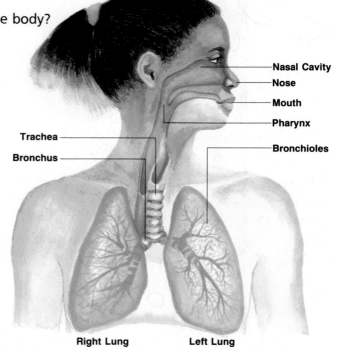

Nasal Cavity
Nose
Mouth
Pharynx
Trachea
Bronchus
Bronchioles
Right Lung   Left Lung

Finding Out More...

1. If a person stops breathing, artificial respiration is necessary to save the person's life. Contact your local Red Cross Chapter to find out information about mouth-to-mouth resuscitation and other lifesaving techniques.

2. **Organize:** Pneumonia, bronchitis, emphysema, and tuberculosis are four respiratory diseases. Find out the symptoms, causes, and treatments for each of these diseases.

3. **Research:** At the top of the windpipe is an organ called the larynx. Using library references, find out why the larynx is important. Write your findings in a report.

4. **Model:** Using the diagrams in this unit as guides, make models of the respiratory system and the urinary system. Label each structure on your models.

REGULATION AND BEHAVIOR

CONTENTS

STUDY HINT Before beginning Unit 15, write the title of each lesson on a sheet of paper. Below each title, write a short paragraph explaining what you think each lesson is about.

What is the nervous system?

▶ Identify the function of the nervous system. ▶ Name the parts that make up the nervous system.

TechTerms

▶ **axon:** fiber that carries messages away from a nerve cell

▶ **dendrite** (DEN-dryt): fiber that carries messages from other neurons to the nerve cell body

▶ **neuron** (NOOR-ahn): nerve cell

The Nervous System Do you belong to an athletic team? Does your team have a captain? Most teams do. The captain gets information and makes important decisions. The team carries out the captain's orders. Your body works in a similar way. The nervous system is the captain of your body. The nervous system controls all of your body's activities.

The nervous system is shown in Figure 1. The nervous system is made up of the brain, the spinal cord, and nerves. Nerves carry information to the spinal cord and brain. Other nerves then carry messages from the brain and spinal cord to the muscles and glands in your body. The muscles and glands carry out the orders of the brain and spinal cord.

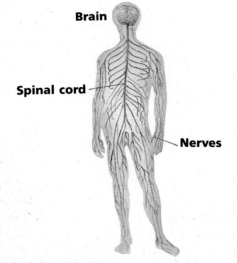

Figure 1 The nervous system

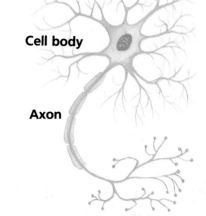

Figure 2 Structure of a neuron

||||▶ *Identify:* Name the parts of the nervous system.

The Central Nervous System Your brain is the control center of your body. The brain is made up of a mass of nervous tissue. The brain is protected by your skull. The spinal cord is made up of many nerves that extend down your back. The spinal cord is protected by the backbone. The brain and spinal cord make up the central nervous system.

||||▶ *Name:* What structure protects the spinal cord?

Nerve Cells Thirty one pairs of nerves branch out from your spinal cord. These nerves branch many times and extend to all parts of your body. Each of the nerves in your body is made up of nerve cells called **neurons** (NOOR-ahns). Neurons are the largest cells in your body. In fact, one neuron in your leg may be as long as 1 meter.

The job of a neuron is to carry messages. Messages travel through a neuron in only one direction. You can see the structure of a neuron in Figure 2. The **dendrites** (DEN-dryts) carry messages toward the nerve cell body, or center of the neuron. The **axon** carries messages away from the cell body. The cell body contains the nucleus of the neuron. It also contains most of the cytoplasm.

||||▶ *Describe:* Describe the structure of a neuron.

LESSON SUMMARY

▶ The nervous system controls all of the body's activities.

▶ The nervous system is made up of the brain, the spinal cord, and nerves.

▶ The brain and spinal cord make up the central nervous system.

▶ A neuron is a nerve cell.

▶ Neurons are made up three parts: a cell body, an axon, and dendrites.

CHECK *Answer the following questions.*

1. What is the job of the nervous system?

2. What are the parts of the nervous system?

3. What parts make up the central nervous system?

4. What is a neuron?

5. How many pairs of nerves branch out from the spinal cord?

6. What are the parts of a neuron?

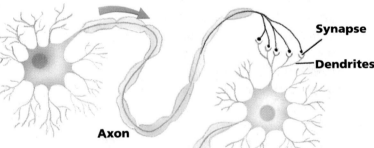

Synapse

Dendrites

Axon

APPLY *Complete the following.*

▶ 7. **Infer:** What is the job of nerve tissue?

▲ 8. **Model:** Draw and label a neuron. Draw arrows on the diagram to show the direction a message travels through the neuron.

9. **Predict:** Unlike other body cells, nerve cells cannot reproduce themselves. What might happen if all of the nerve cells in your hand were destroyed?

InfoSearch..

Read the passage. Ask two questions that you cannot answer from the information in the passage.

Synapses The axon of one nerve cell normally does not touch the dendrites of the next nerve cell. Usually, there is a small gap between the two cells. This gap is called a synapse (SIN-aps). Chemicals released by the axon help to carry messages across the synapse.

SEARCH: Use library references to find answers to your questions.

TECHNOLOGY AND SOCIETY

MAGNETIC RESONANCE IMAGING

Less than 15 years ago, a doctor who suspected that a patient had a tumor would often have to operate. Only by looking inside the body could the doctor find out whether a tumor was present. X-rays can take clear pictures of the bones, but soft organs, such as the brain or liver, do not show up as well.

In 1977, a new technique called magnetic resonance (REZ-un-uns) imaging, or MRI, was first tried on humans. MRI involves the use of magnets and radio waves to form pictures of internal organs. For example, using MRI, doctors can take pictures of the blood vessels of the brain. Doctors also can pinpoint the exact location of tumors of soft organs. This helps doctors plan treatment more effectively. Because MRI can detect tumors of organs such as the kidneys, liver, and brain, it is valuable in detecting cancer.

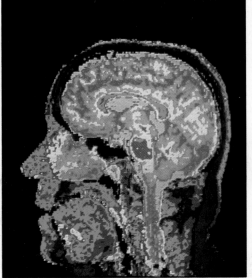

What are the parts of the brain?

Objective ▶ Identify and describe the functions of the three parts of the brain.

TechTerms

- ▶ **cerebellum** (ser-uh-BELL-um): part of the brain that controls balance and body motion
- ▶ **cerebrum** (suh-REE-brum): large part of the brain that controls the senses and thinking
- ▶ **medulla** (muh-DULL-uh): part of the brain that controls heartbeat and breathing rate

The Brain The main job of the brain is to receive and interpret messages. These messages may come from inside or outside your body. Your brain responds to the messages and then controls all of your body's activities. For example, movement, thinking, breathing, and sleeping all are controlled by your brain.

The brain often is called the control center of the body. The brain is made up of three main parts. These parts are the cerebrum (suh-REE-brum), the cerebellum (ser-uh-BELL-um), and the medulla (muh-DULL-uh). Each part of the brain performs a different function.

▶ **Identify:** What is the main job of the brain?

The Cerebrum The largest part of the brain is the **cerebrum.** You can see in Figure 1 that the cerebrum makes up more than two thirds of the brain. The cerebrum has two main jobs. One job of the cerebrum is to interpret information from the sense organs. Your sense organs are your eyes, ears, nose, tongue, and skin. A second job of the cerebrum is to control thinking. Your cerebrum helps you learn, remember, and make decisions.

▶ **Identify:** What are two jobs of the cerebrum?

The Cerebellum The part of the brain located at the back of the brain is the **cerebellum.** The cerebellum controls all body movements. The cerebellum also helps to maintain balance. As your body changes position, the cerebellum receives messages. Then the cerebellum sends messages out to your muscles. The muscles work to help you keep your balance.

▶ **Locate:** Where is the cerebellum located?

The Medulla The smallest part of the brain is the **medulla.** It is made up of nerves from the cerebrum and cerebellum. These nerves form a thick stalk at the base of the skull.

The medulla connects the brain to the spinal cord. The medulla controls digestion, breathing, and heartbeat rate. It also controls the activities of many glands and muscles.

▶ **Classify:** Does the medulla control voluntary or involuntary body activities?

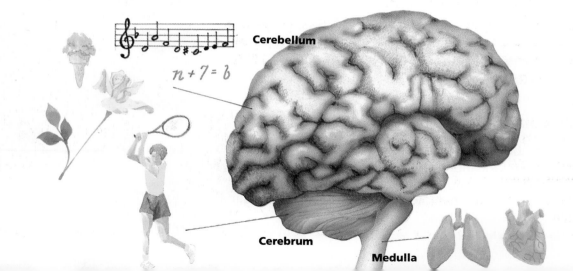

Cerebellum

Cerebrum

Medulla

LESSON SUMMARY

▶ The main job of the brain is to receive and interpret messages from inside and outside the body.

▶ The brain is the control center of the body.

▶ The cerebrum is the large part of the brain that controls the senses and thinking.

▶ The cerebellum is the part of the brain that controls body motion and balance.

▶ The medulla is the small part of the brain that connects the brain and spinal cord and controls heartbeat rate, digestion, and breathing.

CHECK *Find the sentence in the lesson that answers each question. Then, write the sentence.*

1. What are the three parts of the brain?
2. What is the largest part of the brain?
3. Where is the cerebellum located?
4. Where is the medulla located?
5. What is the main job of the brain?
6. What are the sense organs?

APPLY *Identify the part of the brain that controls each of the actions described.*

7. You memorize someone's telephone number.
8. You walk to the store.
9. You breathe faster when you run.
10. You smell smoke.
11. You taste butter in your food.
12. Your heart rate increases as you ride your bicycle.
13. You begin to fall, but then regain your balance.

Skill Builder.................................

▶ **Inferring** When you infer, you form a conclusion based upon fact. Think about the functions of the brain. Then, identify how your body might be affected if the cells making up the medulla of your brain were damaged. Describe how a loss of function of the medulla would affect you.

ACTIVITY

MODELING THE BRAIN

You will need a piece of cardboard, clay of three colors, toothpicks, tape, paper, and a pencil.

▲ 1. **Model:** Using the drawing on page 290 as a guide, draw an outline of the brain on the cardboard. Be sure to include the cerebrum, medulla, and cerebellum in your drawing.

2. Place a piece of clay on top of your cardboard outline. Use the outline to mold the clay into the shape of one part of the brain.

3. Repeat Step 2 for each of the other parts of the brain. Use a different color clay for each part.

4. Cut three rectangles from the paper. Label these cerebrum, medulla, and cerebellum. Use tape to attach each rectangle to a toothpick. Label the parts of your model by placing each toothpick in the clay.

Questions

1. **a. Observe:** Which part of the model is the largest? **b.** Why?

2. **Relate:** If you were to add to your model, what part of the nervous system would come next?

3. To which part of the nervous system does the medulla connect?

Objectives ▶ Define reflex. ▶ Relate reflexes to the stimuli that cause them.

TechTerms

▶ **reflex:** automatic response to a stimulus

▶ **reflex arc:** path of a message in a reflex

▶ **response:** action caused by a stimulus

▶ **stimulus** (STIM-yuh-lus): any action that causes a response

Stimuli and Responses Do you jump at a sudden loud noise? Does your mouth water when you smell food? Do you pull your hand away quickly if you touch something hot? Loud sounds, the smell of food, and heat all are examples of stimuli (STIM-yu-ly). A **stimulus** is any action that causes you to react in some way. The reaction to a stimulus is called a **response**.

▶ **Relate:** How are stimuli and responses related?

Reflexes Some responses are simple. You cannot control them. They happen without you thinking about what you are doing. An automatic response to a stimulus is called a **reflex.**

A reflex usually is a response that protects you in some way. For example, when dust gets into your nose, you sneeze. Dust is a stimulus. Sneezing is the response. Sneezing helps to prevent harmful substances from entering your lungs.

Figure 1 Tearing is a reflex

When dirt gets into your eyes, you blink. Your eyes may also water, or form tears. Blinking your eyes and forming tears are your body's way of protecting your eyes from harmful substances. Sneezing, blinking, and tearing are examples of reflexes.

▶ **Apply:** Is answering a ringing telephone an example of a reflex? Explain.

A Reflex Arc Reflexes usually occur very quickly. One reason reflexes occur so quickly is that they do not involve the brain. Reflexes are controlled by the spinal cord. You can trace the path of a reflex in Figure 2.

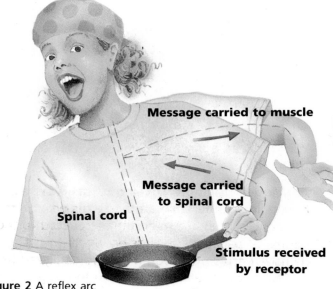

Message carried to muscle

Message carried to spinal cord

Spinal cord

Stimulus received by receptor

Figure 2 A reflex arc

When a reflex takes place, nerves and muscles work together. The stimulus is received by special nerve cells called receptors (ree-SEP-tors). Other nerve cells carry a message to the spinal cord. Another nerve carries a message from the spinal cord to a muscle or a gland. A muscle causes you to move some part of your body. A gland may make your mouth or eyes water. The message travels along a path formed by nerve cells. This path is called a **reflex arc.**

◉ **Observe:** What is the stimulus in Figure 2? What is the response?

LESSON SUMMARY

▶ Any action that causes a response is a stimulus.

▶ An automatic response to a stimulus is called a reflex.

▶ Most reflexes help to protect the body from harm.

▶ Reflexes are controlled by the spinal cord.

▶ The path of a message in a reflex is called a reflex arc.

CHECK *Write true if the statement is true. If the statement is false, change the underlined term to make the statement true.*

1. A reflex is a <u>voluntary</u> response.
2. Reflexes usually work to <u>protect</u> you.
3. The path of a message in a reflex action is called a <u>reflex arc</u>.
4. Any reaction to a stimulus is a <u>reflex</u>.
5. Any change that causes a response is a <u>reflex</u>.
6. Stimuli are received by special nerve cells called <u>receptors</u>.

APPLY *Complete the following.*

7. **Relating Cause and Effect:** Identify the stimulus and the response in each statement.
    a. You step on a tack and pull your foot away.
    b. You open the door after hearing a knock.
    c. You blink when a bright light is shined in your eyes.
    d. Food in your throat causes you to cough.

Designing an Experiment..............

The pupil of your eye controls how much light enters your eye. The pupil changes size in response to different amounts of light.

Design an experiment to solve the problem.

PROBLEM: How does the size of the pupil change as the amount of light in a room changes?

Your experiment should:

1. List the materials you would need.
2. Identify safety precautions that should be followed.
3. List a step-by-step procedure.
4. Describe how you would record your data.

TECHNOLOGY AND SOCIETY

TREATING SPINAL INJURIES WITH BIONICS

A broken or damaged spinal cord can leave a person paralyzed, or unable to move. The nerve signals needed to move muscles can no longer get through the spinal cord to the brain. The degree of paralysis depends on where in the spinal cord the damage occurs. A person might be paralyzed from the neck down or from the waist down. In the past, a person injured in this way could expect to spend his or her life in a wheelchair.

The technology of computers has given some paralyzed patients new hope. Their hope lies in the science of bionics, or computerized replacements for injured body parts. The bionic treatment for a patient unable to walk begins with doctors connecting wires to all of the patient's muscles that are needed for walking. These wires are connected to a computer. Instead of passing through the spinal cord on the way to the brain, the nerve signals pass through the computer. The entire computer can be made so small that a person can easily carry it around.

The science of bionics is still new. As it is further developed, many more paralyzed people will be able to walk and use their arms once more.

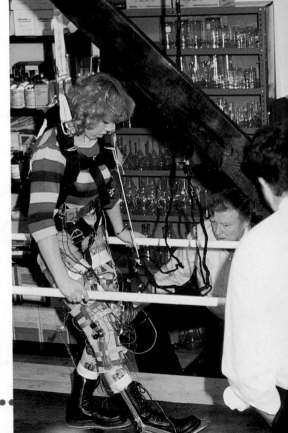

What are sense organs?

Objective ▶ Name the five sense organs and their jobs.

TechTerm

▶ **receptor** (ree-SEP-tor): organ that receives stimuli from the environment

Sense organs Sound, light, and temperature are examples of stimuli. Stimuli are messages. You have special organs that receive stimuli from your environment. These organs are the sense organs. The sense organs are the eyes, nose, skin, ears and tongue.

▶ *Name:* What are the sense organs?

Jobs of the Sense Organs Sense organs work to help you respond to your environment. Sense organs are called **receptors** (ree-SEP-tors) because they receive things. Each sense organ receives certain kinds of messages. The eyes receive pictures of things you look at. These pictures are formed by changes in light. The nose is used for smell. Your skin responds to changes in temperature and pressure. Your skin also responds to pain and touch. Your ears receive sounds from the environment. Your tongue helps you identify different tastes.

▶ *Describe:* What do sense organs do?

Senses Work Together Your senses do not work alone. What do you do when you walk into a darkened room? Your hands reach out to touch things that might be in front of you. Your ears listen for the slightest sound. Your eyes search the dark for some sign of light. Your senses work together to help you.

Your senses help you gather many different kinds of information about your surroundings. In this way, your senses help you to learn. Much that you see, hear, taste, smell, and feel is stored in your mind. This information becomes part of your memory. You remember things when you need to use them. Using information gathered by the senses is important for your safety and for all the things you do.

▶ *Explain:* How do your senses help you to learn?

LESSON SUMMARY

► The sense organs are the eyes, nose, skin, ears, and tongue.

► Sense organs are used to receive stimuli from the environment.

► Your senses work together to help you.

► Your senses help you to learn.

CHECK *Match each sense organ to the kind of message it receives.*

1. pain, temperature, pressure, and touch
2. smell
3. sounds
4. tastes
5. changes in light

   a. tongue
   b. ears
   c. eyes
   d. nose
   e. skin

Complete the following.

6. Sense organs receive stimuli from the _____ .

7. Sense organs are examples of _____ .

8. Your sense organs work _____ to help you.

APPLY *Complete the following.*

9. How do you think the senses work together?

10. What four senses does the skin detect?

11. How does your skin cause your body to respond to cold temperatures?

InfoSearch

Read the passage. Ask two questions that you cannot answer from the information in the passage.

Taste Buds Your tongue is covered with receptors called taste buds. Your taste buds can detect only four tastes: sweet, sour, bitter, and salty. The taste buds that detect each taste are located at different parts of your tongue. Your taste buds cannot taste dry foods. They can only taste foods that are moist. Dry foods do not have a taste. For this reason, foods must be dissolved in water before your taste buds will recognize the food.

SEARCH: Use library references to find answers to your questions.

ACTIVITY

LOCATING TASTE RECEPTORS

You will need colored pencils or markers in four colors, a glass, water, 4 cotton swabs, sugar, lemon juice, table salt, and tonic water.

▲ 1. **Model:** Make a drawing representing your tongue.

2. Wet a cotton swab in the water. Place the cotton swab in the sugar. Touch your tongue with the cotton swab. Observe what part of your tongue detects the sweetness of the sugar. Shade in this part of the tongue in your drawing.

3. Rinse your mouth with water.

4. Repeat Steps 2 and 3 for the table salt.

5. Dip a cotton swab in the lemon juice. Touch the swab to your tongue as you did in Step 2. Rinse your mouth.

6. Repeat Step 5 using the tonic water.

Questions

👁 1. **Observe:** Describe the taste of each substance using the terms salty, bitter, sweet, and sour.

2. **a. Compare:** Look at the models made by two of your classmates. How are your models the same? **b.** How are your models different?

How do you see?

Objective ▶ Name and describe the functions of the parts of the eye.

TechTerms

▶ **cornea** (KOR-nee-uh): clear covering at the front of the eye

▶ **iris** (Y-ris): part of the eye that controls the amount of light entering the eye

▶ **lens:** part of the eye that forms an image on the retina

▶ **pupil** (PYOO-pil): opening in the center of the iris

▶ **retina** (RET-in-uh): part of the eye on which images form.

The Eyes The eyes are the organs of sight. The eyes work by responding to light. In order to see an object, your eyes must respond to light from that object. Different parts of the eye work together to form an image, or a picture from this light. Your brain then receives messages from your eye and interprets what you are looking at.

▶ *Identify:* What are the eyes?

Parts of the Eye Different parts of the eye work together to help you see. Figure 1 shows the parts of the eye. As you read about each part of the eye, locate that part on Figure 1.

▶ **Cornea** The **cornea** (KOR-nee-uh) is a clear covering at the front of the eye.

▶ **Iris** Behind the cornea is a round, colored disk called the **iris** (Y-ris). The iris controls the amount of light entering the eye.

▶ **Pupil** At the center of the cornea is a hole called the **pupil** (PYOO-pil). Light must pass through the pupil to get to the inside of your eye.

▶ **Lens** After light passes through the pupil it passes through the **lens** of the eye. The lens forms an image from the light entering the eye.

▶ **Retina** At the back of the eye is the **retina** (RET-in-uh). The lens of the eye forms images on the retina. The retina responds to the light by generating nerve signals.

▶ **Optic Nerve** Nerve signals formed by the retina are carried to the brain by the optic nerve. The brain then interprets the signals it receives from the optic nerve and tells you what you are seeing.

▶ *Sequence:* List, in the correct order, the parts of the eye through which light passes.

Protecting the Eyes The eyes are protected in many ways. The bones of the face extend in front of the eyes. These bones keep large objects from hitting and damaging the eyes. Eyelids and eyelashes help keep small pieces of matter from entering the eyes. Any particles that do reach the eyes usually are washed away by tears. Tears also keep the eyes from becoming too dry.

▶ *Identify:* How do the bones of the face protect the eyes?

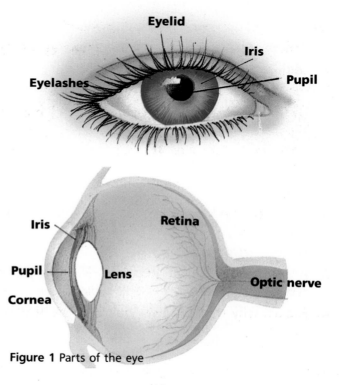

Figure 1 Parts of the eye

LESSON SUMMARY

▶ The eyes are the organs of sight.

▶ Different parts of the eye work together to help you see. The main parts of the eye are the cornea, iris, pupil, lens, retina, and optic nerve.

▶ The eyes are protected by the eyelids, eyelashes, tears, and the bones of the face.

CHECK *Complete the following.*

1. The eyes work by responding to _____ .

2. The clear covering at the front of the eye is the _____ .

3. The size of the pupil is controlled by the _____ .

4. Images are formed on the part of the eye called the _____ .

5. Nerve signals from the eye are carried to the brain by the _____ .

APPLY *Answer the following.*

6. **Observe:** What color are the pupils of your eyes?

7. **Observe:** What color are your irises?

8. What structure of the eye forms images on the retina?

Skill Builder..................................

Applying Definitions Look up the word "lens" in the dictionary. Write the definition in your notebook. How does the function of the lens of your eye compare to the function of lenses made of glass?

Skill Builder..................................

Applying Concepts Many people wear eyeglasses to correct vision problems. Two problems often corrected with eyeglasses are myopia and astigmatism. Use a dictionary or library references to find out what these terms mean. Then, try to find out what features eyeglasses must have to correct these problems.

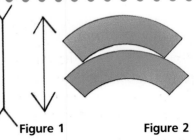

ACTIVITY

ANALYSING OPTICAL ILLUSIONS

You will need a metric ruler and a pencil.

1. **Observe:** Look at Figure 1. Which line appears longer?

2. Look at Figure 2. Which are longer?

3. Look at Figure 3. Which post is tallest?

4. Measure and record the length of each line in Figure 1.

5. Trace the top arc in Figure 2 on a piece of tracing paper. Place the top arc over the bottom arc. Describe your observation.

Questions

1. **a.** Which line looked longer than the other? **b.** How long is each line?

2. Are the arches the same size?

3. **a.** Which post looked shortest? **b.** Which looked tallest? **c.** How tall was each post?

4. Explain why your eye was fooled by the optical illusion in each figure.

Figure 1   Figure 2

How do you hear?

Objective ▶ Describe the jobs of the main parts of the ear.

TechTerms

▶ **cochlea** (KOK-lee-uh): part of the ear that changes vibrations into nerve signals

▶ **eardrum:** sheet of tissue that vibrates when sounds strike it

The Ears The ears are the organs of hearing. The three sections of the ear are the outer ear, the middle ear, and the inner ear. Locate each of these sections on Figure 1.

Each part of the ear has a different job. As you read about the jobs of the parts of the ear, locate each part on Figure 1.

The **outer ear** gathers sounds. The larger the outer ear, the more sound it can gather. Rabbits have larger outer ears than humans. This is one reason why rabbits have better hearing than humans. Between the outer ear and the middle ear is a thin sheet of tissue called the **eardrum.** The eardrum vibrates when sound waves hit it. The ear bones are three small bones in the middle ear. When the eardrum vibrates, it makes the ear bones vibrate also.

A coiled structure called the **cochlea** (KOK-lee-uh) is located in the inner ear. The cochlea receives the vibrations of the ear bones. The cochlea changes vibrations to nerve signals. The nerve signals are sent to the brain. The brain interprets the signals so you can understand what the signals mean.

▶ *State:* What makes the ear bones vibrate?

Sound Waves When someone speaks to you, sound waves are formed. The sound waves are different for each word. These sound waves travel in the air and are gathered by your outer ear. They make your eardrum vibrate. Your ear bones also vibrate. The vibrations are different for each word. The cochlea changes these vibrations into nerve signals. The signals are sent to the brain. The brain figures out what the signals mean.

▶ *Identify:* What happens when signals are sent to the brain?

What You Hear Some sounds are too soft to hear. Their vibrations are not strong enough to make your eardrums vibrate. No signal is sent to the brain. You hear nothing. Other sounds are too high-pitched to hear. The cochlea cannot change these sounds into nerve signals. You do not hear anything.

▶ *Explain:* Why are soft sounds not detected by the ear?

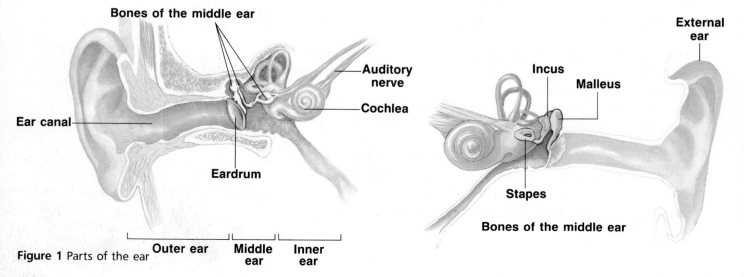

Figure 1 Parts of the ear

Outer ear | **Middle ear** | **Inner ear**

LESSON SUMMARY

▶ The ears are the sense organs of hearing.

▶ Each part of the ear has a different job.

▶ The outer ear gathers sounds.

▶ The ear bones and cochlea are parts of the middle ear.

▶ Some sound waves are too soft or too high-pitched to hear.

CHECK *Complete the following.*

1. The sense organs of hearing are _____ .

2. The three main parts of the ear are _____ .

3. The cochlea receives the vibrations of the ear bones and changes them to _____ .

4. Ear vibrations changed to nerve signals are sent to the _____ .

APPLY *Complete the following.*

5. How do a rabbit's large ears help it to hear?

6. Why can't we hear certain kinds of sounds?

7. What is the brain's role in hearing?

InfoSearch

Read the passage. Ask two questions that you cannot answer from the information in the passage.

Hearing in Animals Many animals have ears that are different from the human ear. Many animals can hear higher-pitched sounds than can humans. Many animals also can hear softer sounds than can humans.

SEARCH: Use library references to find answers to your questions.

Health & Safety Tip

Loud sounds can be damaging to the ears. Many people who work in noisy environments must wear devices to protect their ears. Use library references to find out at what level sounds may damage the ears. Also, find out how you can protect your ears from damage by loud noises.

TECHNOLOGY AND SOCIETY

HEARING AIDS

Anything that stops part of the ear from working can cause deafness. Each part of the ear is needed for good hearing. Sometimes the ear can be damaged but still work a little bit. This is partial deafness. Sometimes deafness is caused by an infection of the ear bones. After the infection, the bones may no longer be able to vibrate.

Deafness sometimes can be corrected by surgery. Damaged parts of the ear often can be repaired or replaced. Materials have been made to replace damaged parts of the ears. Sometimes replacement parts are made from plastic.

Partial deafness can often be corrected by using hearing aids. Hearing aids are like small radios that fit near or into the ear. They make sounds louder, so that even damaged parts of the ear may vibrate.

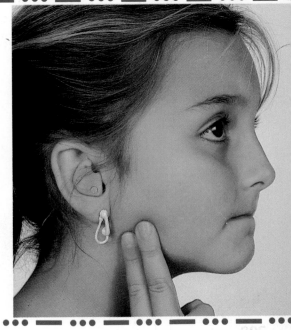

What is the endocrine system?

Objective ► Describe the function of the endocrine system.

TechTerms

► **endocrine** (EN-duh-krin) **gland:** gland that does not have ducts

► **gland:** organ that makes chemical substances used or released by the body

Glands with Ducts A **gland** is an organ that makes substances used or released by the body. Some glands have ducts, or tubes. Substances made by these glands leave the gland through the ducts. Your skin has many sweat glands. Perspiration, or sweat, is made by these glands. Sweat moves from the sweat gland to the surface of the skin by passing through a duct. Salivary (SAL-uh-ver-ee) glands also have ducts. Saliva passes from the salivary glands into the mouth through these ducts.

▐▶ **Name:** What are two glands that have ducts?

Endocrine Glands Some glands do not have ducts. Glands that do not have ducts are called **endocrine** (EN-duh-krin) glands. Substances made by endocrine glands pass from the gland directly into the blood. The blood vessels then carry the substances to the parts of the body where they are needed.

▐▶ **Identify:** How do substances made by endocrine glands get to other parts of the body?

The Endocrine System There are only about ten endocrine glands in the human body. Together, these glands make up the endocrine system. Most of the glands that make up the endocrine system are shown in Figure 1.

👁 **Observe:** Where are the salivary glands located?

Control for Body Functions The job of the endocrine system is to help control bodily functions. Each gland in the endocrine system has a different job. The way each gland works is similar to the way a thermostat and a furnace work. If the temperature in a room goes down, the thermostat signals the furnace to turn up the heat. The "thermostat" of the body is the endocrine gland called the hypothalamus (hy-puh-THAL-uh-mus). If the hypothalamus senses a need for a certain chemical substance in the body, it stimulates an endocrine gland to produce that chemical.

▐▶ **State:** What is the function of the endocrine system?

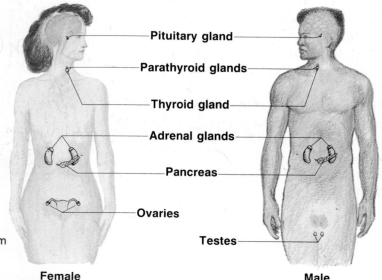

Pituitary gland
Parathyroid glands
Thyroid gland
Adrenal glands
Pancreas
Ovaries
Testes

Figure 1 The endocrine system

Female   Male

LESSON SUMMARY

▶ A gland is an organ that makes substances used or released by the body.

▶ Substances formed by endocrine glands pass directly into the blood.

▶ About ten glands make up the human endocrine system.

▶ The job of the endocrine system is to help control body functions.

CHECK *Write true if the statement is true. If the statement is false, change the underlined term to make the statement true.*

1. An organ that makes substances used or released by the body is called a <u>gland</u>.

2. Glands that do not have ducts are called <u>exocrine</u> glands.

3. The substances made by endocrine glands are carried by <u>ducts</u> to other parts of the body.

4. Sweat glands are <u>endocrine</u> glands.

5. There are about <u>ten</u> endocrine glands in the body.

APPLY *Use the diagram of the endocrine system on page 300 to answer the questions.*

6. **Observe:** What endocrine glands are found only in males?

7. **Observe:** Where is the pituitary gland located?

8. **Identify:** Which glands are found in pairs in the body?

9. The prefix "para-" means "by the side of." Based on this information, where do you think the parathyroid gland is located?

Skill Builder

Using Prefixes A prefix is a word part placed at the beginning of a word. Knowing what the prefix means provides a clue to the meaning of the entire word. Look up the prefixes "endo-" and "exo-" in the dictionary. Write the meanings of these prefixes in your notebook. Next, review what you have learned about endocrine glands and about glands with ducts. Glands with ducts are called exocrine glands. How will knowing the meanings of "endo-" and "exo-" help you remember the two types of glands?

TECHNOLOGY AND SOCIETY

HUMAN GROWTH HORMONE

One hormone produced by the pituitary gland is human growth hormone, or HGH. During childhood, the pituitary gland releases HGH. The amount of HGH released determines how tall a child will grow.

In some children, the pituitary gland does not release enough HGH. As a result, the children's bones and tissues grow very slowly. Without treatment, the children would not grow to normal heights. For many years, children who lacked HGH were treated with injections of the hormone. Unfortunately, there was not enough HGH available to treat all the children suffering from an HGH deficiency.

In 1982, scientists were able to produce HGH in laboratories by genetic engineering. Using genetic techniques, scientists got bacteria to produce large amounts of HGH. As a result, supplies of this hormone greatly increased. Today, a child whose pituitary gland does not produce enough HGH has a good chance of reaching normal height.

Objectives ▶ Define hormones. ▶ Explain some of the jobs of hormones.

TechTerm

▶ **hormone** (HAWR-mohn): chemical substance that regulates body functions

Hormones The chemical substances made by the endocrine glands are called **hormones** (HAWR-mohns). Hormones affect many of your body's functions. For example, growth hormone, which is made by the pituitary (pi-TOO-uh-ter-ee) gland, controls how fast and how much you grow.

▶ *Define:* What is a hormone?

Jobs of Hormones Growth hormone is only one of many hormones made by your body. Hormones may speed up a body function or slow it down. Other hormones do other jobs. Insulin (IN-suh-lin) is a hormone made by the pancreas. Insulin is

needed to keep a balanced amount of sugar in the blood. Still other hormones regulate other glands. Table 1 shows some of the different hormones of the body and what each hormone does.

▶ *Infer:* What do you think would happen if a person had too much growth hormone?

Table 1	Hormones and What They Do	
GLAND	HORMONE	JOB OF HORMONE
Pineal	Melatonin	Helps regulate pituitary gland
Pituitary	Growth hormone	Controls the growth of bones
	ATCH	Controls the release of hormones from the adrenal glands
Thyroid	Thyroxin	Controls rate of body growth
Parathyroid	Parathyroid hormone	Controls the amount of calcium and phosphorus in the blood
Thymus	Thymosin	Controls growth of certain white blood cells
Adrenal	Adrenalin	Controls muscle reaction, blood clotting, and blood pressure
Pancreas	Insulin	Controls blood sugar levels

Reproductive Hormones Some hormones control the development of reproductive organs. The glands that control reproductive organs are different in men and women. Men have glands called testes (TES-teez). The testes produce a hormone called testosterone (tes-TAWS-tur-ohn). Females have glands called ovaries (OH-vuhr-eez). Ovaries produce hormones called estrogen (ES-truh-jun) and progesterone (proh-JES-tur-ohn).

▶ *State:* What hormone does the testes produce?

LESSON SUMMARY

▶ Hormones, produced in endocrine glands, control many of the body's activities.

▶ Each hormone has a specific job.

▶ Ovaries and testes produce hormones that control the development of reproductive organs.

CHECK *Complete the following.*

1. A chemical substance that regulates body functions is called a _____ .

2. Insulin is made by the _____ .

3. The two endocrine glands that produce substances that develop reproductive organs are testes and _____ .

4. Testes are glands found only in _____ .

5. Ovaries are glands found only in _____ .

Answer the following.

6. How are insulin and blood sugar level related?

7. What is the job of growth hormone?

APPLY *Use the table of hormones on page 302 to answer the questions.*

8. **Classify:** Which hormones are found in both males and females and which are found in only one sex or the other?

9. **Infer:** What health problems might someone have if the pituitary gland produced too much or too little growth hormone?

10. **Analyze:** What hormone is needed to release adrenal hormones?

11. **Analyze:** What gland produces thymosin?

InfoSearch

Read the passage. Ask two questions that you cannot answer from the information in the passage.

Thyroxine Thyroxine is the hormone made in the thyroid gland. Thyroxine regulates growth and metabolism. Metabolism refers to the rate at which your body cells use and produce energy. If too little thyroxine is produced, the body metabolism slows down.

SEARCH: Use library references to find answers to your questions.

TECHNOLOGY AND SOCIETY

INSULIN PUMPS

The pancreas releases a hormone called insulin. Insulin helps your body use sugar properly. If the pancreas does not produce enough insulin, sugar stays in the blood instead of passing to the body's cells. When the cells do not get enough sugar, a person may become weak, lose weight, and be very hungry and thirsty. A person suffering from this illness is diabetic.

Diabetes is serious, but it can be treated. There are two kinds of diabetes. One kind can be treated with pills and a weight program. The other kind of diabetes must be treated with daily insulin injections, a carefully balanced diet, and a regular exercise program.

Doctors have not yet found a cure for diabetes. However, they have been trying to make the diabetic's life easier, by searching for alternatives to daily injections. One of these alternatives is the insulin pump. The insulin pump is a machine that releases insulin into the body. One advantage to an insulin pump is that it replaces daily injections. A second advantage to the insulin pump is that it releases insulin into the body only when the body needs it.

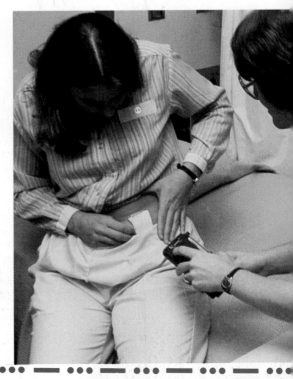

Objective ▶ Describe the two main kinds of behaviors.

TechTerms

▶ **conditioned** (kun-DISH-und) **response:** learned behavior in which a new stimulus causes the same response that an old one did

▶ **innate behavior:** behavior you are born with

▶ **learned behavior:** behavior you practice and learn

Reflexes Everything you do is part of your behavior. There are two main kinds of behavior. One kind of behavior is called inborn, or **innate behavior.** An innate behavior is a behavior you are born with. Reflexes are innate behaviors. Reflexes are automatic responses. You do not learn to make these responses. You cannot control the response. A reflex does not involve any learning or thought. Crying, coughing, swallowing, sneezing, and blinking all are innate behaviors.

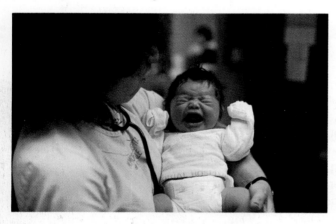

▶ *Identify:* What kind of behavior is a reflex?

Learned Behavior Many of your responses are not automatic. You must learn to make these responses. These responses are called **learned behavior.** You are learning new responses all the time. Learned behavior must be practiced. You also can change learned behavior. Throwing a ball, reading a book, and playing games are examples of learned behaviors.

▶ *Describe:* What are learned behaviors?

Conditioned Responses A **conditioned** (kun-DISH-und) **response** is one kind of learned behavior. In a conditioned response, one stimulus takes the place of another. Ivan Pavlov, a Russian scientist, experimented with this kind of behavior. Pavlov observed that when a dog smells food, its mouth waters. Pavlov decided to ring a bell before he fed the dog. Pavlov repeated this pattern many times. After a while, Pavlov observed something new in the dog's behavior. Whenever the dog heard the bell, its mouth watered, even if there was no food. The dog soon learned that the sound of the bell meant it would be fed. The dog was responding to a new stimulus. The bell took the place of the food. This was a conditioned response.

▶ *Explain:* In Pavlov's experiment, why did the ringing bell cause the dog's mouth to water?

LESSON SUMMARY

▶ A reflex is an innate, or inborn, behavior.

▶ Learned behavior must be practiced.

▶ A conditioned response is a kind of learned behavior.

CHECK *Write true if the statement is true. If the statement is false, change the underlined term to make the statement true.*

1. The two main kinds of behavior are innate behavior and <u>reflexes</u>.

2. A new stimulus causes the same response that an old one did in a <u>conditioned response</u>.

3. Ivan Pavlov studied <u>innate behavior</u> in dogs.

4. A conditioned response is an example of the kind of behavior called <u>learned behavior</u>.

5. Behavior that can be changed is <u>innate</u>.

APPLY *Complete the following.*

6. Which type of behavior, innate or learned, would follow most quickly an accident like stepping on a sharp object?

▶ 7. **Predict:** How would your classmates learn to behave if each time a student was late the teacher awarded that student an "A" for the day?

Health & Safety Tip.........................

In your home, practice what you would do in case of a fire. Practice getting out of your home if a fire were to occur. Be sure you can quickly unlock all doors to the outside. Memorize the phone number for the fire department and the police. If you can respond quickly to a fire, then you have learned a conditioned response that can help you.

Ideas in Action...............................

IDEA: Learning to read is a learned behavior.
ACTION: List ten other activities that you have mastered using learned behavior.

◆▶■◀ PEOPLE IN SCIENCE ▶■◀◆▶■◀◆▶■◀◆▶■◀◆▶■◀◆▶■◀◆▶■◀◆▶■◀◆▶■◀◆▶■◀◆▶■◀
B. F. SKINNER (1904-PRESENT)

B. F. Skinner is an American psychologist. He spent many years studying the way animals learn and demonstrated that the behavior of animals is greatly influenced by learning. During the 1930s, Skinner performed important experiments with rats. One of the experiments involved placing a hungry rat in a special box known as a Skinner box. Inside the box was a lever. Behind the lever was food. When the rat was first put in the box, it ran around nervously. While moving, the rat accidently touched the lever and food dropped into the box. The rat soon learned that pressing the lever caused food to appear.

During the 1950s, Skinner studied how humans learn. He concluded that people learn difficult tasks in a manner similar to that of the rats. Skinner concluded that the easiest way for a person to learn a complex task is to receive rewards after completing each step of the task. He supported the use of teaching machines in the classroom. A teaching machine presents difficult material in a series of small steps. After each step, a student must respond to questions. A teaching machine rewards correct answers.

Objective ▶ Describe the ways in which people learn.

Trial and Error Have you ever borrowed a bunch of keys and then forgotten which one to use? What did you do? You take any key and try it. If it does not open the lock, you try another key. You keep trying the keys until you find the right one. In this way you learn which key to use without being told. This is called trial and error learning. Each time you make an error, you try something else.

▷ *Describe:* What is trial and error learning?

Memory Do you memorize (MEM-uh-rize) things easily? How long does it take you to memorize something? How many of your friends' telephone numbers do you know without looking them up? You memorize things by studying them carefully or by repeating them many times. Memorizing is another way of learning.

▷ *Explain:* How do you memorize things?

Reasoning Reasoning is not the same as trial and error learning or memorizing. Reasoning means solving a problem by thinking about it. You solve the problem by putting together things you have learned in the past. Reasoning can be used to save time in solving problems. Suppose you are given ten keys and asked to open a lock. How can reasoning help you? Look at the shapes of the keys. Look at the shape of the keyhole. You will be able to see that certain keys will not go into the keyhole. You can put them aside. Now you have fewer keys to try. Time is saved.

▷ *Define:* What does reasoning mean?

Reward and Punishment Reward and punishment are often used in learning. This is especially true in training animals. You can teach a dog to stand up by holding food above its head and repeating the words "stand up." When the dog stands up, it is rewarded. It gets the food. After you do this many times, the dog learns to stand up on command, even if there is no food. Other types of behavior can be conditioned by punishment. How can a dog be taught to stop barking a people? What rewards and punishments affect your own behavior?

▷ *Describe:* How can a reward or punishment be a stimulus for a learned behavior?

LESSON SUMMARY

▶ One type of learning is trial and error learning, or trying something new each time you make an error.

▶ You memorize things by studying them carefully or repeating them many times.

▶ Reasoning is a way to solve problems by thinking about past experiences.

▶ Behavior can be conditioned by rewards and punishment.

CHECK *Find the sentence in the lesson that answers each question. Then, write the sentence.*

1. What is trial and error learning?

2. How do you memorize things?

3. What kind of learning helps you save time?

Answer the following.

4. If you can remember your friend's telephone number, what kind of learning have you used?

APPLY *Read each description. Then decide what kind of learning was used.*

5. Dialing each number in the phone book before getting the person you want.

6. Looking up the person's name in the phone book in order to find his or her phone number.

7. Calling the number you have memorized

Answer the following.

8. Explain how you could use reward to change a behavior of your pet.

▶ 9. **Infer:** Why are good study habits important for learning?

Ideas in Action...

IDEA: Most human behavior is learned behavior.
ACTION: Make a list of 10 learned behaviors. Identify each type of behavior as a result of trial and error, memorization, or reasoning.

ACTIVITY

REMEMBERING INFORMATION

You will need 7 index cards, a clock or watch with a second hand, paper, and a pencil.

1. Write a different letter of the alphabet on each of seven index cards.

2. Shuffle the cards and place them face down on a desk.

3. Turn over all 7 index cards. Underneath each letter, write a word that begins with the letter. Try to use words that form a sentence or complete thought.

4. Memorize the sentence. Turn the cards face down.

5. Wait 2 minutes. Then write the sentence on a piece of paper. Underneath the sentence, write the letters that appear on the index cards. Check your answers.

Questions

1. What was the sentence or thought you created from the 7 letters?

2. After waiting 2 minutes, did you correctly identify all 7 letters?

3. What other sentences have you used to help memorize a list of items?

STUDY HINT Before you begin the Unit Challenges, review the TechTerms and Lesson Summary for each lesson in this unit.

TechTerms...

axon (288)	endocrine gland (300)	neuron (288)
cerebellum (290)	gland (300)	pupil (296)
cerebrum (290)	hormone (302)	receptor (294)
cochlea (298)	innate behavior (304)	reflex (292)
conditioned response (304)	iris (296)	reflex arc (292)
cornea (296)	learned behavior (304)	response (292)
dendrite (288)	lens (296)	retina (296)
eardrum (298)	medulla (290)	stimulus (292)

TechTerm Challenges...

Matching *Write the TechTerm that matches each description.*

1. clear covering at the front of the eye
2. part of the ear that changes vibrations into nerve signals
3. organ that receives stimuli from the environment
4. sheet of tissue that vibrates when sounds strike it
5. part of the brain that controls heartbeat and breathing rate
6. nerve cell
7. organ that makes chemical substances used or released by the body
8. learned behavior in which a new stimulus causes the same response that an old one did
9. part of the brain that controls the senses and thinking
10. fiber that carries messages away from a nerve cell

Identifying Word Relationships *Explain how the words in each pair are related. Write your answers in complete sentences.*

1. endocrine gland, hormone
2. axon, dendrite
3. cerebellum, cerebrum
4. pupil, iris
5. reflex, reflex arc
6. response, stimulus
7. lens, retina
8. innate behavior, learned behavior

Content Challenges...

Multiple Choice *Write the letter of the term that best completes each statement.*

1. Coughing is
   a. a learned behavior.   **b.** a conditioned response.   **c.** an innate behavior.   **d.** a stimulus.
2. The lens of the eye forms images on the
   a. cornea.   **b.** pupil.   **c.** iris.   **d.** retina.

3. Insulin is made by the
   a. pituitary gland.   **b.** pancreas.   **c.** ovaries.   **d.** thyroid.
4. The smallest part of the brain is the
   a. medulla.   **b.** cerebrum.   **c.** cerebellum.   **d.** skull.
5. Adrenalin controls
   a. muscle reaction.   **b.** growth.   **c.** the amount of calcium in the blood.   **d.** blood sugar levels.
6. The cochlea is found in the
   a. outer ear.   **b.** middle ear.   **c.** inner ear.   **d.** eardrum.
7. The 'thermostat' of the body is the
   a. cerebrum.   **b.** hypothalamus.   **c.** pituitary gland.   **d.** cerebellum.
8. The nervous system is made up of the brain, the spinal cord, and
   a. muscles.   **b.** glands.   **c.** nerves.   **d.** the skull.
9. The clear covering at the front of the eye is the
   a. retina.   **b.** iris.   **c.** pupil.   **d.** cornea.
10. The largest cells in the body are
    a. axons.   **b.** dendrites.   **c.** hormones.   **d.** neurons.
11. Sneezing and blinking are
    a. receptors.   **b.** reflexes.   **c.** stimuli.   **d.** reflex arc.

True/False *Write true if the statement is true. If the statement is false, change the underlined term to make the statement true.*

1. Estrogen and progesterone are produced by the ovaries.
2. When you solve a problem by thinking about it, you are reasoning.
3. The amount of light entering the eye is controlled by the cornea.
4. The ear responds to light waves.
5. The brain is connected to the spinal cord by the medulla.
6. The path of a message in a reflex is called a conditioned response.
7. Reflexes are controlled by the brain.
8. The brain and spinal cord make up the central nervous system.
9. Growth hormone is made by the pituitary gland.
10. Messages are carried toward the nerve cell body by axons.
11. You are born with learned behaviors.
12. Glands that do not have ducts are called endocrine glands.

Understanding the Features...

Reading Critically *Use the feature reading selections to answer the following. Page numbers for the features follow each question in parentheses.*

1. What does human growth hormone control? (301)
2. **Define:** What is bionics? (299)
3. In what field does B. F. Skinner specialize? (305)
4. **Explain:** Why is MRI valuable in detecting cancer? (289)
5. **Identify:** What are plant hormones called? (303)
6. **Analyze:** Why must a physical therapy assistant be patient and understanding? (293)

Concept Challenges

Interpreting a Table *Use the table to answer the following.*

1. What does thyroxine control?
2. **Classify:** What are two pituitary hormones?
3. **Analyze:** What hormone is needed to regulate the pituitary gland?
4. What gland produces insulin?
5. What minerals are affected by parathyroid hormone?
6. **Hypothesize:** What health problems might a person have if their thymus gland produced too little thymosin?
7. **List:** What three things does adrenalin control?

Table 1	Hormones and What They Do	
GLAND	HORMONE	JOB OF HORMONE
Pineal	Melatonin	Helps regulate pituitary gland
Pituitary	Growth hormone	Controls the growth of bones
	ATCH	Controls the release of hormones from the adrenal glands
Thyroid	Thyroxin	Controls rate of body growth
Parathyroid	Parathyroid hormone	Controls the amount of calcium and phosphorus in the blood
Thymus	Thymosin	Controls growth of certain white blood cells
Adrenal	Adrenalin	Controls muscle reaction, blood clotting, and blood pressure
Pancreas	Insulin	Controls blood sugar levels

Critical Thinking *Answer each of the following in complete sentences.*

1. How do reflexes help an organism survive?
2. **Compare:** How is the eye like a camera?
3. **Contrast:** How is reasoning different from trial and error learning?
4. How is the spinal cord like a tree trunk?
5. How do the nervous system and the endocrine system work together?

Finding Out More

1. Injury to the central nervous system often results from automobile accidents, serious falls, and sports injuries. Make a poster showing some of the ways that people can protect their nervous systems from injury.
2. **Research:** Other parts of the brain are the thalamus, the pons, and the midbrain. Using library references, find out what each of these parts of the brain controls. Organize your findings in a table.
3. Sound intensity is measured in decibels. Find out the safe decibel range for the human ear in an encyclopedia. Then find out the decibel range of ten common noises. Identify any noises that are above the safe range and can damage your ears.
4. Meningitis is a serious disease. Using library references, find out more about meningitis. Write your findings in a report.

HEALTH AND DISEASE

CONTENTS

STUDY HINT Before beginning Unit 16, write the title of each lesson on a sheet of paper. Below each title, write a short paragraph explaining what you think each lesson is about.

16-1 How does the body fight disease?

Objective ▶ Identify the ways your body fights disease.

TechTerms

- ▶ **antibodies** (AN-ti-bahd-eez): substances the body makes to protect itself from disease
- ▶ **white blood cells:** cells that protect the body against disease

Defense Systems Your body is under constant attack by bacteria, viruses, and other harmful substances. However, your body usually can protect itself from illness. Your body has defenses against disease. These defenses include your skin, and the respiratory, digestive, and circulatory systems.

▐▶ *Identify:* What parts of the body defend it from bacteria and viruses?

Skin The skin is a waterproof, germ-proof barrier that covers your body. It is made up of several layers of cells. Working together, the layers of the skin act like a wall to prevent disease-causing substances from entering the body.

▐▶ *Describe:* How does the skin protect the body?

Digestive System Although the skin protects the body from most germs, bacteria and viruses can enter the body through the mouth. The digestive system helps to destroy these substances. Hydrochloric acid in the stomach helps to destroy bacteria and viruses that enter the stomach.

▐▶ *Explain:* How does the digestive system help to protect the body from disease-causing substances?

Respiratory System Bacteria and viruses also can enter the body through the nose. Hairs in the nose filter the air and trap many small particles. Cilia and mucus in the respiratory system also trap germs before they can enter the lungs. The body gets rid of the trapped particles by sneezing and coughing. Sneezing and coughing force mucus and trapped particles out of the body.

▐▶ *Explain:* How does mucus in the nose and windpipe protect the body?

White Blood Cells When germs get past the defenses of the skin, digestive, and respiratory systems, the body's second line of defense goes to work. The **white blood cells** are the body's second line of defense. White blood cells are part of the circulatory system. They travel through the blood in search of bacteria. When white blood cells find bacteria, they surround them. The white blood cells destroy the bacteria by digesting them.

▐▶ *Describe:* How do white blood cells help protect the body?

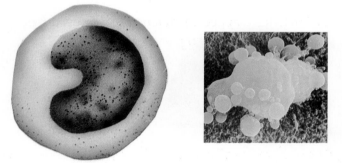

Final Line of Defense If bacteria get through the skin and past the white blood cells, the body still has another line of defense. The body is able to make substances that can destroy the bacteria. These substances are called **antibodies** (AN-ti-bahd-eez). Some antibodies cause bacteria to break up or clump together. Other antibodies make bacterial poisons harmless. Still other antibodies destroy virus particles.

▐▶ *Identify:* What are antibodies?

LESSON SUMMARY

▶ The defense systems of the body include the skin, and the respiratory, digestive, and circulatory systems.

▶ The skin acts like a wall to stop tiny particles from entering the body.

▶ Hydrochloric acid in the digestive system destroys disease-causing organisms.

▶ The hairs in the nose, cilia, and mucus trap small particles, preventing them from entering the body.

▶ White blood cells, produced in the circulatory system, surround and destroy bacteria in the body.

▶ The body produces antibodies that can destroy foreign substances.

CHECK *Write true if the statement is true. If the statement is false, change the underlined term to make the statement true.*

1. Bacteria and <u>viruses</u> can cause disease.

2. The <u>skin</u> allows germs to enter the body.

3. Acid in your <u>blood</u> kills disease-causing organisms.

4. Particles trapped in mucus are removed from the body by <u>sneezing</u> and coughing.

5. White blood cells travel through the body in search of <u>antibodies</u>.

6. The body's first line of defense is <u>antibodies</u>.

APPLY *Complete the following.*

▶ 7. **Infer:** Why is it important to cover your nose with a tissue when you sneeze?

▶ 8. **Infer:** How does putting a bandage over a cut help defend the body against disease?

■ 9. **Hypothesize:** What do you think would happen if germs could enter the skin easily?

Health & Safety Tip..........................

You should always wash a cut thoroughly and apply an antiseptic. An antiseptic is a substance that helps prevent infection. Use a first aid handbook or other references to find out the names of three antiseptics. Describe how each of these antiseptics should be used. Present your findings in a report.

SCIENCE CONNECTION ◆○◇○◆○◇○◆○◇○◆○◇○◆○◇○◆○◇○◆○◇○

PREVENTIVE MEDICINE

One way of keeping healthy is to practice preventive medicine. Preventive medicine means taking steps to prevent sickness. When you keep your body strong and healthy, you are practicing preventive medicine. This means eating a balanced diet and getting the proper amount of rest. Regular exercise also is necessary for a healthy body. Keeping your body clean and free of dirt and germs is another way of preventing disease.

Your doctor and dentist can help you practice preventive medicine. Regular yearly checkups by these professionals can help detect health problems early. The earlier a health problem is caught, the easier it is to treat. This is especially important as you get older.

Good health is not an accident. You must work hard to keep healthy. The way you take care of your body today determines how healthy it will be in the future.

16-2 What is immunity?

Objectives ▶ Define immunity. ▶ Explain the difference between natural immunity and acquired immunity.

TechTerm

▶ **immunity** (im-MYOON-i-tee): resistance to a specific disease

Resisting Disease Antibodies protect the body against foreign substances. After these foreign substances are destroyed, many of the antibodies remain. If the same kind of foreign substances enter the body once again, the remaining antibodies will destroy the substances before they can do any harm. The body has become resistant to these diseases. This resistance to a specific disease is called **immunity** (im-MYOON-i-tee).

▷ *Identify:* What is immunity?

Types of Immunity There are two kinds of immunity. One kind is called natural immunity. The other is called acquired (uh-KWY-urd) immunity. Natural immunity is an immunity that people are born with. In fact, some people have certain kinds of antibodies already in their bodies at birth. Natural immunity is your body's natural defense against certain diseases. Acquired immunity is an immunity that people develop, or acquire, at some time during their lives.

▷ *Compare:* What is the difference between natural and acquired immunity?

Active Acquired Immunity There are two kinds of acquired immunity. One kind is called active acquired immunity. With this kind of immunity, the body resists a certain disease because it has already developed antibodies against the disease. Once you have been exposed to some diseases, your body continues to make the antibodies for that disease. For example, if you have already had chicken pox, you will not get it again.

Your body has developed permanent immunity against chicken pox.

▷ *Explain:* How do you develop active acquired immunity?

Passive Acquired Immunity The second kind of acquired immunity is passive acquired immunity. In this kind of immunity, antibodies are not produced by your body. The antibodies are obtained from somewhere else. You may be injected with antibodies against a certain disease. The person then has a passive acquired immunity against that disease. However, this kind of immunity does not last long. Antibodies that are not made by the body usually are destroyed after a short time.

▷ *Compare:* Which type of acquired immunity lasts the longest?

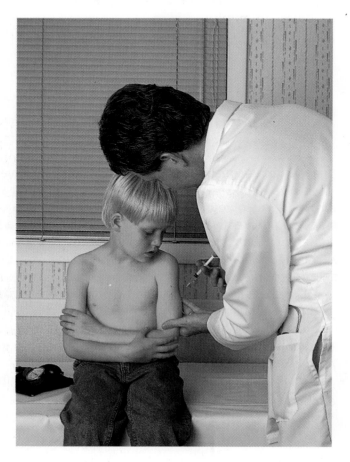

LESSON SUMMARY

▶ Immunity is a resistance to a specific disease.

▶ There are two kinds of immunity—natural immunity and acquired immunity.

▶ Active acquired immunity develops after the body has developed antibodies against a certain disease.

▶ Passive acquired immunity develops when antibodies are injected into the body.

CHECK *Write true if the statement is true. If the statement is false, change the underlined term to make the statement true.*

1. A resistance to disease is called <u>immunity</u>.

2. Immunity a person is born with is called <u>acquired</u> immunity.

3. Immunity a person develops is called <u>natural</u> immunity.

4. Antibodies made by a body remain a <u>short</u> time after the disease it fights is gone.

5. The immunity that results when a body makes antibodies against a disease is an <u>active acquired</u> immunity.

6. Passive acquired immunity is <u>permanent</u>.

APPLY *Complete the following.*

7. **Compare:** How are natural immunity, active acquired immunity, and passive acquired immunity similar?

8. **Relate:** Sometimes a developing baby receives antibodies from its mother. Is this immunity a natural or an acquired immunity? Explain.

InfoSearch..

Read the passage. Ask two questions that you cannot answer from the information in the passage.

Vaccines One way of getting active acquired immunity is through a vaccine. A vaccine is a serum made from a dead or a weakened form of bacteria or viruses. The serum is injected into the body. Vaccines do not cause you to get the disease. The body responds to the vaccine by making antibodies against the disease-causing substance. The antibodies remain in the body and protect it from the disease.

SEARCH: Use library references to find answers to your questions.

▼,▼,▼, LOOKING BACK IN SCIENCE ▼,

EDWARD JENNER'S DISCOVERY OF VACCINATIONS

Modern medicine has become more successful in treating different diseases since the discovery of vaccines. The first vaccine was developed and used in 1796 by Edward Jenner, an English doctor. Jenner noticed that many people who worked near cattle got a mild disease called cowpox. He also noticed that these same people did not get the fatal disease smallpox. To find out why, Jenner tried injecting material from cowpox sores into people. Antibodies against cowpox were produced in these people. Jenner found that these same antibodies were able to destroy the smallpox virus. The people did not get smallpox because the cowpox antibodies protected them against smallpox too.

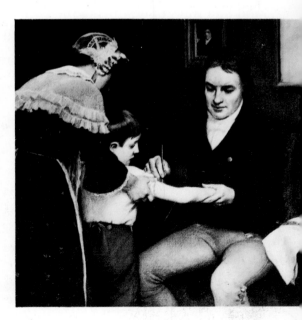

The vaccine Edward Jenner developed was made directly from a disease-causing substance. Today many vaccines are made from chemicals that resemble bacteria and viruses. The human body responds to these artificial vaccines by making antibodies. Diseases that were once widespread, such as polio, have been almost eliminated due to the creation of vaccines.

16-3 What are some bacterial and viral diseases?

Objective ▶ Describe the causes of different types of disease.

Germ Theory Years ago, people believed that a person became ill when evil spirits entered the body. They tried to drive the evil spirits away with magic. About 100 years ago, Louis Pasteur showed that some diseases are caused by bacteria and other microscopic organisms. The idea that diseases are caused by microscopic organisms, or germs, is called the germ theory of disease.

▶ *Explain:* What does the germ theory of disease state?

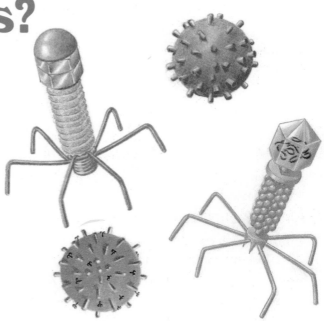

Identifying Germs Pasteur showed that bacteria could cause certain diseases. To fight a disease, it is important to identify the kind of bacteria that causes the disease. Robert Koch studied many of the bacteria that cause diseases in animals and people. He discovered how to grow bacteria outside a living body. This made it possible to study and find ways of destroying the bacteria. Table 1 lists the symptoms of some common diseases caused by bacteria.

Viruses and Disease Some kinds of disease are caused by viruses. The invention of the electron microscope has helped scientists learn about viruses and disease. Scientists have discovered that different kinds of viruses attack different parts of the body. Each virus usually attacks only a certain kind of cell or tissue. For example, the viruses that cause smallpox, measles, and warts seem to attack cells only in the skin. Viruses that cause yellow fever attack cells in the liver. Information about how viruses attack the body helps in the fight against the disease the virus causes. Table 2 lists some common viral diseases and their symptoms.

Table 1 Common Bacterial Diseases	
DISEASE	SYMPTOMS
Diphtheria	Sore throat, fever
Meningitis	fever, headache
Strep throat	sore throat, fever
Tetanus	tightening of body muscles
Toxic Shock Syndrome	fever, low blood pressure

Table 2 Common Viral Diseases	
DISEASE	SYMPTOMS
Chicken pox	skin rash, fever
Hepatitis	jaundice, loss of appetite
Influenza	muscle aches, fever, chills
Measles	pink rash over the body
Mumps	swollen glands, fever

▶ *Describe:* Why is it important to find the germ that causes a disease?

▶ *Describe:* In what way are viruses particular?

316

LESSON SUMMARY

▶ The germ theory of disease states that diseases are caused by microscopic organisms.

▶ Robert Koch discovered how to grow single kinds of bacteria outside a living organism.

▶ The invention of the electron microscope greatly aided the study of viruses and the diseases they cause.

CHECK *Complete the following.*

1. The work of _____ showed that diseases are caused by microscopic organisms.

2. The invention of the _____ helped scientists study viruses.

3. Many kinds of bacteria that cause disease were identified and studied by _____ .

4. The idea that diseases are caused by microscopic organisms is called the _____ .

5. Particular kinds of _____ usually attack only certain types of cells and tissues.

APPLY *Answer the following.*

6. **Infer:** Why do you think the invention of the electron microscope helped in the study of viruses?

Use Tables 1 and 2 on page 316 to answer the following.

7. **Analyze:** If you had a sore throat and fever, what two diseases might you have?

8. **Analyze:** You have swollen glands and a fever. What disease might you have?

9. **Identify:** What are the symptoms of Toxic Shock Syndrome?

InfoSearch

Read the passage. Ask two questions that you cannot answer from the information in the passage.

Food Preservation There are several ways to package and preserve food to prevent spoilage. When foods are canned, they are cooked at high temperatures to kill any bacteria. The foods are then stored in airtight cans or jars. Other foods are preserved by drying. Most of the water is removed from the food. This decreases the chance of spoilage. Raisins are an example of food that has been dried. Another method of food preservation is curing. Food is cooked or dried. Then chemical preservatives such as salt, vinegar, or sugar are added to the food.

SEARCH: Use library references to find answers to your questions.

LOOKING BACK IN SCIENCE

KOCH'S POSTULATES

Bacteriologists are scientists who study bacteria. Many bacteriologists study disease-causing bacteria. They study these organisms to identify the diseases each kind of bacteria causes.

Robert Koch was the first bacteriologist to show that a certain kind of bacterium causes a certain kind of disease. In the late 1800s, he developed a way to determine the relationship between a particular bacterium and a particular disease. This procedure, called Koch's postulates, includes a series of four steps.

The first step is to remove disease-causing bacteria from a diseased animal. The bacteria are then isolated and grown in a laboratory. Next, the laboratory-grown bacteria are injected into a healthy animal.

When the healthy animal becomes sick with the disease, the bacteria are removed from the animal. These bacteria are compared with the original bacteria. If the two bacteria match, bacteriologists assume that the particular bacterium causes the same disease in similar organisms. Modern bacteriologists still follow this same set of steps in their study of bacteria and diseases.

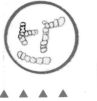

16-4 What are antibiotics?

Objective ▶ Explain how antibiotics are used to fight disease.

TechTerm

▶ **antibiotic** (an-ti-by-AHT-ik): chemical made by a living organism that kills bacteria

Treating Disease Sometimes your body needs help in fighting off disease. Your doctor may prescribe medicine, or you may get an injection. If you are suffering from a disease caused by bacteria, you may be treated with an **antibiotic** (an-ti-by-AHT-ik). An antibiotic is a chemical substance that kills harmful bacteria.

▶ ***Identify:*** What is an antibiotic?

Discovery of Antibiotics The use of antibiotics to treat disease is a fairly recent discovery. In 1929, Alexander Fleming, an English bacteriologist (bak-tir-ee-AHL-uh-jist), was the first person to observe the action of an antibiotic. Fleming was growing bacteria in a dish. He noticed that the bacteria did not grow in a part of the dish where some mold had formed. Fleming guessed that the mold produced a substance that was harmful to bacteria. Thirteen years later, scientists were able to separate this substance from the same kind of mold. The scientists called this substance penicillin (pen-uh-SIL-in). Penicillin was the first antibiotic.

Figure 1 Alexander Fleming

Since the discovery of penicillin, many other antibiotics have been discovered. Most antibiotics are produced from molds. Other antibiotics are made from fungi and bacteria. Certain kinds of plants and animals also produce some types of antibiotics.

Figure 2 Growing Mold

▶ ***Explain:*** Where does penicillin come from?

How Antibiotics Work Antibiotics are not all the same. Different antibiotics have different effects on a particular disease. This means that each antibiotic can be used to fight only certain kinds of organisms. For example, penicillin works only against certain bacteria. The penicillin destroys the bacteria and stops it from reproducing. Penicillin does not work against viruses. No antibiotic can destroy all types of bacteria.

A problem with antibiotics is that they may make some people sick. Antibiotics also can cause an allergic reaction. For example, some people are allergic to penicillin. If they take this antibiotic, a fever or rash may develop. In severe reactions, the person may not be able to breathe. For this reason, it is important that you tell your doctor if you are allergic to any medications.

▶ ***Apply:*** Why is it important to inform your doctor of any drug allergies you might have?

LESSON SUMMARY

▶ Antibiotics are chemical substances that kill harmful bacteria.

▶ Alexander Fleming was the first person to discover the use of antibiotics to treat disease.

▶ Antibiotics are produced from molds, fungi, and bacteria.

▶ Each antibiotic can be used to fight specific bacteria.

▶ Antibiotics may make some people sick or cause an allergic reaction.

CHECK *Find the sentence in the lesson that answers each question. Then, write the sentence.*

1. In Fleming's experiment, where did the bacteria not grow?

2. What was the first antibiotic discovered?

3. What are antibiotics?

4. Do antibiotics work against viruses?

5. What kind of reaction do antibiotics have on some people?

6. What are most antibiotics made from?

7. Who was the first person to observe an antibiotic?

APPLY *Complete the following.*

8. **Hypothesize:** Is it possible for a person taking penicillin for a bacterial disease to get sick with a different type of disease? Explain.

9. **Infer:** Would penicillin be an effective treatment for a person suffering with the measles?

Skill Builder

Building Vocabulary The prefix "anti-" means "against". Use this information to write a definition of the word "antibiotic." Then look up the word in a dictionary to check your definition. Write definitions for the terms "antibodies" and "antiseptic." Use a dictionary to check your definitions.

Health & Safety Tip

If you are allergic to any antibiotic, such as penicillin, you should have some kind of identification on you at all times. Visit a pharmacy to get information about bracelets or charms that contain medical information. It is important to have this kind of information on you in case of an emergency.

⬥▪◀ PEOPLE IN SCIENCE ▶■◀⬥▪◀⬥▪◀⬥▪◀⬥▪◀⬥▪◀⬥▪◀⬥▪◀⬥▪◀⬥▪◀⬥▪◀⬥▪◀⬥▪◀⬥▪◀

JOSEPH LISTER (1827–1912)

Joseph Lister was an English doctor who practiced medicine during the middle of the 19th century. He read about the discoveries of Louis Pasteur regarding bacteria. Lister guessed that there was a connection between bacteria and infections.

During the time that Lister practiced medicine, most people who underwent surgery developed some kind of infection after the surgery. Many of the patients died from these infections. Joseph Lister hypothesized that these infections were caused by bacteria. He thought that if the tools the doctor used during surgery were cleaned, bacteria on the tools would be removed, and infections would decrease. Lister introduced sterilization (ster-uh-li-ZAY-shun) as a means of cleaning surgical instruments. In this method, surgical instruments are placed in boiling water. The heat destroys bacteria. Lister also insisted that doctors carefully wash their hands in antiseptic solutions before operating on patients. When doctors began following Lister's suggestions, the number of post-operative infections declined. The work of Lister saved many lives.

Objective ▶ Identify types of heart disease and their causes.

TechTerms

▶ **atherosclerosis** (ath-ur-oh-skluh-ROH-sis): buildup of fat deposits on artery walls

▶ **coronary** (KOWR-uh-ner-ee) **arteries:** arteries that carry blood and oxygen to the tissues of the heart

▶ **heart attack:** lack of blood and oxygen to a part of the heart

Heart Disease More than half of all deaths in the United States are caused by heart disease. Heart disease affects both the heart and the blood vessels. Some kinds of heart disease are genetic (juh-NET-ik), or passed from parent to offspring. If someone in your family has heart disease, there is a chance that you also may develop heart disease. Other kinds of heart disease are the result of the person's lifestyle and environment. Table 1 lists some of the factors that contribute to heart disease.

Table 1 Factors Contributing to Heart Disease	
▶ age	▶ high blood pressure
▶ family history	▶ obesity
▶ gender	▶ physical inactivity
▶ smoking	▶ high cholesterol levels

▶ *Infer:* What are some ways you can prevent heart disease?

Atherosclerosis In one kind of heart disease, fatty substances coat the inside walls of the arteries. One of these fatty substances is cholesterol (kuh-LES-tuh-rohl). Cholesterol is a fatty substance found in animal products. As fat builds up in an artery, the artery becomes more narrow. This condition is known as **atherosclerosis** (ath-ur-oh-skluh-ROH-sis). The heart must work harder to pump blood through the narrow arteries.

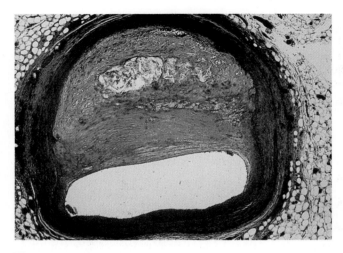

▐▐▐▶ *Explain:* What is atherosclerosis?

Heart Attack Like other body cells, the cells of the heart need food and oxygen. Heart cells also need to get rid of waste materials. The heart has its own system of blood vessels to take care of these needs. This system is called the coronary (KOWR-uh-ner-ee) system. **Coronary arteries** carry oxygen and blood to the heart.

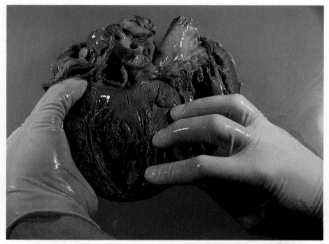

Atherosclerosis may make the coronary arteries narrower. Sometimes a clot forms and completely blocks a coronary artery. This blockage stops blood and oxygen from reaching a part of the heart. The affected part of the heart cannot do its work. This condition is called a **heart attack.** A person having a heart attack usually feels a sharp pain in the chest. Heart attacks often are fatal.

▐▐▐▶ *Describe:* What is a heart attack?

LESSON SUMMARY

▶ Heart disease affects both the heart and the blood vessels.

▶ Atherosclerosis is a condition in which fat deposits build up on the walls of arteries.

▶ Coronary arteries carry blood to the heart.

▶ A heart attack occurs when part of the heart does not receive blood and oxygen.

CHECK *Complete the following.*

1. More than half of all deaths in the United States are caused by _____ .

2. The _____ arteries carry blood and oxygen to all parts of the heart.

3. A blockage in a coronary artery can cause a person to suffer a _____ .

4. A fatty substance that can coat the inside walls of arteries is _____ .

5. A buildup of cholesterol on the artery walls causes the heart to work _____ .

APPLY *Complete the following.*

6. **Hypothesize:** Does the width of an artery increase or decrease as a person ages?

7. **Hypothesize:** What do you think you can do to reduce the chance that you will be affected by atherosclerosis later in life?

8. **Analyze:** Which of the factors listed in Table 1 can be controlled?

9. Pick three factors in Table 1. Explain how you can lower your risk of getting heart disease for each of the three factors chosen.

InfoSearch

Read the passage. Ask two questions that you cannot answer from the information in the passage.

Strokes Atherosclerosis affects all the arteries in the body, not only the coronary arteries. Atherosclerosis also makes the arteries leading to the brain narrower. These narrow arteries may become blocked by a blood clot. The blocked artery prevents blood and oxygen from reaching the brain. A lack of oxygen to the brain is called a stroke. A stroke may cause paralysis. A serious stroke may even cause death.

SEARCH: Use library references to find answers to your questions.

SCIENCE CONNECTION

CHOLESTEROL

Cholesterol is a fatty substance found in meats and other animal products. Cholesterol also is made by the cells of the body. Cholesterol is needed by the body for most chemical reactions to take place. However, many people eat foods that contain a high amount of cholesterol. This increases the level of cholesterol in the blood. This in turn increases the chance that cholesterol will build up on artery walls.

Research suggests that you can reduce your risk of developing heart disease by eating foods that are low in cholesterol. To cut back on cholesterol, limit the amount of red meat, eggs, and whole milk dairy products you eat. However, since these foods contain many valuable vitamins and minerals, they should not be totally eliminated from your diet. Low-cholesterol foods, such as chicken and fish, can be substituted for red meat. Studies have shown that people who eat more fish than red meat have less heart disease than people who eat red meat more often. By developing healthful eating habits now, you can help avoid serious health problems.

What is cancer?

Objectives ▶ Describe the effect of cancer on the body. ▶ Identify some possible causes of cancer.

TechTerms

▶ **benign** (bi-NYN) **tumor:** mass of cells that is usually harmless

▶ **malignant** (muh-LIG-nunt) **tumor:** harmful mass of cells that can spread throughout the body

▶ **tumor** (TOO-mur): mass or lump of cells

Growing Wild Most cells in the body divide and produce new cells. Sometimes the growth of cells goes wild. New cells grow where they are not needed. Their growth becomes rapid and uncontrolled. The new cells crowd nearby cells and rob these cells of their nutrients. They prevent the nearby cells from working in a normal way. The new cells form a mass, or lump, called a **tumor** (TOO-mur).

▶ *Identify:* What is a tumor?

Kinds of Tumors There are two kinds of tumors. One kind grows in one place in the body. It does not spread to other places. This type of tumor often stops growing once it gets to a certain size. Tumors of this kind are called **benign** (bi-NYN) **tumors.** Benign tumors usually are not a serious health problem.

The other kind of tumor spreads to other places in the body. As it spreads, it causes harm to the body. If the growth of the tumor is not stopped, the person eventually may die. This kind of tumor is called a **malignant** (muh-LIG-nunt) **tumor.** The condition the body is in after the spread of malignant tumors is called cancer.

▶ *Describe:* What is a cancer?

How Cancer Spreads Cancer spreads from one part of the body to another part of the body in different ways. In one way, cancer cells break away from the tumor. The cancer cells enter the bloodstream. They travel with the blood to other parts of the body. A new tumor will then begin to grow at the place where the cancer cells come to rest.

▶ *Explain:* How can cancer cells spread through the body?

Possible Causes Scientists are not sure what causes a cancer to start growing. They do know that certain chemicals increase the chance of getting cancer. Studies have shown that exposure to too much sunlight and X rays can cause cancer. Scientists believe that certain viruses and hormones also may cause cancer. Research also has shown that people who smoke cigarettes are more likely to get lung cancer than nonsmokers.

▶ *List:* What are three possible causes of cancer?

Early Signs The earlier cancer is detected, the more likely a person is to survive the disease. Table 1 lists the seven warning signs of cancer. If you ever have any of these warning signs, you should see a doctor immediately.

Table 1   Seven Warning Signs of Cancer
1. A sore that does not heal
2. Unusual bleeding
3. A lump in the breast or other area beneath the skin
4. Constant indigestion or trouble swallowing
5. A nagging cough
6. A change in the size, shape, or color of a wart or mole
7. A change in bowel or bladder habits

▶ *Explain:* Why is it important to detect cancer early?

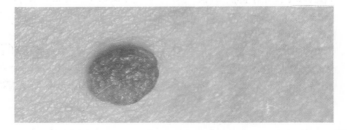

LESSON SUMMARY

- ▶ A tumor is a mass or lump of cells.
- ▶ Benign tumors are harmless growths of cells.
- ▶ The spread of harmful malignant tumors is called cancer.
- ▶ Overexposure to sunlight and X rays, and cigarettes contribute to developing cancer.
- ▶ The chances of curing cancer are best when treatment starts early.

CHECK *Complete the following.*

1. A tumor that does not spread to other parts of the body is a _____ tumor.

2. A cancer is the spread of _____ tumors.

3. Cells from a tumor sometimes travel through the _____ to new places in the body.

4. Nonsmokers are less likely to get cancer than _____ .

5. The earlier cancer is detected, the better a person's chance of _____ .

APPLY *Complete the following.*

▶ 6. **Infer:** Why do X-ray technicians wear lead aprons when taking X rays?

7. Why is it important to wear a sunscreen when out in the sun?

🖥 8. **Hypothesize:** The incidence of lung cancer in women has increased greatly in the past ten years. Why do you think this is so?

Skill Builder.....................................

Building Vocabulary Look up the words "benign" and "malignant" in a dictionary. Determine whether these words are antonyms or synonyms. Write two sentences for each term that describe different kinds of tumors.

Health & Safety Tip.........................

One way to prevent cancer is to detect it early. You should look over your entire body for abnormalities, such as growths or sores. Use library references or first-aid books to find out about self-examination. Practice self-examination every month. The more familiar you are with these practices, the better chance you have of finding an abnormality and getting it treated early.

TECHNOLOGY AND SOCIETY ▪▪▪ ▪▪▪ ▪▪▪ ▪▪▪ ▪▪▪ ▪▪▪ ▪▪▪ ▪▪▪ ▪▪▪ ▪▪▪ ▪▪▪ ▪▪▪

CANCER DETECTION AND TREATMENT

Different forms of cancer are detected in different ways. Skin cancer is the easiest form of cancer to cure if it is detected early. Noting a change in a mole or skin mark is a common method of detecting skin cancer. Breast cancer can sometimes be detected by examining one's breasts for a lump or thickening. Breast X rays, or mammograms, also can detect the presence of tiny cancerous tumors.

Cancer treatment varies with each different form of cancer. Sometimes the tumor is surgically removed from the body. This treatment often works when the tumor is found early, before it has spread. Sometimes the tumor cannot be removed without damaging the organ it is growing in. Doctors then use radiation to kill the cancer cells. Powerful beams of light called laser beams have been used to destroy cancer cells. Certain drugs also have been found to slow down the growth of tumors. The problem with these drugs is that they affect normal cells as well as cancerous cells. This causes harm to the body. The best way to avoid cancer is to live a lifestyle aimed at prevention.

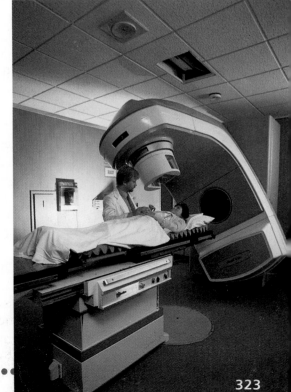

Objective ► Describe the causes and symptoms of AIDS.

TechTerms

► **AIDS:** viral disease that attacks a person's immune system

► **immune system:** body system made up of cells and tissues that help a person fight disease

Immune System Your body is constantly fighting off germs and disease-causing organisms. The first line of defense against bacteria and viruses are structures located at body openings. For example, hairs in your nose work to filter particles from air taken in through the nose. However, some disease-causing substances manage to get past these hairs. Your body's second line of defense must then work to destroy the foreign substances. This second line of defense is called the **immune system.** The immune system is made up of cells and tissues that fight disease.

▌▶ Identify: What is the immune system?

HIV Virus Viruses cause disease. They are selective in the cells or tissues that they attack. A particular virus called the HIV virus attacks a person's immune system. The illness caused by the HIV virus is called Acquired Immune Deficiency Syndrome or **AIDS.** AIDS is a viral disease that kills white blood cells in a person's immune system. The person loses the ability to fight disease. For this reason, people with AIDS easily get diseases that most healthy people can fight off.

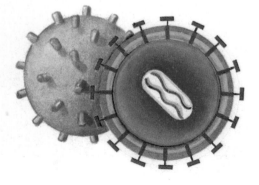

These diseases often are fatal for a person with AIDS.

▌▶ Describe: What effect does the HIV virus have on the body?

Transmission of AIDS It is not easy to become infected with AIDS. AIDS cannot be transmitted by casual contact. It cannot travel through air, food, or water. People with AIDS have the HIV virus in their blood and body fluids. In order to contract AIDS, you must exchange bodily fluids with an infected person. The virus can enter the bloodstream by sexual contact with someone who has AIDS. The HIV virus also can enter the bloodstream of intravenous drug users who use contaminated needles. Another way the HIV virus enters the bloodstream is through a blood transfusion of infected blood.

▌▶ Explain: How is the HIV virus transmitted?

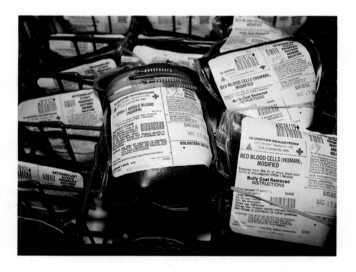

AIDS Treatment There is no known cure for AIDS at this time. However, scientists are working on ways to treat AIDS patients. Doctors have developed ways to strengthen the immune systems of people with AIDS. Scientists have been able to prolong the life of a person with AIDS, but there is no long-term cure.

► Infer: Why do you think AIDS research is so important?

LESSON SUMMARY

▶ The immune system is made up of cells and tissues that help the body fight disease.

▶ AIDS is a viral disease that kills white blood cells in a person's immune system.

▶ AIDS cannot be transmitted by casual contact.

▶ At present, there is no long-term treatment or cure for AIDS.

CHECK *Complete the following.*

1. The body's first line of defense against disease are structures located at _____ .

2. The immune system is made up of cells and _____ that fight disease.

3. The HIV virus attacks the body's _____ .

4. People with AIDS have the HIV virus in their _____ and body fluids.

5. Is there a cure for AIDS?

6. Is it easy to contract AIDS?

APPLY *Complete the following.*

▶ 7. **Infer:** The leading cause of death among AIDS patients is an infection of the lungs. How is the HIV virus responsible for such infections?

8. Could a person with AIDS transmit the virus to another person sitting in the same room?

▶ 9. **Infer:** Some large cities distribute clean needles to drug users. Why do you think they do this?

▶ 10. **Infer:** How does making the immune system stronger lengthen the life of a person infected with AIDS?

InfoSearch

Read the passage. Ask two questions that you cannot answer from the information in the passage.

Blood Screening Blood transfusions often are used during surgical procedures. The person undergoing surgery is given blood that was previously taken from a blood donor. Without the blood transfusion, the person undergoing surgery would die, due to loss of blood. In 1985, a method of testing donor blood for the HIV virus became available. The presence of certain antibodies in the blood indicates that it is contaminated with HIV. Most hospitals use this test to be sure that the donor blood is safe for transfusions.

SEARCH: Use library references to find answers to your questions.

ACTIVITY

INVESTIGATING THE SPREAD OF AIDS

You will need paper, pencil, and a calculator (optional).

1. Study Table 1.
2. Rank the states in order from the state with the most reported AIDS cases to the fewest reported AIDS cases.

Questions

1. a. Which state has the most reported AIDS cases?
   b. Which state has the fewest reported AIDS cases?

2. **Calculate:** What is the total number of reported AIDS cases in Table 1?

▶ 3. **Analyze:** Do you think AIDS is a nationwide problem?

Table 1 Reported AIDS Cases in 10 U.S. States Through April 1990*	
STATE	TOTAL CASES
California	25,549
Florida	11,439
Georgia	3429
Illinois	3922
Maryland	2483
Massachusetts	2790
New Jersey	8790
New York	29,335
Pennsylvania	3640
Texas	9058

*Source: GMHC

How do some drugs affect the body?

Objective ▶ Describe the effects of some drugs on the body.

TechTerms

▶ **addiction** (uh-DIK-shun): uncontrollable dependence on a drug

▶ **depressant** (di-PRES-unt): drug that slows down the central nervous system

▶ **hallucinogen** (hul-LOO-suh-nuh-jen): drug that causes a person to see, hear, smell, and taste things in a strange way

▶ **stimulant** (STIM-yuh-lent): drug that speeds up the central nervous system

Drugs A drug is a chemical substance that causes a change in the body. Aspirin and antacids are drugs that can be bought in a store. You do not need a doctor's prescription for these drugs. Other drugs, such as antibiotics, can be bought only with a doctor's prescription.

▶ *Define:* What is a drug?

Depressants Drugs are grouped according to the effects they have on the body. **Depressants** (di-PRES-unts) are drugs that slow down the central nervous system. They slow down the heartbeat and breathing rate. Large amounts of depressants can cause a person to go into a coma, or even die.

The most widely abused depressant is alcohol. Barbiturates (bar-BICH-uhr-its) are depressants commonly used in sleeping pills and sedatives.

Narcotics (nar-KOT-iks) are depressants used as pain killers. If barbiturates or narcotics are used over a period of time, the user may develop an **addiction** (uh-DIK-shun) to the drug. An addiction is an uncontrollable dependence on a drug. People can develop both physical and mental addictions to drugs. The addicted person cannot stop using the drug without going through a period of sickness.

▶ *Describe:* What effect do barbiturates have on the body?

Stimulants Some drugs speed up the action of the central nervous system. These drugs are called **stimulants** (STIM-yuh-lents). In many ways, the effects of stimulants are opposite those of depressants. Stimulants speed up a person's heartbeat and rate of breathing. Cocaine is a commonly abused stimulant. Crack is a purified form of cocaine that is extremely dangerous. You may be surprised to learn that caffeine also is a stimulant. Caffeine is found in coffee, tea, cola, and chocolate.

▶ *Contrast:* How does the effect of a stimulant differ from the effect of a depressant?

Hallucinogens Some drugs change the way a person receives information through the senses. For example, they cause a person to see, hear, smell, and taste things in a strange way. These drugs are called **hallucinogens** (hul-LOO-suh-nuh-jenz). Hallucinogens often make a person feel panicked or threatened. For this reason, people who take hallucinogens often are dangerous to themselves and others. LSD and marijuana are two commonly abused hallucinogens.

▶ *Explain:* What effect do hallucinogens have on the body?

LESSON SUMMARY

- A drug is a chemical substance that causes a change in the body.

- Depressants slow down the action of the central nervous system.

- People who use drugs, such as depressants, can become addicted to the drug.

- Stimulants speed up the action of the central nervous system.

- Hallucinogens change the way a person receives information through the senses.

CHECK *Answer the following.*

1. What are two drugs that you can buy without a doctor's prescription?

2. What effect do depressants have on the body?

3. What group of depressants often are used as pain killers?

4. What effect do stimulants have on the body?

5. What kind of drug is marijuana?

6. What is the most abused depressant drug?

APPLY *Complete the following.*

7. What effect does coffee have on the central nervous system?

8. **Relate:** Which drug, a barbiturate or cocaine, would most likely make a person feel wide awake? Explain your answer.

Health & Safety Tip

The vapors from certain common household products can have a harmful effect on the body when inhaled. Sniffing large amounts of airplane cement, cleaning fluids, and paint thinners can cause damage to the tissues of the heart, liver, and kidneys. Sniffing such substances may cause unconsciousness, drowsiness, and even death. You should be sure to work with these products only in well-ventilated areas.

Skill Builder

Classifying Use library references to find out information about the following drugs: amphetamines, heroin, morphine, tranquilizers, PCP, and nicotine. Then classify each drug as either a stimulant, a depressant, or a hallucinogen. Present your findings in a report.

ACTIVITY

READING DRUG LABELS

You will need 3 different over-the-counter drug or medication labels or packaging, paper and a pencil.

1. Examine one of your drug or medication labels. Write the name of the drug or medication and its use.

2. Find the dosage instructions and record the dosage.

3. Find the expiration date and record it.

4. Record any special warnings or precautions.

5. Repeat Steps 1–4 for each drug or medication.

Questions

1. What does the expiration date tell you about the drug?

2. What information is given in the dosage directions?

3. **Hypothesize:** Why is the recommended dosage of a drug for an adult different from that for a child?

4. **Observe:** What warning statement is found on all labels?

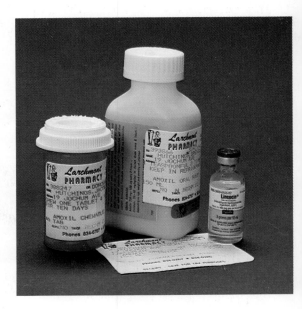

How does alcohol affect the body?

Objective ▶ Describe the effects alcohol has on the body.

TechTerms

- ▶ **alcoholic** (al-kuh-HOWL-ik): person who is dependent on alcohol
- ▶ **cirrhosis** (suh-ROH-sis): liver disorder caused by damaged liver cells

Ethyl Alcohol You have learned that a drug is a substance that causes a change in the body. One of the most commonly abused drugs is ethyl (ETH-ul) alcohol. Ethyl alcohol is the alcohol that is in drinks, such as beer, wine, and whiskey. Alcohol is a drug because it causes a change in the body. Most people think that alcohol is a stimulant. However, alcohol is a depressant. It slows down the action of the central nervous system.

▶ **Explain:** What effect does alcohol have on the body?

Alcohol and the Brain The amount of alcohol present in the bloodstream is called Blood Alcohol Concentration (BAC). The effect of alcohol on the body increases as the Blood Alcohol Concentration increases. When people drink alcohol, their body systems slow down. They think and react more slowly. Their movements become clumsy.

▶ **Infer:** Why do you think it is dangerous for a person who has been drinking to drive a car?

Alcohol and the Body Alcohol affects many organs of the body. If a person does not drink very much, or very often, the effects of alcohol wear off after a period of time. However, if the drinking continues over a period of time, the body can be harmed. The liver can be greatly affected by alcohol. Large amounts of alcohol taken over a long period of time slowly destroy the liver tissues. The liver then loses its ability to carry out its functions. This condition is called **cirrhosis** (suh-ROH-sis). Cirrhosis of the liver often leads to death.

▶ **Identify:** What is cirrhosis of the liver?

Alcoholics Alcohol is addictive for some people. These people drink more and more alcohol as time goes by. They become dependent on alcohol. A person who cannot control his or her drinking of alcohol is called an alcoholic (al-kuh-HOWL-ik). Alcoholism is a disease because it causes harmful changes in body organs. Many alcoholics get help from groups such as Alcoholics Anonymous or AL-ATEEN.

▶ **Describe:** What is an alcoholic?

LESSON SUMMARY

► Ethyl alcohol is a depressant drug.

► Alcohol slows down the working of the body systems.

► Cirrhosis is a liver disorder caused by destroyed liver cells.

► An alcoholic is a person who is dependent upon alcohol.

CHECK *Find the sentence in the lesson that answers each question. Then, write the sentence.*

1. What kind of alcohol is in alcoholic drinks?

2. What effect does alcohol have on body systems?

3. How does a person act after drinking alcohol?

4. What effect does drinking over a long period of time have on the body?

5. What is a person who is dependent on alcohol called?

APPLY *Answer the following.*

6. **Relate:** A person drinks a glass of wine every night before dinner. Is this person an alcoholic? Explain your answer.

Health & Safety Tip......................

When two different drugs are taken together, their effects can be increased greatly. Therefore, even a single alcoholic drink should not be taken with any other kind of drug. Even a harmless drug, such as an aspirin, can change the way alcohol affects a person. To be safe, alcohol should never be mixed with other drugs.

Skill Builder...................................

Researching Pregnant woman who drink have a high risk of giving birth to babies with fetal alcohol syndrome. Fetal alcohol syndrome can cause low birth weight, and both physical and mental disabilities. Use library resources to find out more about fetal alcohol syndrome. Present your findings in a report.

ACTIVITY

COMPARING BLOOD ALCOHOL CONCENTRATION

You will need graph paper and a pencil.

1. Carefully read the table.

2. Use the information in the table to make a bar graph.

Questions

1. **Observe:** At what blood alcohol concentration does the drinker have difficulty thinking?

2. **Observe:** About how much alcohol is present in the bloodstream of a person who has had 4 drinks in two hours?

3. **Analyze:** Is it possible to "drink yourself to death"? Explain your answer.

Table 1 Blood Alcohol Concentration (BAC) and its Effects		
*DRINKS PER HOUR	BAC (PERCENT)	EFFECTS
1	0.02–0.03	Feeling of relaxation
2	0.05–0.06	Slight loss of coordination
3	0.08–0.09	Loss of coordination, trouble talking and thinking, legal intoxication
4	0.11–0.12	Slowed reaction time; lack of judgement
7	0.20	Difficulty thinking; loss of motor skills
14	0.40	Unconsciousness; vomiting may occur
17	0.50	Deep coma; if breathing ceases, death
* 1 drink equals 12 oz beer, 4 oz wine, or 1 oz whisky		

Objective ▶ Explain the effects of tobacco on the body.

TechTerm

▶ **nicotine** (NIK-uh-teen): stimulant found in tobacco

Tobacco Tobacco contains more than 100 chemical substances. When tobacco burns, many new substances are formed. All of these substances are taken into the body with the tobacco smoke. One of the harmful substances in tobacco smoke is **nicotine** (NIK-uh-teen). Nicotine is a stimulant drug. Tars and carbon monoxide are other harmful substances found in tobacco smoke. When inhaled, tar coats the lining of the lungs. Carbon monoxide is a poisonous gas that may cause dizziness and headaches.

▶ *Identify:* What are three harmful substances found in tobacco smoke?

Effect on Breathing The air tubes in the lungs are lined with cells. These cells have tiny hairs called cilia. The cilia beat back and forth to push mucus from the lungs toward the throat. Foreign substances trapped in the mucus get pushed out of the lungs. Smoking causes the cilia to stop working. Without the action of the cilia, mucus collects in and blocks the air tubes. This trapped air causes the air sacs in the air tubes to break. With fewer air sacs at work, less oxygen flows into the lungs with each breath. Less carbon dioxide is released from the body with each exhalation. The smoker must breathe harder and more often to get enough oxygen.

▶ *Explain:* What effect does smoking have on the cilia in the air tubes of the lungs?

Illnesses When the cilia stop beating, foreign substances remain trapped in the lungs. There, these substances may cause infection and disease. Smokers are more likely to get heart disease, cancer, and emphysema (em-fuh-SEE-muh) than nonsmokers. Smokers have a higher death rate from these diseases than do nonsmokers. Compare the healthy lung on the left in the picture with the smoker's lung on the right.

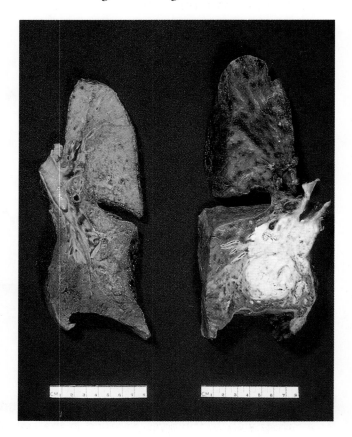

Smoking also places a great strain on the heart. With fewer air sacs in the lungs, the blood gets less oxygen. The heart must work harder to supply enough fresh oxygen for body cells. Nicotine also makes the blood vessels become narrower. The heart must work harder to pump the blood through the smaller openings. This puts an even greater strain on the heart.

▶ *Describe:* What effect does nicotine have on blood vessels?

LESSON SUMMARY

- ▶ Tobacco smoke contains many different kinds of harmful chemicals, including nicotine, tar, and carbon monoxide.
- ▶ Smoking causes the cilia lining the air tubes in the lungs to stop working.
- ▶ Smoking increases the chance of getting certain serious illnesses.
- ▶ Smoking puts a strain on the heart and blood vessels.

CHECK *Complete the following.*

1. Nicotine, carbon monoxide, and _____ are examples of harmful substances found in tobacco smoke.

2. When cilia in the air tubes stop beating, _____ collects and blocks the tubes.

3. Nicotine makes blood vessels become _____ .

4. Smoking puts a great strain on the _____ .

APPLY *Complete the following.*

▶ 5. **Infer:** What effect would cigarette smoking have on a person's ability to exercise?

6. **Hypothesize:** Nicotine is an addictive drug. Why do you think it may be difficult for a person to quit smoking?

InfoSearch

Read the passage. Ask two questions that you cannot answer from the information in the passage.

Sidestream Smoke Research has shown that cigarette smoke has a harmful effect on nonsmokers as well as smokers. For example, children of smokers are twice as likely to have respiratory problems as children of nonsmokers. This is caused by the exposure to sidestream smoke from a burning cigarette. In recent years, smoking has been banned from many public places. This was done in an attempt to protect nonsmokers from sidestream smoke. For example, smoking is restricted in many restaurants, offices, airplanes, and trains.

SEARCH: Use library references to find answers to your questions.

ACTIVITY

CLASSIFYING CIGARETTE ADVERTISEMENTS

You will need magazines, newspapers, paper and pencil.

1. Find several advertisements for cigarettes.
2. Find several advertisements against smoking.
3. Read the advertisements carefully.
4. List the warnings contained in the advertisements.
5. **a.** Make a list of the pictures found in the cigarette advertisements. **b.** Make another list of the pictures which discourage people from smoking.

Questions

1. Did you find more advertisements for or against smoking?
2. How many different warning statements were in the advertisements?
3. **a. Analyze:** Which companies or organizations sponsor the pro-smoking advertisements? **b.** Which organizations sponsor the anti-smoking advertisements?

STUDY HINT Before you begin the Unit Challenges, review the TechTerms and Lesson Summary for each lesson in this unit.

TechTerms................................

addiction (326)
AIDS (324)
alcoholic (328)
antibiotic (318)
antibodies (312)
atherosclerosis (320)
benign tumor (322)

cirrhosis (328)
coronary arteries (320)
depressant (326)
hallucinogen (326)
heart attack (320)
immune system (324)

immunity (314)
malignant tumor (322)
nicotine (330)
stimulant (326)
tumor (322)
white blood cells (312)

TechTerm Challenges................................

Matching *Write the TechTerm that best matches each description.*

1. cells that protect the body against disease
2. stimulant drug found in tobacco
3. resistance to a certain disease
4. liver disorder caused by damaged liver cells
5. buildup of fat deposits on artery walls
6. person who is dependent on alcohol
7. arteries that carry blood and oxygen to the tissues of the heart
8. kind of drug that causes a person to see, hear, mell, and taste things in a strange way
9. lack of blood and oxygen to a part of the heart
10. uncontrollable dependence on a drug
11. mass or lump of cells
12. body system made up of cells and tissues that help a person fight disease
13. viral disease that attacks a person's immune system

Identifying Word Relationships *Explain how the terms in each pair are related. Write your answers in complete sentences.*

1. benign tumor, malignant tumor
2. depressant, stimulant
3. antibiotic, antibodies
4. coronary arteries, atherosclerosis
5. alcohol, cirrhosis
6. immunity, immune system

Content Challenges................................

Multiple Choice *Write the letter of the term that best completes each statement.*

1. All of the following are stimulants EXCEPT
   a. nicotine. **b.** tar. **c.** caffeine. **d.** cocaine.

2. Drugs that slow down the central nervous system are
   a. hallucinogens. **b.** depressants. **c.** barbiturates. **d.** stimulants.

3. The immunity a person is born with is called
   a. active immunity. **b.** acquired immunity. **c.** passive immunity. **d.** natural immunity.

4. AIDS is a disease of the
   a. circulatory system. **b.** immune system. **c.** digestive system. **d.** respiratory system.

5. All of the following body systems help prevent infection EXCEPT the
   a. respiratory system.   b. digestive system.   c. circulatory system.   d. nervous system.

6. AIDS is a disease that is NOT transmitted by
   a. infected blood.   b. casual contact.   c. contaminated needles.   d. sexual contact.

7. Hepatitis usually is caused by a
   a. virus.   b. bacterium.   c. fungus.   d. protozoan.

8. Drugs that cause people to see things that do not exist are
   a. sleeping pills.   b. barbiturates.   c. depressants.   d. hallucinogens.

9. The stimulant drug found in tobacco is
   a. tar.   b. nicotine.   c. carbon monoxide.   d. carbon dioxide.

10. The risk factor for heart disease that you can control is
    a. age.   b. gender.   c. physical inactivity.   d. family history.

11. Cirrhosis of the liver is a disease caused by
    a. hepatitis.   b. alcohol.   c. tobacco.   d. depressants.

12. The body's first line of defense against disease is the
    a. mouth.   b. white blood cells.   c. skin.   d. nose.

13. All of the following are caused by viruses EXCEPT
    a. strep throat.   b. chicken pox.   c. measles.   d. mumps.

14. All of the following are caused by bacteria EXCEPT
    a. influenza.   b. meningitis.   c. tetanus.   d. toxic shock syndrome.

15. Crack is a form of
    a. heroin.   b. cocaine.   c. LSD.   d. opium.

Completion *Write the term that best completes each statement.*
1. Alcohol is classified as a _____ drug.

2. Nicotine is classified as a _____ drug.

3. When the body has developed an uncontrollable dependance on the drug, it has become _____ .

4. Vaccines are used to acquire _____ immunity.

5. Active acquired immunity lasts a _____ time.

6. The germ theory was first stated by _____ .

7. Substances produced by the body to fight disease are called _____ .

8. Most antibiotics are produced from _____ .

9. The kind of tumor that may be harmless is a _____ tumor.

10. When there is a blood clot in the brain, it is called a _____ .

Understanding the Features...

Reading Critically *Use the feature reading selections to answer the following. Page numbers for the features are shown in parentheses.*

1. **Identify:** What are some ways you can practice preventive medicine? (313)

2. How has the discovery of vaccines helped in the prevention of disease? (315)

3. **List:** What are the four steps in Koch's Postulates? (317)

4. How has the work of Joseph Lister saved lives? (319)

5. **List:** What are some low-cholesterol foods that can be substituted for red meat? (321)

6. **Identify:** What are some of the ways cancers are treated? (323)

Concept Challenges..

Interpreting a Table *Use the following table to answer the questions.*

*DRINKS PER HOUR	BAC (PERCENT)	EFFECTS
Table 1		**Blood Alcohol Concentration (BAC) and its Effects**
1	0.02–0.03	Feeling of relaxation
2	0.05–0.06	Slight loss of coordination
3	0.08–0.09	Loss of coordination, trouble talking and thinking, legal intoxication
4	0.11–0.12	Slowed reaction time; lack of judgement
7	0.20	Difficulty thinking; loss of motor skills
14	0.40	Unconsciousness; vomiting may occur
17	0.50	Deep coma; if breathing ceases, death

* 1 drink equals 12 oz. beer, 4 oz. wine, or 1 oz whisky

1. What is the percent of alcohol in the blood-stream when you are considered legally intoxicated?
2. What is the effect of one drink in one hour on the body?
3. What is the percent of alcohol in the body when the body goes into a coma?
4. **Calculate:** How many ounces of wine is there in 7 'drinks'?
5. **Infer:** Why do you think it is dangerous to drive after 4 drinks?
6. **Calculate:** If a person had a drink that contained two ounces of whisky, what effect would it have on the body?
7. What is the BAC percent when vomiting or unconsciousness may occur?
8. What does BAC stand for?

Critical Thinking *Answer each of the following in complete sentences.*

1. Why is it dangerous for drug users to share hypodermic needles?
2. **Infer:** A person has smoked cigarettes for over ten years. Do you think it is too late for this person to quit smoking? Explain your answer.
3. The surgeon general has issued a warning statement that must appear on all containers of alcoholic beverages. Do you agree or disagree with this labeling?
4. **Infer:** Why do you think dentists often wear face masks and rubber gloves when examining patients?
5. **Infer:** How do you think a drug addiction affects a person's life? Explain your answer.

Finding Out More..

1. Caffeine is an addictive drug that may be associated with heart disease and certain kinds of cancers. Use library references to find out some of the common food or drink items that contain caffeine. Research the dangers of caffeine addiction, as well as the symptoms of caffeine withdrawal.
2. Many schools now have chapters of SADD (Students Against Drunk Drivers). Investigate the process of setting up a chapter at your school. Then as a class, set up a chapter at your school.
3. Immunization against disease is important to good health. Use library references to find out the different immunizations that are recommended. Visit or call your doctor and find out if your immunizations are up to date.
4. Obtain information about cancer prevention from the local chapter of the American Cancer Society. Create a classroom display of the pamphlets and other information your receive.

REPRODUCTION AND DEVELOPMENT

CONTENTS

STUDY HINT As you read each lesson in Unit 17, write the lesson title and lesson objective on a sheet of paper. After you complete each lesson, write the sentence or sentences that answer each objective.

17-1 What is meiosis?

Objectives ▶ Define gamete. ▶ Describe the type of cell division that produces gametes.

TechTerms

▶ **gamete** (GAM-eet): reproductive cell
▶ **meiosis** (my-OH-sis): cell division that produces gametes

Body Cells Each body cell of an animal contains the same number of chromosomes. For example, each body cell in a bullfrog contain 26 chromosomes. Fruit flies have 8 chromosomes in each body cell. The body cells of humans contain 46 chromosomes.

▶ *Identify:* How many chromosomes are in each human body cell?

Gametes Sperm cells and egg cells are reproductive cells. Reproductive cells also are called **gametes** (GAM-eets). Gametes develop from special cells in the body. During the formation of gametes, the number of chromosomes changes. It is cut in half. Each gamete contains only half as many chromosomes as a body cell. The process by which gametes form is called **meiosis** (my-OH-sis). The cells formed by meiosis have half the number of chromosomes as body cells.

Nucleus of fruit fly body cell

Nucleus of fruit fly gamete

▶ *Compare:* What is the relationship between the number of chromosomes in body cells and gametes?

Meiosis Meiosis begins when the nucleus of a cell breaks down. The pairs of chromosomes in the cell split apart and move to opposite ends of the

cell. The cell divides into two daughter cells. Each daughter cell is a gamete. Each gamete contains half the number of chromosomes of the original cell.

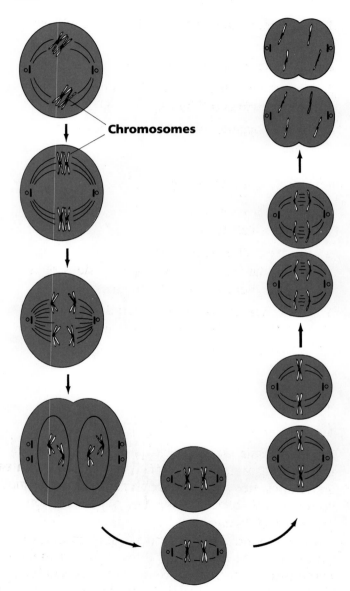

Chromosomes

The body cell of a fruit fly contains 8 chromosomes. After meiosis, there are only 4 chromosomes in each gamete. Your body cells contain 46 chromosomes. After meiosis, there are only 23 chromosomes in each gamete that forms.

▶ *Explain:* Why do gametes contain half the number of chromosomes as body cells?

LESSON SUMMARY

▶ Each body cell of an animal contains the same number of chromosomes.

▶ Reproductive cells are called gametes.

▶ Gametes are produced by meiosis.

▶ A gamete contains half the number of chromosomes of a body cell.

CHECK *Complete the following.*

1. All human body cells contain _____ chromosomes.

2. Sperm cells and egg cells are _____ cells.

3. The process that reduces the number of chromosomes in a cell by one-half is _____ .

4. Meiosis occurs only in the formation of _____ .

APPLY *Complete the following.*

5. How many chromosomes does a human egg cell contain? Why?

6. **Relate:** What would happen if the chromosomes did not move to opposite sides of the nucleus before the cell divided?

7. **Compare:** How are mitosis and meiosis similar? Hint: review the process of mitosis discussed in Lesson 4-6.

8. **Analyze:** Which of the gametes shown would be produced by an organism with 14 chromosomes in its body cells?

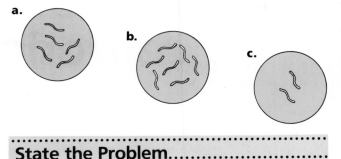

....
State the Problem ....

Study the illustrations. Then, state the problem.

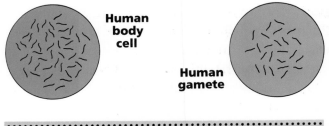

Human body cell

Human gamete

....
Skill Builder ....

▲ *Organizing Information* When you organize information, you put the information in some kind of order. Divide a sheet of paper into two columns labeled "Mitosis" and "Meiosis." Beneath each heading, list all the facts you have learned about each process.

◆○◆ **SCIENCE CONNECTION** ◆○◆○◆○◆○◆○◆○◆○◆○◆○◆○◆○◆○◆○◆

NONDISJUNCTION

During meiosis, chromosome pairs separate to form gametes. Usually, a gamete receives one chromosome from each pair. Sometimes, however, the chromosome pairs do not separate correctly. This condition is called nondisjunction (nahn-dis-JUNK-shun).

When nondisjunction occurs, both chromosomes in a pair go to the same gamete. As a result, one gamete has too many chromosomes and the other gamete has too few. If either of these gametes unites with a normal gamete during fertilization, the organism does not develop properly.

There are several conditions that result from nondisjunction. One of these conditions is Down's syndrome. The body cells of a person with Down's syndrome contain 47 chromosomes instead of 46. People with Down's syndrome can lead productive lives, but they usually have some degree of mental retardation.

What are the parts of the female reproductive system?

Objective ▸ Describe the organs of the female reproductive system.

TechTerms

▸ **cervix** (SUR-viks): narrow end of the uterus

▸ **ovaries** (OH-vuhr-eez): main organs of the female reproductive system

▸ **oviduct** (OH-vuh-dukt): long tube between the ovary and the uterus

▸ **uterus** (YOOT-ur-us): organ in which an embryo develops

▸ **vagina** (vuh-JY-nuh): birth canal

Reproductive Systems Most systems of the body are the same in males and females. This is not true of reproductive systems. The male and female reproductive systems are quite different. All the organs of the female reproductive system are located inside a female's body. In the male reproductive system, some organs are inside the body, while other organs are located outside the body. The main job of both the male and female reproductive systems is to produce offspring.

▸ *Compare:* How is the reproductive system different from other body systems?

Ovaries The **ovaries** (OH-vuhr-eez) are the main organs of the female reproductive system. Ovaries are egg-shaped structures. One ovary lies on each side of the female's body.

The ovaries contain two different kinds of cells. One kind of cell produces hormones. The other kind of cell produces eggs. Eggs are female sex cells.

▸ *Describe:* What do the ovaries produce?

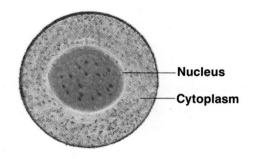

Figure 1 An egg cell

Pathway of Eggs You can see the structure of the female reproductive system in Figure 2. A long tube called an **oviduct** (OH-vuh-dukt) connects each ovary to the **uterus** (YOOT-ur-us). The uterus is a hollow organ with thick, muscular walls. It is inside this organ that an embryo develops.

As the uterus extends downward, it becomes narrower. The narrow end of the uterus is called the **cervix** (SUR-viks). The cervix connects the uterus to the **vagina** (vuh-JY-nuh). The vagina is the opening through which a baby passes during birth. The vagina also is called the birth canal.

▸ *Explain:* Why is the vagina called the birth canal?

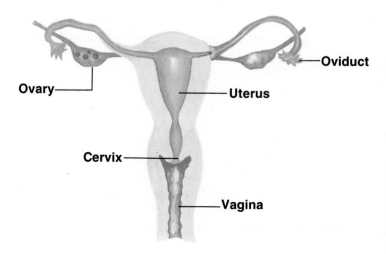

Figure 2 The female reproductive system

LESSON SUMMARY

▶ All the parts of the female reproductive system are located inside the female's body.

▶ The main organs of the female reproductive system are the ovaries.

▶ The ovaries produce hormones and eggs.

▶ An embryo develops inside the uterus.

▶ The vagina is the passageway through which a baby moves during birth.

CHECK *Answer the following.*

1. What are the main organs of the female reproductive system?

2. What are female sex cells called?

3. What is the name of the long tube that connects an ovary to the uterus?

4. What is the narrow end of the uterus?

5. What is another name for the vagina?

APPLY *Answer the following.*

6. **Relate:** Through which structures of the female reproductive system does a baby pass?

7. **Describe:** Where are the female reproductive organs located?

8. **Infer:** The walls of the uterus are flexible, and can expand and contract. How does this help a developing embryo?

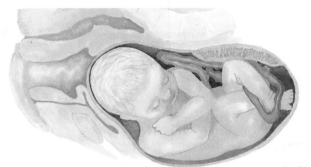

9. **Relate:** What happens to the size of the uterus as the embryo becomes larger?

10. **Infer:** Why is the vagina also called the birth canal?

11. **Applying definitions:** The term "ova" means "egg." A duct is a tube. Relate these definitions to the function of the oviduct.

Skill Builder

Making a Diagram A diagram is a way to organize information. Diagrams also illustrate relationships between the parts of a group. Make a diagram that shows how the parts of the female reproductive system are related. Label the parts of your diagram.

SCIENCE CONNECTION

ECTOPIC PREGNANCY

The joining together of a sperm cell and an egg cell is called fertilization (fur-tul-ih-ZAY-shun). Fertilization usually occurs in the oviducts. The fertilized egg then travels through the oviduct to the uterus. Once inside the uterus, the egg attaches to the wall of the uterus. Here the fertilized egg develops into an embryo.

Sometimes, a fertilized egg attaches to a structure outside the uterus. This condition is called an ectopic (ek-TAHP-ik) pregnancy. Ectopic pregnancies can occur in the oviducts, in the abdomen, or in the ovaries.

An ectopic pregnancy in the oviducts is called a tubal ectopic pregnancy. A tubal ectopic pregnancy is the most common ectopic pregnancy. This kind of ectopic pregnancy can be caused by a swelling of the oviducts. It also can be caused by scar tissue. As the embryo develops in the oviducts, it blocks the blood supply. This blockage causes severe pain in the pregnant woman.

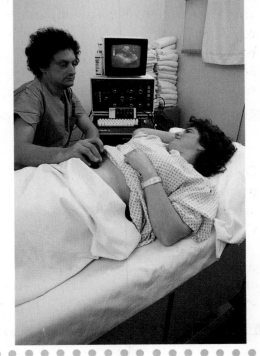

What are the parts of the male reproductive system?

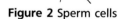

Figure 2 Sperm cells

Objective ▶ Describe the structures of the male reproductive system.

TechTerms

▶ **scrotum** (SKROH-tum): pocket of skin that protects and holds the testes

▶ **testes** (TES-teez): main organs of the male reproductive system

▶ **testosterone** (tes-TAWS-tuhr-ohn): hormone produced in the testes

▶ **urethra** (yoo-REETH-ruh): tube that carries sperm to the outside of the male's body

The Testes The main organs of the male reproductive system are the **testes** (TES-teez). The testes are egg-shaped structures located outside the body. The testes rest in a pocket of skin called the **scrotum** (SKROH-tum). The organs of the male reproductive system are shown in Figure 1.

▶ *Identify:* What are the main organs of the male reproductive system?

Testosterone Two kinds of cells are located in the testes. One type of cell produces **testosterone** (tes-TAWS-tuhr-ohn). Testosterone is a hormone that controls the development of secondary sex characteristics. In a male, these characteristics include a deeper voice and the growth of body hair. Testosterone also causes male sex cells to develop inside the testes. Male sex cells are called sperm.

▶ *Describe:* What substances are produced inside the testes?

The Urethra The **urethra** (yoo-REETH-ruh) is part of the excretory system. Urine made in the kidneys travels from the bladder to the outside of the body through the urethra.

In males, the urethra performs another job. It is the passageway through which sperm leaves a male's body. For this reason, the urethra is considered to be a part of the male reproductive system.

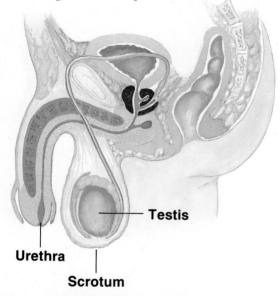

Testis

Urethra

Scrotum

Figure 1 The male reproductive system

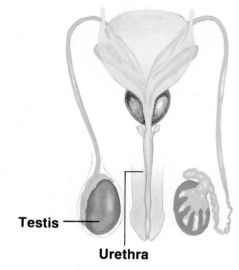

Testis

Urethra

▶ *Explain:* Why is the urethra part of two different body systems in males?

LESSON SUMMARY

▶ The testes are the main organs of the male reproductive system.

▶ Testosterone is a male hormone that triggers the development of secondary sex characteristics.

▶ The urethra is part of the excretory system.

▶ In males, both urine and sperm leave the body through the urethra.

CHECK *Complete the following.*

1. The main organs of the male reproductive system are the _____ .

2. The testes are suspended within a pocket of skin called the _____ .

3. One kind of cell in the testes produces the hormone _____ .

4. Male sex cells are called _____ .

5. In males, both sperm and urine leave the body through the _____ .

6. Two secondary sex characteristics in males are a deeper _____ and the growth of body hair.

7. The testes produce both testosterone and _____ .

APPLY *Complete the following analogies.*

8. **Relate:** Ovaries are to females as _____ are to males.

9. Eggs are to females as _____ are to males.

Answer the following.

10. **Compare:** How is the shape of the testes similar to the shape of the ovaries?

11. **Name:** To what two body systems does the urethra belong?

12. **Define:** What is a hormone? Use the glossary of this text for help if necessary.

13. **Define:** What is the scrotum?

Skill Builder

▲ *Organizing Information* When you organize information, you put the information in some kind of order. A table is one way to organize information. Make a table that describes the parts of the male reproductive system. Include a description of each part in the table.

SCIENCE CONNECTION ◆○◆○◆○◆○◆○◆○◆○◆○◆○◆○◆○◆○◆○◆○◆

SEXUALLY TRANSMITTED DISEASES

A disease that can be passed from one person to another is called a contagious (kun-TAY-jus) disease. Some contagious diseases are spread by pathogens (PATH-uh-junz), or germs, in the air. Other contagious diseases are spread by pathogens contained in food or water.

A few contagious diseases are spread only through sexual contact with an infected person. These diseases are called sexually transmitted (trans-MIT-ed) diseases. The pathogens that cause sexually transmitted diseases can live only inside a human body. When outside the body, these pathogens quickly die. As a result, sexually transmitted diseases are not spread by casual contact.

Gonorrhea and syphilis are two sexually transmitted diseases. These diseases usually are treated with penicillin or other antibiotics. Many cases of gonorrhea and syphilis are cured through early diagnosis and treatment.

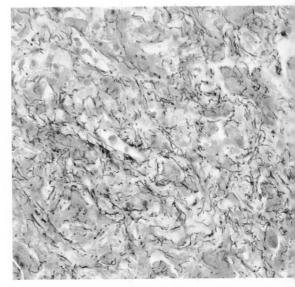

Figure 3 Syphilis

What is menstruation?

Objective ► Describe the menstrual cycle.

TechTerms

► **menstrual** (MEHN-struhl) **cycle:** monthly cycle of change that occurs in the female reproductive system
► **menstruation** (men-stroo-WAY-shun): process by which blood and tissue from the lining of the uterus break apart and leave the body
► **ovulation** (oh-vyuh-LAY-shun): release of a mature egg from the ovary

Menstrual Cycle When a female is born, her body contains all the egg cells she will ever have. However, the eggs are not mature, or fully developed. The eggs do not begin to mature until the female reaches puberty (PYOO-bur-tee). Puberty generally begins between the ages of 10 and 14. In females, puberty is marked by the beginning of the **menstrual** (MEHN-struhl) **cycle.** The menstrual cycle is a monthly cycle of change that occurs in the female reproductive system.

▶ *Identify:* What is the menstrual cycle?

Ovulation The menstrual cycle occurs every 28 to 32 days. It is triggered by the release of hormones in the female reproductive system. One hormone causes an egg to mature in an ovary. Another hormone causes the walls of the uterus to thicken and the supply of blood to the uterus to increase. Yet another hormone triggers **ovulation** (oh-vyuh-LAY-shun). Ovulation occurs when a mature egg leaves an ovary and travels into an oviduct.

▶ *Explain:* What causes ovulation?

Menstruation After an egg is released from the ovary, it travels through an oviduct. If the egg does not meet sperm in the oviduct, it begins to break apart. The amount of hormones in the reproductive system decreases. This causes the thickened walls of the uterus to also break apart. About 14 days after ovulation, **menstruation** (men-stroo-WAY-shun) occurs. Menstruation is the process by which blood and tissue from the uterus break apart and leave the body.

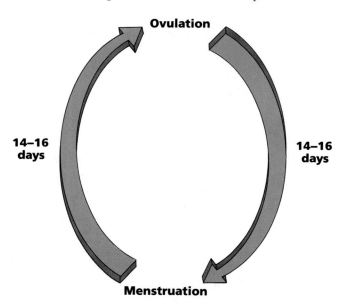

Ovulation

14–16 days   **14–16 days**

Menstruation

Soon after menstruation, a new egg begins to mature in the ovary. The menstrual cycle is repeated. The cycle is continuously repeated well into adulthood. For most females, the menstrual cycle continues until around age 50.

▶ *Identify:* At about what age does the menstrual cycle stop occurring?

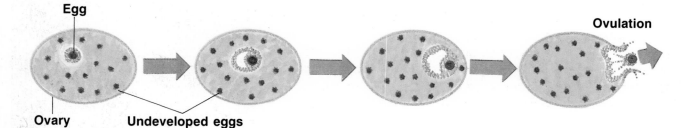

Egg

Ovulation

Ovary   **Undeveloped eggs**

LESSON SUMMARY

▶ The menstrual cycle is a series of changes in the female reproductive system that occur about once a month.

▶ During ovulation, a mature egg leaves an ovary and travels to the oviduct.

▶ The process by which blood and tissue lining leave the uterus is called menstruation.

CHECK *Complete the following.*

1. A female is born with all the _____ cells she will have in her lifetime.

2. The series of monthly changes in a female's reproductive system is called the _____ .

3. Ovulation is triggered by the release of _____ into parts of the reproductive system.

4. During _____, blood and tissue from the uterus are released from the body.

5. A mature egg moves from an ovary into an oviduct during _____ .

APPLY *Complete the following.*

6. **Hypothesize:** Could a female over age 50 continue to have a menstrual cycle? Explain.

7. **Compare:** What effect do hormones have on the development of sex cells in the male and female reproductive systems?

8. **Sequence:** The pictures below show how an egg changes before ovulation. Place the pictures of the changing egg into the correct order.

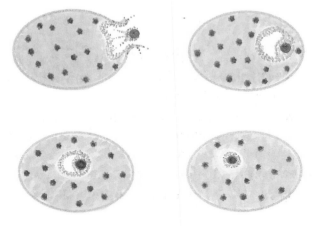

Skill Builder

Diagraming When you organize information, you put the information in some type of order. A diagram is one way to organize information. Draw a diagram of the female reproductive system. Indicate the path a mature egg takes through the system during menstruation.

SCIENCE CONNECTION

HORMONE THERAPY

In order to work properly, the body needs a certain amount of hormones. Sometimes, a person's body does not produce enough of one or more hormones. This is called a hormone deficiency. Certain diseases, such as diabetes, are caused by a hormone deficiency.

Hormone therapy is a way of treating people suffering from a hormone deficiency disease. In hormone therapy, patients are given doses of hormones that their bodies lack. As hormone levels become normal, the symptoms of the disease lessen. The patient's body works properly. However, hormone therapy does not cure the disease.

Addison's disease is a hormone deficiency disease. A person with Addison's disease does not have a proper amount of hormones made by the adrenal glands. As a result, the person may experience a loss of body fluids and muscle weakness. If left untreated, the disease can be fatal. However, through hormone therapy, the person can live a normal life.

17-5 How does fertilization take place?

Objective ▶ Describe how a sperm cell and egg cell join to form a zygote.

TechTerms

- ▶ **fertilization** (fur-tul-i-ZAY-shun): joining of one sperm cell and one egg cell
- ▶ **zygote** (ZY-gote): fertilized egg produced by fertilization

Sperm Cells Sperm cells are microscopic. The largest part of a sperm cell is a round part called the head. The head contains the cell nucleus. A sperm cell also has a long tail. The motion of the tail helps the sperm cell move. This helps the sperm cell reach the egg.

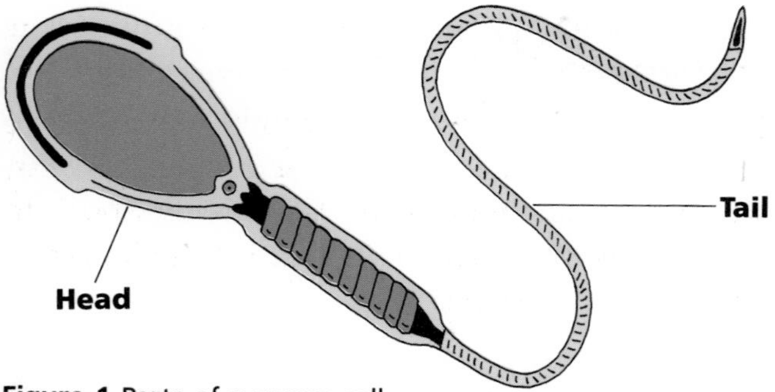

Figure 1 Parts of a sperm cell

▶ ***Describe:*** How does a sperm cell move?

Mature Egg During ovulation, a mature egg is released from an ovary. The egg passes into the oviduct. Tiny hairs line the walls of the oviduct. The motion of these hairs move the egg through the oviduct.

▶ ***Explain:*** What causes an egg to move through the oviduct?

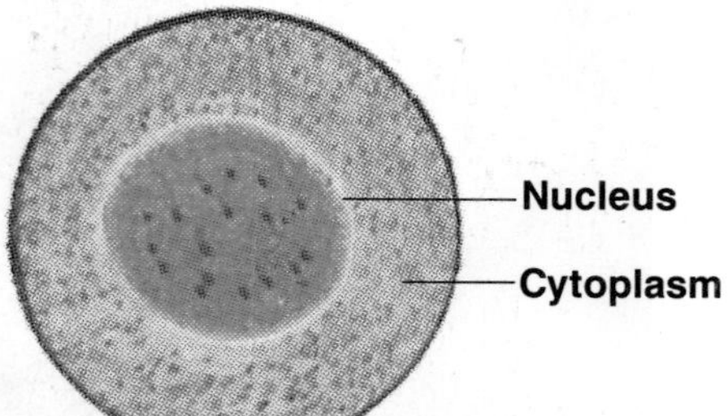

Fertilization Sperm cells enter the female reproductive system through the uterus. The sperm cells move across the uterus and into the oviduct. Of the millions of sperm cells that enter the uterus, only a few may reach the oviduct.

During ovulation, an egg leaves an ovary and enters the oviduct. If a sperm cell meets an egg cell in the oviduct, **fertilization** (fur-tul-i-ZAY-shun) can occur. Fertilization is the joining of one sperm and one egg. An egg can be fertilized only during ovulation.

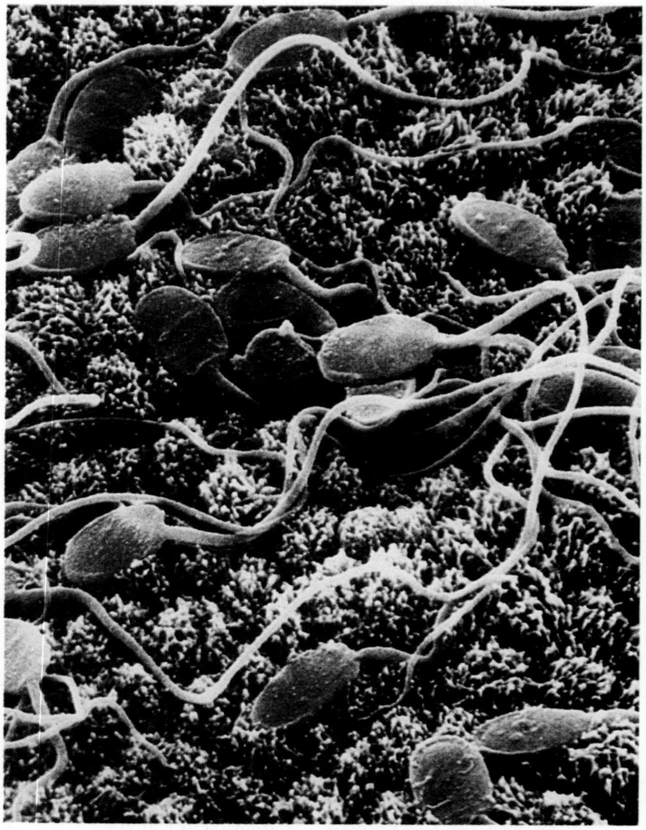

▶ ***Identify:*** What is fertilization?

Zygote During fertilization, the nuclei of a sperm cell and an egg cell join together. The new cell that results from fertilization is called a **zygote** (ZY-gote). A zygote is a fertilized egg. The zygote travels down the oviduct and enters the uterus.

▶ ***Explain:*** What causes a zygote to form?

LESSON SUMMARY

▶ Sperm cells have a head that contains the cell nucleus, and a tail that helps the sperm cell move.

▶ A mature egg is released from the ovary, and passes through the oviduct.

▶ Only a few sperm cells reach the oviduct.

▶ Fertilization can occur only during ovulation.

▶ The new cell produced by fertilization is called a zygote.

CHECK *Complete the following.*

1. A sperm cell moves due to the motion of its _____ .

2. During _____, a mature egg leaves an ovary and travels into the oviduct.

3. The joining of a sperm cell and an egg cell is called _____ .

4. The new cell produced by fertilization is called a _____ .

5. Fertilization occurs in the _____ of a female.

APPLY *Complete the following.*

6. **Relate:** During which stage of a female's menstrual cycle can fertilization occur? Why?

7. **Contrast:** How is the motion of sperm cells and egg cells different?

Answer the questions about the diagram.

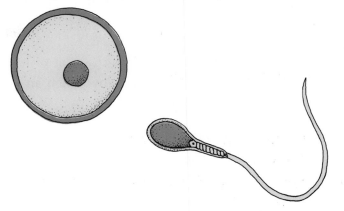

8. **Analyze:** The sperm cell shown is a human gamete. How many chromosomes does the sperm cell contain?

9. How many chromosomes does the human egg cell contain? Explain your answer.

Health & Safety Tip..........................

The length of a woman's menstrual cycle can be affected by outside factors such as illness, stress, weight loss, or weight gain. Use a health book to find out two other factors that affect the length of a woman's menstrual cycle.

TECHNOLOGY AND SOCIETY

IN VITRO FERTILIZATION

In order for fertilization to occur, a mature egg must first be released into the oviduct. Certain disorders of the female reproductive system prevent ovulation. Females suffering with these disorders are unable to release eggs contained in their ovaries. As a result, fertilization by natural means is impossible.

In recent years, scientists have developed medical procedures to help the fertilization process. One technique is called *in vitro* fertilization. In this procedure, eggs are removed from a woman's ovaries. The eggs are taken to a laboratory and mixed with sperm cells. Fertilization, or joining of egg and sperm cells, occurs in the laboratory. The fertilized eggs are then placed inside the woman's uterus. Chances are that one of the fertilized eggs may develop into a normal fetus. *In vitro* fertilization has produced a number of healthy babies. These babies sometimes are called "test tube babies."

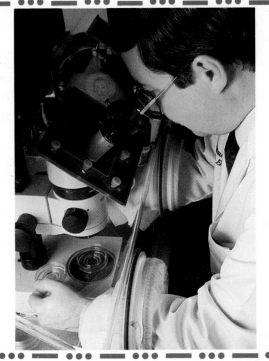

How does a human embryo develop?

Objective ► Describe the process by which an embryo develops into a fetus.

TechTerms

► **amnion** (AM-nee-on): fluid-filled membrane that surrounds an embryo
► **embryo** (EM-bree-oh): hollow ball of cells formed by cell division of the zygote
► **fetus** (FEET-us): developing baby
► **placenta** (pluh-SEN-tuh): organ through which an embryo receives nourishment and gets rid of wastes
► **umbilical** (um-BIL-ih-kul) **cord:** connects the embryo to the placenta

Early Changes After fertilization, the zygote divides by mitosis. Two cells are formed. These cells are attached to one another. Both of these cells divide to form four attached cells. This cell division continues until a hollow ball of cells is formed. The hollow ball of cells attaches itself to a wall of the uterus. This mass of cells is now called an **embryo** (EM-bree-oh). All of the tissues and organs of the body form from the cells in the embryo.

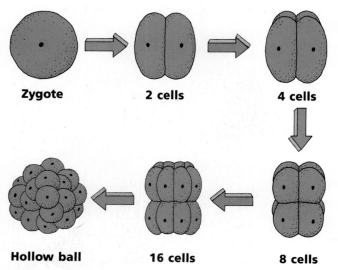

Zygote   **2 cells**   **4 cells**

Hollow ball   **16 cells**   **8 cells**

▌▶ *Identify:* What is an embryo?

Needs of the Embryo Tissues that surround the embryo develop into a thick flat structure called the **placenta** (pluh-SEN-tuh). The placenta is an organ through which the embryo receives nourishment. The embryo also gets rid of wastes through the placenta.

The embryo is attached to the placenta by the **umbilical** (um-BIL-ih-kul) **cord.** The umbilical cord is a thick, ropelike structure. One kind of blood vessel in the umbilical cord carries nourishment from the placenta to the embryo. A different kind of blood vessel carries wastes from the embryo to the placenta.

▌▶ *Describe:* How does an embryo receive nourishment?

Protection The embryo is surrounded by a clear, fluid-filled sac. This sac is called the **amnion** (AM-nee-on). The fluid inside the sac cushions and protects the developing embryo.

▌▶ *Explain:* How does the amnion help the developing embryo?

Fetus After about eight weeks, the embryo begins developing a heart, brain, and nerve cord. Eyes and ears also begin to form. When bone forms in the organism's skeleton, it is called a **fetus** (FEET-us). The fetus continues to grow and develop inside the uterus. Finally, about nine months after fertilization, a baby is born.

▌▶ *Describe:* When is a developing organism called a fetus?

LESSON SUMMARY

► An embryo is a hollow ball of cells attached to a wall of the uterus.

► An embryo receives nourishment and rids itself of wastes through the placenta.

► The umbilical cord connects an embryo to the placenta.

► The amnion cushions and protects the developing embryo.

CHECK *Complete the following.*

1. After fertilization, a zygote divides by _____ .

2. All the tissues and organs of the body form from the cells of the _____ .

3. The umbilical cord connects the embryo to the _____ .

4. When the skeleton of an embryo is filled with bone, the embryo is called a _____ .

5. The embryo receives nourishment and gets rid of wastes through the _____ .

APPLY *Complete the following.*

6. **Relate:** Identify the stages of development that occur after an egg is fertilized.

7. Why does one type of blood vessel bring nutrients to the embryo while another removes its waste products?

Ideas in Action..................................

IDEA: Parts of a chicken egg contain structures for a developing embryo much like the structures that surround a human embryo.

ACTION: Study the parts of an uncooked egg. Identify structures of the egg that resemble structures that surround a developing embryo.

ACTIVITY

GRAPHING CHANGES IN FETAL DEVELOPMENT

You will need graph paper and a pencil.

1. Study the information in the Data Table.

2. Make a bar graph that shows how the length of a baby changes as it develops.

3. Find the difference between the shortest and longest lengths in the Data Table. The difference is the total length that needs to be represented on the vertical axis. Use this information to determine the length one box on the axis represents.

4. Label each column along the horizontal axis with the time it represents.

5. Fill in the column above each time label with a bar whose top ends at the appropriate length line.

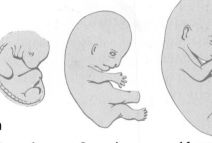

5 weeks   9 weeks   14 weeks

Questions

1. How long is the embryo at the end of 9 weeks of development?

2. Between which weeks does the developing baby have the greatest increase in length?

3. **Analyze:** Does the length of a developing baby increase in a set pattern? Explain.

Data Table:	Growth of a Developing Baby	
TIME (WEEKS)	MASS (g)	LENGTH (cm)
5	0.5	1.3
9	10	5
14	113	18
26	907	38
38	3180	51

What are the stages of human development?

Objective ▶ Identify the stages of the human life cycle.

Life Cycle A developing baby goes through a series of stages before birth. After birth, a human also goes through a series of stages. The stages of development are called a life cycle. There are five stages in the human life cycle.

▶ *Identify:* What is a life cycle?

Early Years The earliest stage of human life is called infancy. Infancy begins at birth and ends at age 2. This stage of life is marked by a rapid increase in size. The muscles and nerves of the infant also develop quickly. Mental skills develop and the infant begins to interact with its surroundings. By age 2, most infants are able to walk and speak a few words.

Childhood usually is defined as between ages 2 and 12. During childhood, muscle development allows more complex activities. Children become more independent and are able to feed and dress themselves. Mental abilities also increase. Most children learn to read and write during this stage of development.

▶ *Describe:* What are some of the changes a person goes through during childhood?

Adolescence Between the ages of 11 and 14, most young people go through a period of rapid physical change. This stage is called adolescence (ad-ul-ES-uns). The beginning of adolescence is called puberty (PYOO-bur-tee). During puberty, the sex organs develop. These organs release hormones that cause growth spurts. Adolescents of both sexes develop the ability to reproduce.

▶ *Explain:* What causes rapid growth spurts during puberty?

Later Years Adulthood is the stage at which the physical growth of the human body is complete. Muscle development and coordination reach their peak during early adulthood. Between the ages of 30 and 50, muscle tone often decreases.

Old age, or the beginning of the aging process, occurs at different times in different people. People who have exercised regularly and eaten a balanced diet all of their lives may not show signs of aging until their late 70s or early 80s. Old age usually is marked by a decline in muscle strength. Sense organs such as the eyes and ears may not work as well. The bones of older adults often become brittle and can break more easily.

▶ *Describe:* What effect do diet and exercise have on the aging process?

LESSON SUMMARY

▶ The stages of development in humans is called the human life cycle.

▶ The earliest stage of human life is called infancy.

▶ The second stage in the human life cycle is childhood.

▶ During adolescence, a person goes through a period of rapid physical change.

▶ Adulthood is the stage at which the physical growth of the body is complete.

▶ Old age is the last stage of the human life cycle.

CHECK *Complete the following.*

1. The human life cycle consists of _____ stages.

2. The earliest stage of human development is called _____ .

3. The period between ages 2 and 12 is generally called _____ .

4. A person develops the ability to reproduce during the stage called _____ .

5. The stage during which the physical growth of the human body is complete is _____ .

APPLY *Complete the following.*

6. **Relate:** Usually, a very young infant does not cry in the presence of a stranger, but a 9-month old will scream for a parent. What does this behavior indicate about the development of the infant?

7. **Compare:** What is a general statement that compares muscle development during infancy with that of old age?

8. **Relate:** At what stage in the life cycle is a person most likely to first put together a jigsaw puzzle?

Skill Builder

🔺 *Organizing Information* When you organize information, you put the information in some kind of order. A table is one way of organizing information. Make a table with five columns. Label the columns with the stages of the human life cycle. Then list the traits of each stage under the appropriate heading.

CAREER IN LIFE SCIENCE

DAY CARE WORKER

Day care workers supervise groups of infants or young children. They carry out educational plans under the supervision of a certified teacher. With a group of infants, this means feeding and changing the babies on schedule. It also means creating an environment that is stimulating to the infants. In groups of young children, the plan may call for reading a book, instructing a craft, or supervising free play.

Day care workers generally work in a center for young children. A day care worker must be familiar with the general stages of emotional, social, physical, and intellectual skills that a child goes through. Day care workers also should be patient and understanding. To be a day care worker you need a high school diploma. Many day care workers also have a two-year degree in Early Childhood education.

For more information, write to the Day Care and Child Development Council of America, 1401 K St., N.W., Washington, D.C. 20005.

STUDY HINT Before you begin the Unit Challenges, review the TechTerms and Lesson Summary for each lesson in this unit.

TechTerms..

amnion (346)	menstrual cycle (342)	testes (340)
cervix (338)	menstruation (342)	testosterone (340)
embryo (346)	ovaries (338)	umbilical cord (346)
fertilization (344)	oviduct (338)	urethra (340)
fetus (346)	ovulation (342)	uterus (338)
gamete (336)	placenta (346)	vagina (338)
meiosis (336)	scrotum (340)	zygote (344)

TechTerm Challenges..

Matching *Write the TechTerm that matches each description.*
1. process by which gametes form
2. hollow organ in which an embryo develops
3. process by which blood and tissue from the uterine lining breaks apart and leaves the body
4. clear, fluid-filled sac that protects the developing embryo
5. developing baby
6. organ through which an embryo receives nourishment and gets rid of waste
7. hollow ball of cells formed by cell division of the zygote
8. fertilized egg
9. main organs of the male reproductive system
10. long tube between an ovary and the uterus
11. thick, ropelike structure

Fill in *Write the TechTerm that best completes each statement.*
1. Reproductive cells also are called _____ .
2. The main organs of the female reproductive system are the _____ .
3. The pocket of skin that surrounds and protects the testes is called the _____ .
4. Sperm is carried to the outside of the body through the _____ .
5. The release of a mature egg from the ovary is called _____ .
6. The joining of one sperm cell and one egg cell is called _____ .
7. The start of the _____ marks the beginning of puberty in females.
8. The hormone that controls the development of male characteristics is _____ .
9. The _____ often is called the birth canal.
10. The _____ connects the uterus to the vagina.

Content Challenges..

Multiple Choice *Write the letter of the term that best completes each statement.*
1. In females, puberty begins with the start of
   a. diffusion. **b.** fertilization. **c.** menstruation. **d.** pregnancy.
2. Muscular strength begins to decrease during
   a. infancy. **b.** adolescence. **c.** adulthood. **d.** old age.

3. The release of an egg is called
   a. fertilization.   b. ovulation.   c. menstruation.   d. pregnancy.

4. The structure that performs the jobs of excretion, respiration, and circulation for the developing fetus is called the
   a. amnion.   b. uterus.   c. birth canal.   d. placenta.

5. Fertilization in humans takes place in the
   a. vagina.   b. uterus.   c. ovary.   d. oviduct.

6. Sperm are stored in the
   a. urethra.   b. vas deferens.   c. epidymis.   d. testes.

7. Sperm and testosterone are produced in the
   a. epidymis.   b. scrotum.   c. testes.   d. urethra.

8. The ovaries produce all of the following EXCEPT
   a. eggs.   b. progesterone.   c. estrogen.   d. testosterone.

9. The most rapid physical changes take place in humans during
   a. infancy.   b. adolescence.   c. adulthood.   d. old age.

10. The narrow end of the uterus is called the
    a. vagina.   b. cervix.   c. oviduct.   d. ovary.

True/False *Write true if the statement is true. If the statement is false, change the underlined term to make the statement true.*

1. Once bone replacement is complete, the embryo is called a <u>fetus</u>.
2. The <u>urethra</u> carries material between the embryo and the placenta.
3. The process by which blood and tissue leave the uterus is called <u>fertilization</u>.
4. Reproductive cells are sometimes called <u>gametes</u>.
5. Gametes are formed by <u>mitosis</u>.
6. The growth process in humans ends at <u>adolescence</u>.
7. Eggs are <u>female</u> sex cells.
8. An egg can be fertilized by <u>many</u> sperm.
9. The clear fluid-filled membrane surrounding the embryo is called the <u>placenta</u>.
10. A <u>zygote</u> undergoes cell division until a hollow ball of cells forms.
11. People develop the ability to reproduce during <u>adulthood</u>.
12. Humans reproduce by <u>asexual</u> reproduction.

Understanding the Features..

Reading Critically *Use the feature reading selections to answer the following. Page numbers for the features follow each question in parentheses.*

1. **Define:** What is nondisjunction? (337)
2. **Infer:** Why is an ectopic pregnancy dangerous? (339)
3. **Name:** What are three sexually transmitted diseases? (341)
4. How is hormone therapy useful in treating Addison's disease? (343)
5. **Analyze:** How do 'test-tube babies' get their name? (345)
6. **Infer:** Why do day care workers need to be patient and understanding? (349)

Interpreting a Diagram *Use the following diagram of meiosis to complete the following.*

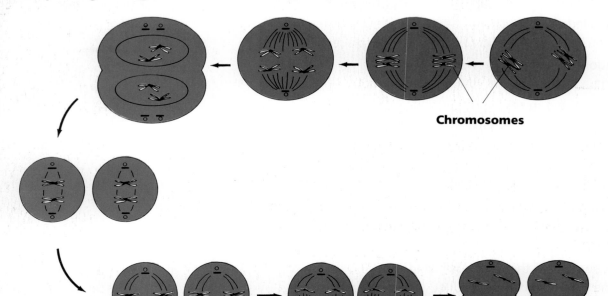

Chromosomes

👁 **1. Observe:** How many cells are formed when meiosis is complete?
👁 **2. Observe:** How many chromosomes are in each cell when meiosis is complete?
3. In what stage does the nucleus break down?
4. Describe: What is happening in Step 7?

Critical Thinking *Answer each of the following in complete sentences.*

1. Compare: How are the functions of the testes and ovaries similar?

▶ **2. Infer:** Why do you think it is important for a pregnant woman to avoid alcohol, tobacco, and drugs?

3. Identify: How do you think you can slow down the effects of aging?

4. Hypothesize: What would happen if the placenta became detached from the fetus?

▶ **5. Infer:** Why do you think the regular monthly changes in females is called a cycle?

6. Contrast: What is the difference between meiosis and mitosis?

1. Find out some of the dangers that are associated with teenage pregnancy. Use library references to find out any complications for the mother as well as for the child. Present your findings in a report.

2. During the 1950s, doctors prescribed a drug called thalidomide for morning sickness. Research what effect thalidomide had on the developing fetus. Report your findings to the class.

3. Use library references to research some of the ways technology is being used to help couples

who have trouble conceiving. Write a report on one of these techniques. Include a description of the technique, the costs involved, and the success rate of the technique. Present your findings in an oral report.

4. Use library references to make a time line showing the different stages in a developing fetus. Break the time line into trimesters. Include when organs are formed, when the arms and legs form, when the heart and brain form, and when the fetus begins to rapidly gain weight.

HEREDITY AND GENETICS

CONTENTS

STUDY HINT As you read each lesson in Unit 18, write the topic sentence of each paragraph in the lesson on a sheet of paper. After you complete each lesson, compare your list of topic sentences to the Lesson Summary.

What is heredity?

Objective ► Explain why offspring have some of the traits of their parents.

TechTerms

► **genetics** (juh-NET-iks): study of heredity

► **heredity** (huh-RED-uh-tee): passing of traits from parents to offspring

► **inherited** (in-HER-uh-ted) **traits:** traits that are passed from parents to their offspring

► **traits:** characteristics

Traits Suppose you were asked to make a list of ten characteristics that describe your appearance. Your list might include: brown eyes, black hair, 5 feet tall. It is unlikely that another student would list the same exact characteristics. This is because each person is unique. The characteristics you might include on your list are called **traits.** Traits are characteristics of an organism.

▐▐▐▶ *Define:* What are traits?

Identifying Traits Do you look like your mother, your father, or a mixture of both your parents? Children often look like their parents or grandparents in some way.

There is a reason for this. During fertilization, male and female sex cells join together. Each of these sex cells contains material that affects the development of the offspring. Traits that are passed from parents to their offspring are called **inherited** (in-HER-uh-ted) **traits.** Eye color is an example of an inherited trait.

► *Infer:* How was your eye color determined?

Heredity The passing of traits from parents to their offspring is called **heredity** (huh-RED-uh-tee). The first person to study heredity was Gregor Mendel, an Austrian monk. Mendel studied how traits are passed from one generation to the next. The field of biology that studies heredity is called **genetics** (juh-NET-iks). Gregor Mendel often is called the "father of genetics."

▐▐▐▶ *Describe:* What is the relationship between traits and genetics?

LESSON SUMMARY

► Traits are the characteristics of an organism.
► Traits that are passed from parents to their offspring are called inherited traits.
► The passing of traits from parents to offspring is called heredity.

CHECK *Complete the following.*

1. Characteristics of an organism are called _____ .

2. An example of an inherited trait is _____ .

3. Inherited traits are passed from _____ to their offspring.

4. The first scientist to study inherited traits was _____ .

5. The modern science of heredity is called _____ .

APPLY *Answer the following.*

► 6. **Infer:** Why do brothers and sisters often have similar traits?

► 7. **Analyze:** Two children have the same last name, the same birthdate, and the same inherited traits. How are the children related?

► 8. **Infer:** Why do you think the study of genetics is important?

InfoSearch

Read the passage. Ask two questions that you cannot answer from the information in the passage.

Human Genetics Human genetics is a difficult field to study. One difficulty is that the life span of humans is long compared with the life span of other animals. As a result, a scientist could not study all of the offspring produced in many generations of one family. Another difficulty is that humans produce fewer offspring than other animals. It is hard to compare traits with fewer offspring to observe. Scientists often study the genetics of plants and animals to learn more about human genetics.

SEARCH: Use library references to find answers to your questions.

◄◊►◄ PEOPLE IN SCIENCE ►◄◊►◄◊►◄◊►◄◊►◄◊►◄◊►◄◊►◄◊►◄◊►◄◊►◄

GREGOR MENDEL (1822–1884)

In the mid-1800s, a young monk named Gregor Mendel took care of a monastery garden in Czechoslovakia. Mendel noticed that the pea plants growing in the garden had a variety of characteristics. Some plants had tall stems. Other plants had short stems. The seeds of some plants were round; others were wrinkled. Mendel wondered what caused the plants to have these different characteristics.

Mendel began to experiment with the pea plants. Because pea plants grow and reproduce quickly, Mendel was able to observe many generations of the pea plants. From his experiments, Mendel concluded that traits were passed from the parent plants to their offspring. The results of Mendel's experiments were published in 1866. However, it was not until 1900 that scientists began to analyze Mendel's research. Since that time, many new discoveries have been made about heredity. However, Mendel's original hypothesis still forms the basis of modern genetics.

What are genes and chromosomes?

Objective ▶ Describe how genes and chromosomes are involved in heredity.

TechTerms

▶ **chromosomes** (KROH-muh-sohms): thread-like structures in the nucleus of a cell that control heredity

▶ **genes** (JEENS): parts of a chromosome that control inherited traits

Threads of Life Fine, threadlike structures are located in the nucleus of a cell. These threadlike structures are called **chromosomes** (KROH-muh-sohms). Chromosomes control heredity. Chromosomes sometimes are called the "threads of life" because they control heredity during cell division. During cell division, each chromosome makes a copy of itself. A pair of identical chromosomes is formed. Each new daughter cell receives one chromosome from each of the pairs.

▶ **Define:** What are chromosomes?

Chromosomes and Reproduction In both asexual and sexual reproduction, chromosomes are passed from parent to offspring. During asexual reproduction, each daughter cell receives its chromosomes from the single parent. During sexual reproduction, the daughter cell receives chromosomes from each parent cell. All new organisms contain chromosomes from both of its parents.

▶ **Explain:** How does an organism produced by sexual reproduction get its chromosomes?

Chromosomes and Genes Along each chromosome, there are many dark bands. Each band is a small part of the chromosome called a **gene.** A gene is a part of the chromosome that controls inherited traits. Each gene affects a different trait. Genes determine your height, eye color, hair color, and many other characteristics. Each gene is located at a certain place on the chromosome. Genes also control the life processes of your cells.

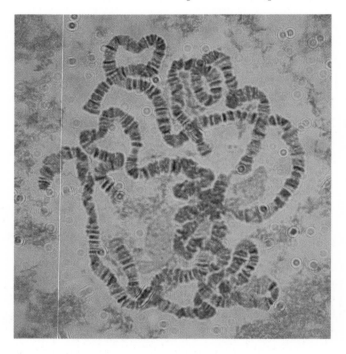

▶ **Identify:** What do genes control?

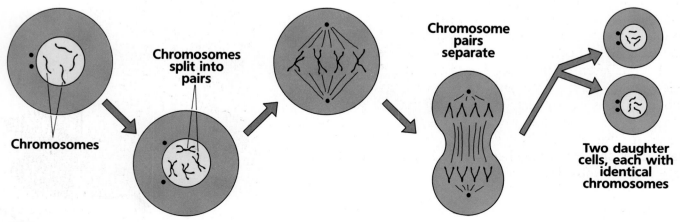

Chromosomes

Chromosomes split into pairs

Chromosome pairs separate

Two daughter cells, each with identical chromosomes

LESSON SUMMARY

▶ Chromosomes are threadlike structures in the nucleus of a cell that control heredity.

▶ An organism receives chromosomes from its parent or parents.

▶ Genes control inherited traits.

CHECK *Answer the following.*

1. Where are chromosomes located?

2. Where does an organism produced by sexual reproduction receive its chromosomes from?

3. Where are genes located?

4. What are the life processes of cells controlled by?

5. What kind of traits are controlled by genes?

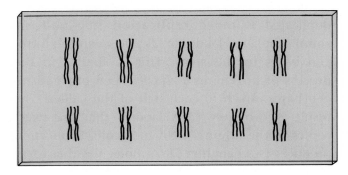

APPLY *Answer the following.*

6. **Infer:** Which organism would more closely resemble its parent or parents, one produced by asexual reproduction or one produced by sexual reproduction? Why?

7. **Analyze:** Suppose the chromosomes in a cell did not double to form pairs of identical chromosomes during cell division. What effect would this have on the daughter cells?

Ideas in Action

IDEA: Your inherited traits are controlled by the genes that you received from your parents. Eye color, eye shape, hair color, nose shape, and skin color all are inherited traits.

ACTION: Divide a piece of paper into four columns. Label the first column "Traits." In the first column, list ten inherited traits. Label the second column "Me," the third column "Mother," and the fourth column "Father." Then in each of the columns, describe the traits for each individual. Circle the traits that you share with one or both of your parents. Analyze your findings to see which parent you most closely resemble.

◆▷■◁ PEOPLE IN SCIENCE ▷■◁◇▷■◁◇▷■◁◇▷■◁◇▷■◁◇▷■◁◇▷■◁◇▷■◁◇▷■◁◇▷■◁

BARBARA MCCLINTOCK (1902-1992)

Barbara McClintock is an American plant geneticist (juh-NET-uh-sist). For more than 60 years, she has studied the inherited traits of corn. Dr. McClintock was especially interested in the color of corn kernels. By careful experimentation, she identified the genes that control this trait.

In 1931, Dr. McClintock made a startling discovery. She found that some genes "jump," or change position on a chromosome. The "jumping" gene caused changes in the traits controlled by the genes next to the place where it landed. For example, when a "jumping" gene landed next to the gene that controlled kernel color, the kernels were speckled.

Dr. McClintock announced her results in 1951. Most scientists at this time did not take her work seriously. They thought genes always remained in a definite position on a chromosome. It was not until the 1970s that other scientists began to take note of her conclusions. In 1983, more than 50 years after she announced her discoveries, Dr. McClintock received the Nobel Prize.

How do chromosomes carry traits?

Objective ▶ Describe the chemical makeup of chromosomes.

TechTerms

- ▶ **DNA:** chemical contained in chromosomes
- ▶ **replication** (rep-luh-KAY-shun): process by which DNA is duplicated

Composition of Chromosomes Many years of research were needed before scientists learned that chromosomes contained genes. Scientists then wondered what chemical makes up chromosomes. Their question was answered with the discovery of **DNA.** DNA stands for deoxyribonucleic (dee-oks-ee-ry-boh-noo-KLEE-ik) acid. DNA is the chemical that makes up chromosomes.

▶ **Define:** What is DNA?

Structure of DNA A molecule of DNA looks like a twisted ladder. The sides of the DNA ladder are made up of sugars and phosphates. The steps of the DNA ladder are made up of four kinds of nitrogen bases.

The four nitrogen bases that make up a DNA molecule are adenine (AD-uh-neen), guanine (GWAH-neen), thymine (THY-meen), and cytosine (SYT-oh-seen). Scientists use the capital letters A, G, T, and C to represent each of the four nitrogen bases. For example, A stands for adenine. Each step on the DNA ladder is made up of two bases. The bases always join together in certain pairs. Base A always pairs with base T. Base G always pairs with base C.

▶ **Describe:** How are nitrogen bases arranged in a DNA molecule?

Genetic Code A single DNA molecule, or ladder, can have thousands of steps. The number and arrangement of these steps form a genetic "code."

This code determines the kind of gene that forms. Different genes determine different kinds of inherited traits of an organism.

▶ **Explain:** What determines the inherited traits of an organism?

Replication During cell division, each chromosome doubles to form a pair of identical chromosomes. Molecules of DNA in the parent chromosome also double. The process by which DNA is duplicated is called **replication** (rep-luh-KAY-shun). The DNA ladder breaks between the nitrogen bases in the steps. This is similar to the process of unzipping a zipper. Then other nitrogen bases attach to each half of the ladder. The result is two new DNA ladders that are exact copies of the original DNA molecule. Replication produces two daughter DNA molecules that carry the same genetic code as the parent molecule.

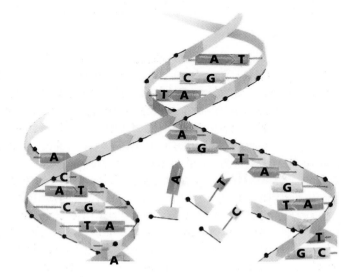

▶ **Define:** What is replication?

LESSON SUMMARY

- Chromosomes are made up of DNA.
- A molecule of DNA contains sugars, phosphates, and four nitrogen bases.
- The four nitrogen bases always join together in certain pairs.
- The genetic code controls inherited traits.
- DNA duplicates itself in a process called replication.

CHECK *Find the sentence in the lesson that answers each question. Then, write the sentence.*

1. What is the main substance in chromosomes?
2. What are the steps of a DNA ladder made up of?
3. What base does base G always pair with?
4. What does the genetic code determine?
5. How does DNA duplicate itself?

APPLY *Answer the following.*

6. **Predict:** Suppose DNA in the chromosomes of a cell did not replicate. What effect would this have on the daughter cells?

7. **Hypothesize:** Two developing human embryos contain the same genetic code. How will these offspring resemble one another?

Use the art showing replication on page 358 to answer the following.

8. **Observe:** What are the sides of the DNA ladder made up of?

9. **Observe:** Can base C ever pair with base T? Explain your answer.

InfoSearch.....................................

Read the passage. Ask two questions that you cannot answer from the information in the passage.

Mutations Sometimes a DNA molecule does not replicate itself exactly. The chromosomes do not pair correctly. Such changes in genes and chromosomes are called mutations (myoo-TAY-shuns). Most mutations are harmful to organisms. However, some mutations make an organism more suited to its environment. The mutation can help the organism survive. Many scientists think that helpful mutations lead to the development of new kinds of organisms.

SEARCH: Use library references to find answers to your questions.

LOOKING BACK IN SCIENCE

DISCOVERING THE STRUCTURE OF DNA

The structure of DNA was discovered in 1953 by an American biologist, James Watson, and a British scientist, Francis Crick. Watson and Crick built a model of a DNA molecule. Their model showed how the atoms in a DNA molecule resemble a twisted ladder. Watson and Crick also discovered that the order of nitrogen bases in the steps of the DNA molecule determine the particular traits controlled by the genes.

The discovery of Watson and Crick is considered one of the greatest scientific breakthroughs of the twentieth century. However, Watson and Crick were not the only scientists working with DNA. Watson and Crick based their discovery on the work of another British scientist, Rosalind Franklin. Franklin had taken many X-ray photographs of DNA crystals. Based on these photographs, Franklin predicted the shape and composition of a DNA molecule.

Why can offspring differ from their parents?

Objective ▶ Explain the difference between dominant and recessive traits.

TechTerms

▶ **dominant** (DOM-uh-nuhnt) **gene:** stronger gene that always shows itself

▶ **hybrid** (HY-brid): having two unlike genes

▶ **pure:** having two like genes

▶ **recessive** (ri-SES-iv) **gene:** weaker gene that is hidden when the dominant gene is present

Genes and Traits Gregor Mendel hypothesized that inherited factors produced certain traits. These factors are now called genes. Mendel based his hypothesis on the pea plants he was studying. Some of the plants were tall. Others were short. Mendel decided there must be a gene for tallness and another gene for shortness. The trait of height for a pea plant is determined by two genes, one from each parent. Each trait of an organism is determined by at least one gene from each parent.

▶ *Describe:* How are traits determined?

Pure Plants Mendel found that one kind of tall pea plant always had tall offspring. These plants have two tall genes, one from each parent. Mendel called these plants, **pure** tall. An organism is pure if it has two like genes for a given trait. Mendel found that some short pea plants always have short offspring. These plants have two short genes, one from each parent. Mendel called these plants pure short.

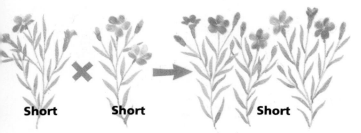

Short   **Short**   **Short**

▶ *Identify:* What genes does a short pea plant always have?

Hybrids Mendel wondered what the offspring of one pure tall parent and one pure short parent would look like. He cross-pollinated a pure tall pea plant and pure short pea plant. All of the offspring were tall. They had received a tall gene from the tall parent and a short gene from the short parent. They had one tall gene and one short gene and grew up tall. Mendel called these plants hybrids (HY-bridz). Organisms that have two unlike genes for a trait are **hybrid** for that trait.

▶ *Contrast:* What is the difference between a pure tall plant and a hybrid tall plant?

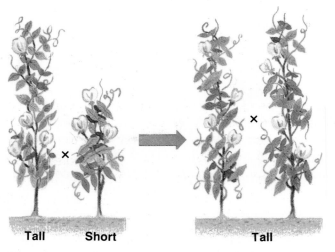

Tall   **Short**   **Tall**

Dominant Genes Mendel found that in a hybrid, one gene always shows itself. The other gene is hidden. Mendel called the gene that shows itself the **dominant** (DOM-uh-nuhnt) **gene.** He called the hidden gene a **recessive** (ri-SES-iv) **gene.** The tall gene is dominant in pea plants. The short gene is recessive. That is why plants with one tall gene and one short gene are always tall.

▶ *Compare:* What is the difference between a dominant and a recessive gene?

LESSON SUMMARY

▶ Each trait is determined by at least one gene from each parent.

▶ An organism with two like genes for a trait is called pure for that trait.

▶ An organism with two unlike genes for a trait is called hybrid for that trait.

▶ A dominant gene always shows itself over a recessive gene.

CHECK *Complete the following.*

1. A pea plant with two tall genes is called _____ .

2. A hybrid tall pea plant has one tall gene and one _____ gene.

3. The gene that always shows itself is called the _____ gene.

4. The recessive gene for height in pea plants is _____ .

5. All short pea plants have two _____ genes.

APPLY *Answer the following.*

6. **Relate:** What combinations of genes could a tall pea plant have?

7. **Predict:** Could a short pea plant be hybrid? Explain your answer.

8. One parent is pure for brown hair. The other parent is pure for blond hair. Brown hair is the dominant trait. What color hair will the off-spring have?

9. **Infer:** Why do you think scientists use symbols to represent dominant and recessive genes?

Skill Builder ....

Researching Gregor Mendel often is called the "Father of Genetics." Use library references to find information about Gregor Mendel's life. Make a poster that illustrates Mendel's work. Diagram and illustrate his experiments and the results. Present your poster to the class.

Skill Builder ....

Defining There are three basic principles of genetics that are based on Mendel's studies. The three principles are: The Principle of Dominance, The Principle of Segregation, and the Principle of Independent Assortment. Use library references to define each of these terms.

ACTIVITY

MODELING THE STRUCTURE OF DNA

You will need a pair of scissors, tracing paper, and construction paper in white, black, green, red, blue, and yellow.

1. Trace the DNA parts shown on tracing paper.

2. Trace and cut out 12 white sugar molecules and 21 black phosphate molecules. Label the molecules.

3. Trace and cut out 3 green T's, 3 red A's, 3 yellow G's, and 3 blue C's. Label the bases.

4. Construct the sides of the DNA molecule with the sugars and phosphates.

5. Complete the DNA molecule by fitting the base pairs along the sides of the molecule.

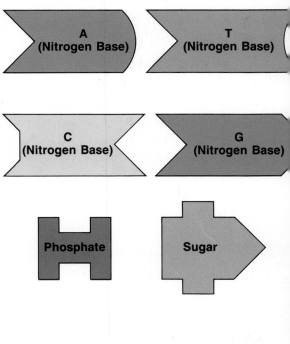

Questions

1. **a.** Which nitrogen base always pairs with C? **b.** Which base always pairs with T?

2. **a.** What chemicals make up the sides of the DNA molecule? **b.** Which chemicals make up the rungs?

3. **Infer:** Do all DNA models look exactly the same? Explain.

How do genes combine in offspring?

Objective ▶ Identify what causes differences in the traits of parents and their offspring.

TechTerm

▶ **Punnett square:** chart that shows possible gene combinations

Gene Symbols Organisms have at least two genes for every trait. They receive at least one gene from each parent. Symbols are used to represent the combinations of genes. A capital letter represents a dominant trait. A lower case letter represents a recessive trait. In humans, the gene for brown eyes is dominant. The symbol for this gene is B. The gene for blue eyes is recessive. The symbol for this gene is b.

▶ *Explain:* What does a capital letter show about a gene?

Predicting Traits Suppose a father has pure brown eyes. His genes are BB. Suppose a mother has pure blue eyes. Her genes are bb. What color eyes will the offspring have? One way to predict the eye color trait is to use a **Punnett square.** A Punnett square is a chart used to show possible gene combinations.

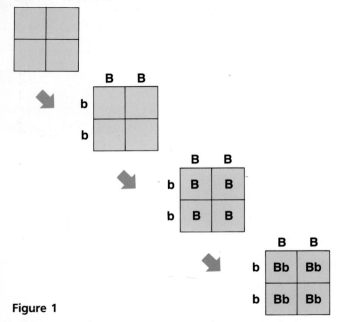

Figure 1

You can use a Punnett square to predict the eye color of the offspring in the above example. Draw a box with four squares in it, as shown in Figure 1. Write the genes for brown eyes (BB) from the father across the top. Write the genes for blue eyes (bb) from the mother down the side. Now fill in each square with a gene from the father, and a gene from the mother. As you can see, all the offspring will be hybrid dominant.

👁 *Observe:* What color eyes will all the offspring shown in Figure 1 have?

Combining Hybrids What happens when both parents are hybrid for a trait? A Punnett square also can be used to predict the genetic makeup of their offspring. Figure 2 shows that one-fourth of the offspring will be BB, or pure dominant. One-half will be Bb, or hybrid dominant. One-fourth will be bb, or pure recessive. This means that three-fourths of the offspring will have brown eyes. Only one-fourth will have blue eyes.

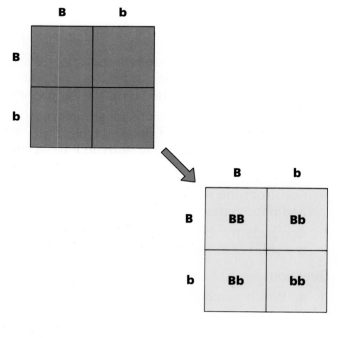

Figure 2

▶ *List:* What are the three kinds of traits that the offspring of hybrid parents can have?

LESSON SUMMARY

▶ A capital letter represents a dominant gene, and a lower case letter represents a recessive gene.

▶ A Punnett square is a chart used to show possible gene combinations.

▶ When one parent is pure dominant and the other is pure recessive, the offspring are all hybrid dominant.

▶ When both parents are hybrid for a trait, one-fourth of the offspring are pure dominant, one-half are hybrid dominant, and one-fourth are pure recessive.

CHECK *Complete the following.*

1. A capital letter always represents a _____ gene.

2. The recessive gene for blue eye color in humans is shown by the symbol _____ .

3. When one parent is pure dominant and the other parent is pure recessive, their offspring are all hybrid _____ .

4. If both parents are hybrid then one-half of their offspring will have _____ genes.

5. A person with blue eyes always has the gene combination _____ .

APPLY *Use the Punnett square and information to answer the following. A pure white (bb) guinea pig is crossed with a hybrid black (Bb) guinea pig. There are four offspring.*

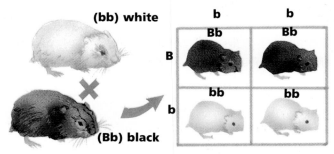

6. How many of the offspring will be black?

7. How many of the offspring will be white?

8. Are any of the offspring hybrid?

InfoSearch

Read the passage. Then ask two questions that cannot be answered from the information in the passage.

Height In humans, height is affected by many genes. None of these genes is dominant. Therefore, the height of a person depends upon the total effect of all the genes for height. This results in a wide range of heights among humans.

SEARCH: Use library references to find answers to your questions.

ACTIVITY

INVESTIGATING THE EFFECT OF CHANCE ON HEREDITY

You will need a penny, a nickel, masking tape, a marker, paper, and a pencil.

1. Place a piece of tape on both sides of each coin.
2. Label one side of each coin with an F to represent the gene for unattached ear lobes. Label the other side of each coin with an f to represent the gene for attached ear lobes.
3. Copy the Data Table on a sheet of paper.
4. Flip the two coins 50 times each. Record the gene combination that results from each flip by placing a dash in the appropriate column on your paper.
5. After 50 flips, total the number of dashes in each column.

Data Table: Gene Pairs		
FF	Ff	ff
1. 2. 3. 4. 5.		

Questions

1. **a.** Which trait is dominant?   **b.** Which trait is recessive?
2. What were the results of the coin flips for each pair of genes?

How are traits blended?

Objective ▶ Describe what occurs when a dominant gene is not present in a gene pair.

TechTerm

▶ **blending:** combination of genes in which a mixture of both traits shows

Eye Color In humans, the gene for brown eyes is dominant. The gene for blue eyes is recessive. Yet some people have eyes that are not brown or blue. They have hazel or green eyes. This is caused by genes for eye color that are neither dominant nor recessive. When these genes combine, a mixture of both traits shows. This kind of gene combination is called **blending.** A person's skin color and hair color also can result from the blending of genes.

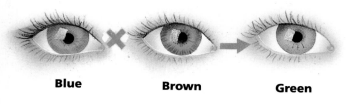

Blue **Brown** **Green**

▐▶ *Define:* What is blending?

Melanin Scientists have learned that four pairs of genes control skin color. None of these genes is dominant and each affects skin color equally. These genes control the amount of melanin found in the skin. Melanin is a chemical substance that gives skin its color. The more melanin found in the skin, the darker the color.

▢*Hypothesize:* Why are no two skin colors exactly alike?

Blending in Chickens There is a kind of chicken that has a gene for black feathers and a gene for white feathers. Neither of these genes is dominant over the other. The capital letter B stands for black feathers. The capital letter W stands for white feathers. A chicken that is BB, or pure black, has black feathers. A chicken that is WW, or pure

white, has white feathers. When a pure black chicken and a pure white chicken are mated, all of the offspring have the gene combination BW. These hybrids are neither black nor white. They are grey. Blending has taken place.

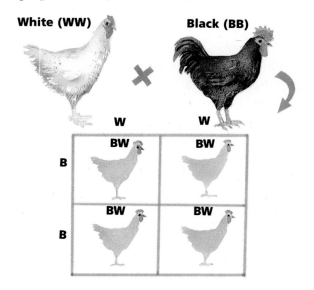

▐▶ *Describe:* What causes a chicken to have grey feathers?

Blended Genes Blended genes do not disappear. They show up again when hybrids are mated. If two hybrid grey chickens are mated, one-fourth of the offspring are BB, or pure black. One-half of the offspring are BW, or hybrid grey. One-fourth of the offspring are WW, or pure white.

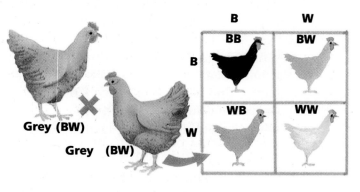

▐▶ *Relate:* What kinds of gene combinations can the offspring of two hybrid grey chickens have?

LESSON SUMMARY

▶ Blending is a gene combination in which a mixture of traits shows.

▶ Four different pairs of genes control skin color.

▶ There is no dominant gene in blending.

▶ When hybrids with blended traits are mated, the pure traits show again in some of the offspring.

CHECK *Answer the following.*

1. What are green eyes in humans caused by?

2. What color feathers will a chicken with the gene combination WW have?

3. What is the gene combination for a chicken with grey feathers?

Complete the following.

4. A person who is very fair-skinned has a _____ amount of melanin.

5. If two hybrid grey chickens are mated, one-half of their offspring will have the gene combination _____ .

6. Chickens with a gene combination of BW will be _____ in color.

7. The offspring of two hybrid grey chickens can have feathers that are colored black, _____, or grey.

APPLY *Answer the following.*

8. **Infer:** Why are capital letters used to represent all the genes for feather color in certain types of chickens?

9. **Hypothesize:** Why does blending occur only in hybrids?

Skill Builder

▶ ***Predicting*** Some traits in animals are determined by blending. For example, neither of the genes for tail length in cats is dominant. Use a Punnett square to predict the tail length in the kittens if two short-tailed cats (LN) are crossed. Let L represent the gene for long tail and N represent the gene for no tail.

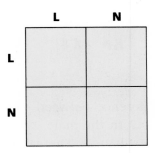

	L	N
L		
N		

◆◆◆◆ CAREER IN LIFE SCIENCE ◆◆◆◆◆◆◆◆◆◆◆◆◆◆◆◆◆◆◆◆◆◆◆◆◆◆◆◆◆◆

GENETICS COUNSELOR

Do you enjoy working in a science laboratory? Do you find genes and inherited traits interesting? If you do, you may enjoy a career as a genetics counselor. A genetics counselor studies inherited diseases. Some genetics counselors work in laboratories. They investigate how certain diseases are passed from parents to offspring. Other genetics counselors meet with parents who are concerned about passing on certain genes to their children. The counselors provide the parents with support and information about inherited diseases.

To become a genetics counselor, you need a four-year degree. Most genetics counselors also have a master's degree in either genetic counseling, nursing, or social work. Genetics counselors work mainly in medical centers and teaching hospitals. For more information, write to the March of Dimes, Birth Defects Foundation, 1275 Mamaroneck Avenue, White Plains, NY 10605.

How is sex determined?

Objective ▶ Explain how chromosomes in a sperm cell determine the sex of offspring.

X and Y All human traits are determined by 23 pairs of chromosomes. Look at the 23 pairs of chromosomes below.

Do you see the difference between the two chromosomes in pair 23? In the male, the chromosomes of this pair are not alike. These different chromosomes are called X and Y. Male cells have one X chromosome and one Y chromosome. Female cells have two X chromosomes. The X and Y chromosomes determine the sex of an organism.

▶ *Identify:* Which chromosomes determine sex?

Sperm Cells Male cells have an XY pair of chromosomes. During meiosis, each sperm cell receives only one chromosome from each pair. One-half of the sperm cells receive an X chromosome. The other half receive a Y chromosome. Every sperm contains either an X or a Y chromosome.

▶ *Explain:* Why do some sperm cells have an X chromosome while other sperm cells have a Y chromosome?

Egg Cells Female cells have an XX pair of chromosomes. During meiosis, each egg cell receives

one chromosome from each pair of chromosomes. Therefore, each egg cell receives one X chromosome. All egg cells contain X chromosomes.

▶ *Relate:* Why do egg cells contain only one X chromosome?

Sex of Offspring The sex of offspring is controlled by chromosomes in male sperm cells. As a result, all children inherit their sex from their fathers. During fertilization, if the egg cell is fertilized by a sperm cell carrying an X chromosome, the fertilized egg will have two X chromosomes (XX). It will develop into a female. If the sperm is carrying a Y chromosome, the zygote will have one X chromosome and one Y chromosome (XY). It will develop into a male. Half the sperm cells carry the X chromosome. The other half carry the Y chromosome. Therefore, over a large number of births, half will be female and half will be male.

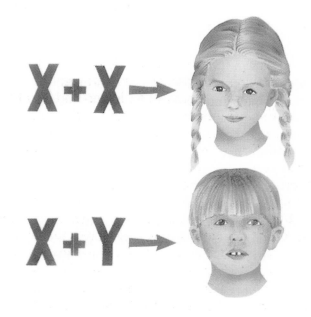

▶ *Identify:* Which parent is responsible for the sex of the offspring?

LESSON SUMMARY

▶ Male cells are XY, while female cells are XX.

▶ Each sperm cell has either an X or a Y chromosome.

▶ Each egg cell always has one X chromosome.

▶ When a sperm cell fertilizes an egg cell, the zygote becomes a female if it is XX or a male if it is XY.

CHECK *Write true if the statement is true. If the statement is false, change the underlined term to make the statement true.*

1. The sex of offspring is controlled by chromosomes in the <u>female</u>.

2. A <u>female</u> cell contains two X chromosomes.

3. An <u>egg</u> cell can contain either an X or a Y chromosome.

4. A zygote that has two X chromosomes will develop into a <u>male</u>.

5. A zygote that has an X and a Y chromosome will develop into a <u>male</u>.

APPLY *Answer the following.*

6. **Hypothesize:** Which are more alike, sperm cells or egg cells?

7. **Calculate:** Last year, 250 babies were born at a certain hospital. Estimate the number of males born there during the year. Explain your answer.

....
InfoSearch.......................................

Read the passage. Ask two questions that you cannot answer from the information in the passage.

Sex Chromosomes In birds, moths, butterflies, and some fish, the sex chromosomes are the reverse of human sex chromosomes. In these animals, the male has two identical (XX) chromosomes, and the female has two different ones (XY). In these animals, it is the egg cell, not the sperm cell, that determines the sex of the offspring.

SEARCH: Use library references to find answers to your questions.

TECHNOLOGY AND SOCIETY

AMNIOCENTESIS

Amniocentesis (am-nee-oh-sen-TEE-sis) is a test performed during a woman's pregnancy. The test involves inserting a needle through the woman's abdomen and into her uterus. Some of the amniotic fluid that surrounds the fetus is removed through the needle. The amniotic fluid contains cells that have been shed by the fetus. Doctors can learn about fetal chromosomes and enzyme levels by studying the cell cultures.

More than 175 different genetic disorders can be identified by amniocentesis. In addition, the test reveals the sex of the fetus. This knowledge is important when looking for inherited diseases that are passed to one sex only. Amniocentesis usually is recommended for pregnant women who are 35 or older. Women over 35 have the greatest risk of having a baby with an abnormal number of chromosomes. Down's syndrome is one genetic disorder that is caused by an abnormal number of chromosomes. Almost half of the babies born with Down's syndrome have mothers who are over 35 years old.

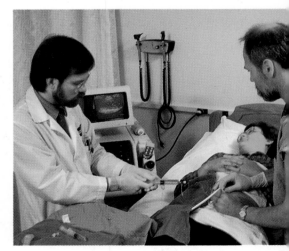

What are sex-linked traits?

Objective ▶ Describe how certain traits are inherited along with sex.

TechTerm

▶ **sex-linked traits:** traits that are controlled by the sex chromosomes

Inherited Traits Some traits are inherited along with sex. These traits are controlled by the X and Y sex chromosomes. Traits that are inherited along with sex are called **sex-linked traits.**

Like all other chromosomes, the sex chromosomes carry genes. The X chromosome carries many genes, but the Y chromosome carries few genes. Most of the genes for sex-linked traits are found on the X chromosome.

▶ **Define:** What are sex-linked traits?

Sex-linked Disorders Some hereditary disorders are controlled by the sex chromosomes. Two sex-linked disorders are hemophilia (hee-moh-FIL-ee-uh) and colorblindness. Hemophilia is a disorder in which the blood does not clot properly. People with hemophilia can easily bleed to death. Even a small cut can prove to be life-threatening. Colorblindness is a disorder in which a person cannot see the difference between certain colors.

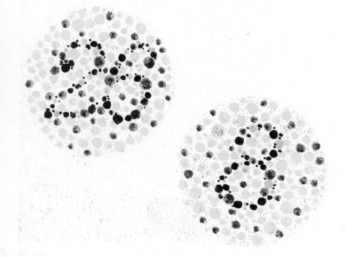

Sex-linked disorders are found more often in men than in women. This occurs because the genes for most sex-linked disorders are recessive. In women there are two X chromosomes. Both X chromosomes contain genes for these traits. While one of these chromosomes may have the gene for the disorder, the other X chromosome usually has a normal gene. Since the gene for the disorder is recessive, a female with each type of gene still appears normal.

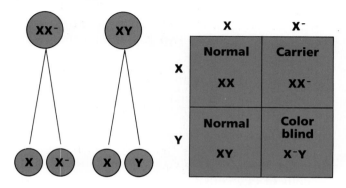

▶ **Explain:** Why does a female with a gene for a sex-linked disorder appear normal?

Carriers Woman who have one normal gene and one gene for a sex-linked disorder are said to be carriers of the disorder. Although they do not show the disorder, they can pass the gene for it to their children. Men who have a sex-linked disorder also are called carriers, because they can pass the disorder to their offspring.

When the sex chromosomes of a female carrier separate during meiosis, half the eggs get the normal chromosome and half get the chromosome for the disorder. If an egg cell containing the X chromosome for the disorder is fertilized by a sperm containing the normal X chromosome, the daughter will be normal. If an egg containing the X chromosome for the disorder is fertilized by a sperm with a Y chromosome, the son will have the disorder.

▶ **Explain:** Why do more men have sex-linked disorders?

LESSON SUMMARY

▶ Traits that are inherited with the sex chromosomes are called sex-linked traits.

▶ Most of the genes for sex-linked traits are carried on the X chromosome.

▶ Hemophilia and colorblindness are two sex-linked disorders.

▶ Most of the genes for sex-linked traits are recessive.

▶ Females can be carriers of genes for sex-linked disorders without actually having the disorder.

▶ Half the eggs of a female carrier of a sex-linked disorder get a normal chromosome and half get a chromosome for the disorder.

CHECK *Write true if the statement is true. If the statement is false, change the underlined term to make the statement true.*

1. Most of the genes for sex-linked traits are found on the X chromosome.

2. The genes for most sex-linked disorders are <u>dominant</u>.

3. Traits that are inherited along with sex are called <u>sex chromosomes</u>.

4. A female who has a gene for a sex-linked disorder but does not show the disorder is a <u>carrier</u>.

5. A male who has a normal Y chromosome and an X chromosome for a disorder <u>will not</u> have the disorder.

APPLY *Answer the following.*

▶ 6. **Infer:** How can a female have the sex-linked disorder of hemophilia?

7. **Describe:** What kind of gene pair is found in the female carrier of a sex-linked disorder?

State the Problem

State the problem for this experiment.

A person once tried to develop a breed of mice without tails. He cut off the tails of mice before letting them mate. He repeated this for many generations. The results were always the same. All the offspring were born with normal tails.

LOOKING BACK IN SCIENCE

HEMOPHILIA IN THE ROYAL FAMILY

Hemophilia is a sex-linked disorder. Since it is a recessive trait, males generally have the disease. Females generally are carriers who do not show the disease. A famous carrier of hemophilia was Queen Victoria of England. Three out of her four offspring inherited the gene for hemophilia. Two of Queen Victoria's daughters were carriers of hemophilia. One of her sons was a hemophiliac. These offspring passed the trait along to each of their own children. Four of Queen Victoria's granddaughters were carriers. Three of her grandsons were hemophiliacs. As her grandchildren married members of other royal families, the disorder continued to spread. As a result, hemophilia was quite common among European royalty during the nineteenth century.

What are some inherited diseases?

Objective ▶ Identify some inherited diseases.

TechTerm

▶ **inherited disease:** disease caused by an inherited gene

Inherited Diseases Sometimes a gene is defective or abnormal. The abnormal gene stops the body from working properly. An abnormal gene can be passed from the parents to their offspring. The gene can cause disease in the offspring. A disease that is caused by an inherited abnormal gene is called an **inherited disease.**

▷ *Define:* What is an inherited disease?

Sickle-Cell Anemia Sickle-cell anemia is an inherited disease that mainly affects people of African descent. The red blood cells of a person with sickle-cell anemia are not shaped normally. Rather than being rounded, these red blood cells are shaped like a sickle or crescent. Because of their abnormal shape, sickle cells are easily trapped in blood vessels. They clog the blood vessels and block the flow of blood. The clogged blood vessels may cause severe pain and, in some cases, an early death.

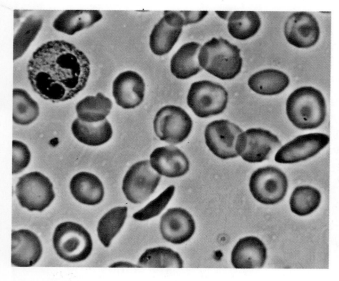

The gene for sickle-cell anemia is recessive (s). The gene for normal red blood cells is dominant (S). People who inherit two recessive genes (ss) for the trait have the disease. People who inherit one dominant and one recessive gene (Ss) for the trait are carriers of the disease.

▷ *Explain:* People with one recessive gene for the sickle-cell anemia trait do not show the disease. Why?

Tay-Sachs Disease Tay-Sachs is a disease that affects mainly Jewish children with Eastern European ancestry. An abnormal gene stops the child's body from producing an enzyme that breaks down fat. As a result, the fat gathers in the brain cells. This can cause brain damage and death.

The gene for Tay-Sachs disease also is recessive. A person who inherits two recessive genes for this trait will have the disease. A person who inherits one recessive gene and one dominant gene is a carrier.

▷ *Relate:* What body function does the recessive gene for Tay-Sachs disease affect?

PKU PKU, or phenylketonuria (fen-il-keet-uh-NYOOR-ee-uh), is another kind of inherited disease. People with PKU are missing an important enzyme that is made in the liver. As a result, they cannot break down certain chemicals in food. The chemicals build up in the body and may cause brain damage and mental retardation.

▷ *Describe:* What effect does the disease PKU have on the body?

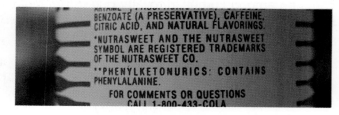

LESSON SUMMARY

▶ An inherited disease is caused by an abnormal gene that is inherited.

▶ Sickle-cell anemia is an inherited disease in which red blood cells are irregularly shaped.

▶ The gene for sickle-cell anemia is recessive.

▶ Tay-Sachs disease is an inherited disease that stops the body from producing an enzyme that breaks down fat.

▶ The gene for Tay-Sachs disease is recessive.

▶ PKU is an inherited disease that prevents the body from breaking down certain chemicals found in foods.

CHECK *Find the sentence in the lesson that answers each question. Then, write the sentence.*

1. What are inherited diseases caused by?

2. How do sickle cells differ from normal red blood cells?

3. Who is considered a carrier of Tay-Sachs disease?

4. What is PKU?

Complete the following.

5. A person with Tay-Sachs disease lacks an enzyme that breaks down _____ .

6. A person with PKU cannot break down _____ found in foods.

7. Sickle cells clog the blood vessels and block the flow of _____ .

8. A defective gene can be passed from _____ to their offspring.

APPLY *Answer the following.*

9. **Compare:** How are the effects of PKU and Tay-Sachs disease similar?

10. **Hypothesize:** Are the offspring of parents who are both pure dominant for normal red blood cells in danger of getting sickle-cell anemia? Why?

InfoSearch

Read the passage. Ask two questions that you cannot answer from the information in the passage.

Treating PKU PKU is not a curable disease. However, if the disease is detected early enough, it can be treated. All newborn babies are given a simple blood test that checks for the presence of PKU. A child who has PKU is put on a special diet. Foods high in protein must be avoided. By following the special diet, the child can grow and live a normal life.

SEARCH: Use library references to find answers to your questions.

ACTIVITY

PREDICTING TAY-SACHS DISEASE

You will need paper and a pencil.

1. Copy the chart shown on a piece of paper.

2. Predict the gene combinations each child will inherit from its parents. Fill in the gene combinations on the chart.

Questions

1. Are the parents pure dominant, pure recessive, or hybrid dominant for the trait?

2. What gene combination is found in a carrier for the trait?

3. How many of the offspring have Tay-Sachs disease?

4. What is the gene combination found in normal offspring?

5. **Analyze:** What is the chance that parents who are both carriers of the Tay-Sachs trait will produce a child who has the disease?

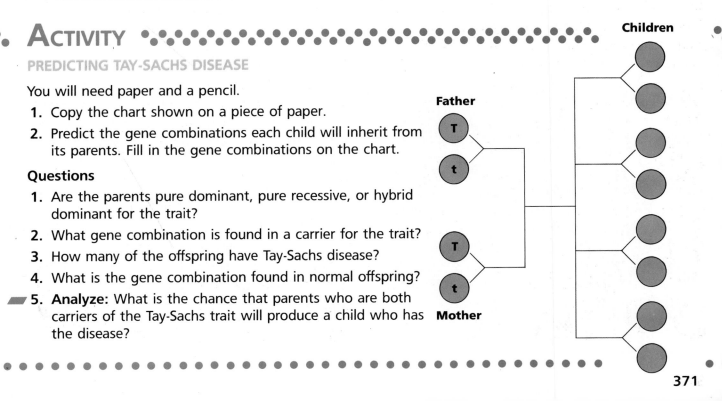

Can the environment affect inherited traits?

Objective ▶ Understand how the living conditions of an organism affect the way it develops.

Genes and Traits Genes control many of your traits. This is true for all living things. Your hair color and skin color are determined by genes. So is your sex. The size and color of flowers also are determined by genes. However, other factors can affect traits. The environment affects the traits of living things in many ways.

▶ *Identify:* What affects the traits of living things besides genes?

Conditions in the Environment The right environment is important for the development of certain traits. Food, air, water, and sunlight all are parts of the environment. Organisms need all of these things to grow and survive. Green plants need sunlight to develop and grow properly. Plants that do not get enough sunlight will be smaller and weaker than normal. When the environment is not right for an organism, certain traits many not develop at all.

Living things are affected by a shortage of food. A young organism may not develop properly if it does not get the right food. If the food supply increases, development may become normal again. Living things can develop normally only when conditions in the environment are normal.

Malnourished children

▶ *Infer:* What might happen to a green plant if its environment is not right?

Environment and Genes Plants that are grown in poor soil are small in size. They produce only a few fruits. Adding nutrients to the soil changes the environment. The plants grow larger and produce more fruit. The size of a plant and the amount of fruit it produces are traits. The genes for these traits are not affected by the environment. Only the development of the trait is affected by the environment. A change in the environment does not change the genes. A change in the environment can help the gene to do its work. A change in the environment also can prevent the gene from doing its work.

▶ *Describe:* What effect does the environment have on the genes in a plant?

LESSON SUMMARY

▶ The environment affects the traits of certain things.

▶ The right environment is important for the development of living things.

▶ Living things develop normally when the conditions in the environment are normal.

▶ A change in the environment does not change the genes.

CHECK *Complete the following.*

1. The traits of all living things are controlled by _____ .

2. Food, air, _____, and sunlight are all parts of the environment.

3. The environment affects the development of _____ .

4. A change in the environment of an organism has no effect on the _____ of the organism.

5. A young organism does not develop normally if its environment does not supply enough of the right _____ .

6. Living things develop normally when conditions in the _____ are normal.

APPLY *Answer the following.*

7. **Infer:** A green plant was stored in a school closet during the summer vacation. When students returned to the school in the fall, they found the plant had died. What probably caused the plant to die?

8. Gardeners often add fertilizer to soil. Does the fertilizer affect the genes of plants growing in the soil?

9. **Hypothesize:** Sunlight helps the body to synthesize vitamin D. What would happen to a person if they were never exposed to sunlight?

Designing an Experiment

Design an experiment to solve the problem.

PROBLEM: How does the amount of water affect the growth and development of a plant?

Your experiment should:

1. List the materials you would need.

2. Identify safety precautions that should be followed.

3. List a step-by-step procedure.

4. Describe how you would record your data.

◇◆◇◆ SCIENCE CONNECTION ◆◇◆◇◆◇◆◇◆◇◆◇

PLANT FERTILIZERS

Have you ever "fed" your plants? Although plants can make their own food, they need nutrients from the soil to carry on their food-making processes. For example, plants need phosphorus, potassium, nitrogen, sulfur, and magnesium. Sometimes soil does not have enough of these substances. This results in a need for plant food to be added to the soil. Plant food contains the nutrients a plant needs to grow.

The scientific name for plant food is fertilizer (fur-TUL-i-zur). Some fertilizers come directly from living things. Other fertilizers are mixtures of chemicals. Chemical fertilizers are produced in factories. The most widely used chemical fertilizers contain nitrogen, phosphorus, and potassium. These are the three plant nutrients in which soil is most frequently deficient. The amount of these nutrients needed by a plant varies according to the kind of plant and the type of soil in which the plant is growing.

18-11 How is genetics used to improve living things?

Objective ► Explain the different methods of controlled breeding.

TechTerms

- **controlled breeding:** mating organisms to produce offspring with certain traits
- **hybridization** (hy-brid-ih-ZAY-shun): mating two different kinds of organisms
- **inbreeding:** mating closely related organisms
- **mass selection:** crossing plants with desirable traits

Controlled Breeding People have raised animals and planted crops for thousands of years. Over time, they have learned that some plants produce better crops than others. They saw that some animals had more desirable traits than others. People learned to mate animals with desirable traits. Mating organisms to produce offspring with certain traits is called **controlled breeding.** The offspring produced usually had the same desirable traits as their parents.

▶ *Define:* What is controlled breeding?

Mass Selection Plant growers use a process called **mass selection** to produce plants with certain traits. In mass selection, plants with desirable traits are crossed. Seeds produced by these plants are collected and planted. New plants develop from these seeds. If the new plants have the same desirable traits, their seeds are collected and planted. The process goes on for many generations of the plants. The result of mass selection is a new kind of plant with certain desirable traits.

▶ *Describe:* What is the purpose of mass selection?

Inbreeding Another way of producing organisms with favorable traits is by **inbreeding.** In-breeding is the mating of closely related organisms. Offspring produced by inbreeding have genes that are very similar to their parents' genes. For example, inbreeding is used to breed racehorses, whose desirable trait is speed.

▶ *Compare:* How do the genes of offspring produced by inbreeding compare with the parents' genes?

Hybridization Sometimes two organisms with different kinds of genes are crossed. The offspring show traits of both parents. The mating of two different kinds of organisms is called **hybridization** (hy-brid-ih-ZAY-shun). A mule is the result of hybridization. Its father is a donkey and its mother is a horse.

When two different organisms are crossed, the offspring usually shows the best traits of both parent organisms. For example, when a rye plant and a wheat plant are crossed, they produce triticale (trit-i-KAY-lee). Triticale is more nutritious than either wheat or rye.

▶ *Explain:* How is a mule produced?

374

Mating organisms with certain traits to produce offspring with those same traits is called controlled breeding.

In mass selection, plants with desirable traits are crossed until a new variety of plant is developed.

Inbreeding is the mating of organisms with similar genes.

In hybridization, two different kinds of organisms are mated.

A hybrid organism usually shows the best traits of both parents.

CHECK *Find the sentence in the lesson that answers each question. Then, write the sentence.*

1. What is the purpose of controlled breeding?

2. How are new plants with desirable traits produced?

3. What is the mating of closely related organisms called?

4. What is hybridization?

5. How are the genes of offspring produced by inbreeding related to their parents' genes?

APPLY *Complete the following.*

6. **Compare:** How are mass selection, inbreeding, and hybridization alike?

7. **Relate:** Could inbreeding and hybridization be considered opposite processes? Why?

▶ 8. **Infer:** Why is inbreeding dangerous to the organism?

9. Would you rather have an inbred dog as a pet or a hybrid dog as a pet? Explain your answer.

InfoSearch.....................................

Read the passage. Ask two questions that you cannot answer from the information in the passage.

Problems with Inbreeding When inbreeding is allowed to continue for a number of generations, undesirable traits may begin to show. For example, inbreeding of plants may cause the crop yield to become smaller and smaller. The plants themselves may become smaller and weaker. Hereditary diseases in animals usually are very rare. However, inbreeding of animals may cause these diseases to appear much more often.

SEARCH: Use library references to find answers to your questions.

CAREER IN LIFE SCIENCE

ANIMAL BREEDER

Animal breeders study the genes that control certain traits in animals. They use this information to breed animals and produce offspring with desirable traits. Animal breeders also identify the traits that are desirable in different kinds of animals. For example, a thoroughbred racehorse is bred for speed. A workhorse is bred for strength. A cow may be bred to produce more milk.

Most animal breeders work directly with animals outdoors. Other animal breeders do research in a laboratory. In order to become an animal breeder, you need a high-school diploma. You should take courses in agriculture and biology. A four-year degree in animal science often is necessary. For more information, write to your county or state department of agriculture.

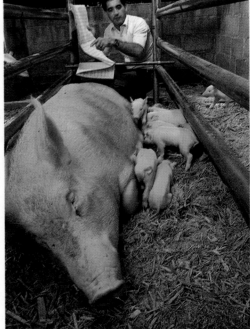

What is genetic engineering?

Objective ▶ Describe one method used to produce new DNA.

TechTerms

- ▶ **genetic engineering** (juh-NET-ik en-juh-NEER-ing): methods used to produce new forms of DNA
- ▶ **gene splicing** (SPLYS-ing): moving a section of DNA from the genes of one organism to the genes of another organism

Changing DNA Until recently, new forms of life could only be invented in books and movies. Today, however, advances in modern genetics have made it possible for new forms of bacteria and other simple organisms to be made in the laboratory. This is the result of a new technology called **genetic engineering** (juh-NET-ik en-juh-NEER-ing). Genetic engineering is a process by which new forms of DNA are produced.

▧▶ *Define:* What is genetic engineering?

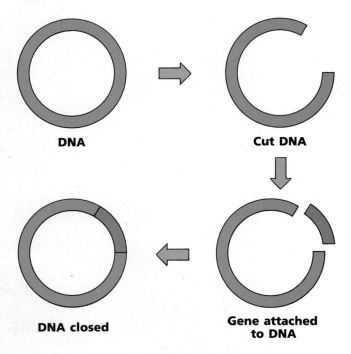

DNA   **Cut DNA**

DNA closed   **Gene attached to DNA**

Gene Splicing One method of genetic engineering is called **gene splicing** (SPLYS-ing). Gene splicing is a process in which a section of DNA from one organism is transferred to another organism.

Gene splicing takes place in three steps. First a DNA chain is temporarily opened up. This is done with certain enzymes. New genes from another organism are then added, or spliced, into the DNA. Finally, the DNA chain is closed.

▧▶ *Describe:* What occurs during gene splicing?

Benefits Genetic engineering has many benefits. New genes have been added to the DNA of certain bacteria. These bacteria can then produce substances that otherwise could only be made by the human body. Human insulin needed by diabetics is an example of such a substance. Bacteria can produce insulin after human genes are spliced into the DNA of bacteria.

Scientists hope that someday genetic engineering can be used to correct genetic disorders. This may involve adding normal genes to cells that are missing these genes. Another use of genetic engineering could be to improve certain traits in plants and animals used for food.

▧▶ *Explain:* How can bacteria produce human insulin?

Disadvantages Some scientists are concerned about possible dangers in working with new forms of DNA. There is some concern that experiments with the DNA of bacteria might produce a disease-causing organism for which there is no cure. To help prevent this from happening, the federal government has set up special rules that must be followed in experiments using genetic engineering.

▧▶ *Identify:* What is a possible danger in working with new forms of DNA?

LESSON SUMMARY

▶ Genetic engineering is a process used to produce new forms of DNA.

▶ Gene splicing is a form of genetic engineering in which part of a DNA chain from one organism is added to a DNA chain from a different organism.

▶ Genetic engineering can produce substances normally made only by the human body.

▶ Scientists hope genetic engineering may be used to correct genetic disorders in the future.

▶ There are possible dangers involved in working with new forms of DNA.

CHECK *Complete the following.*

1. Methods used to produce new forms of DNA are called _____ .

2. A process in which DNA from one organism is added to the DNA of another organism is called _____ .

3. Certain kinds of genetically engineered bacteria can produce _____ , which is needed by diabetics.

4. Scientists hope that someday genetic engineering can be used to correct _____ .

5. A danger with genetic engineering is the possibility of producing a _____ organism for which there is no known cure.

APPLY *Answer the following.*

6. **Relate:** PKU is a genetic disorder that causes the lack of a certain enzyme. How might genetic engineering someday be used to correct this disorder?

7. **Infer:** Through genetic engineering, scientists have been able to place a gene from a firefly into the DNA of a plant. What effect do you suppose this gene has on the plant?

Skill Builder ....

▲ *Modeling* When you model, you use a copy or an imitation of an object to help explain or understand something. Make a model showing the three stages in gene splicing. Use one colored piece of construction paper to represent the original gene. Use another colored piece of construction paper to represent the new section. Make one model for each step in the process. Write an explanation of what is happening under each step.

TECHNOLOGY AND SOCIETY

INTERFERON

Genetic engineering is being used to produce hormones and other important body chemicals. One such substance is called interferon (in-tur-FEER-ahn). Until recently, interferon could only be produced in human body cells. The body produced the chemical naturally. When a virus enters a cell, the cell produces interferon. The interferon leaves the infected cell and alerts other body cells that a virus is present. The other cells of the body respond by making chemicals to fight the virus.

For many years, scientists thought that interferon could be used to fight serious diseases such as cancer. However, there was not enough interferon available to test this hypothesis. As a result of gene splicing, bacterial cells can now be made to produce human interferon. Scientists now have enough of this substance to carry out their experiments.

STUDY HINT Before you begin the Unit Challenges, review the TechTerms and Lesson Summary for each lesson in this unit.

TechTerms..

blending (364)
chromosomes (356)
controlled breeding (374)
DNA (358)
dominant gene (360)
gene splicing (376)
genes (356)
genetic engineering (376)

genetics (354)
heredity (354)
hybrid (360)
hybridization (374)
inbreeding (374)
inherited disease (370)
inherited traits (354)

mass selection (374)
Punnett square (362)
pure (360)
recessive gene (360)
replication (358)
sex-linked traits (368)
trait (354)

TechTerm Challenges..

Matching *Write the TechTerm that matches each description.*

1. moving a section of DNA from the genes of one organism to the genes of another organism
2. field of biology that studies heredity
3. mating of closely related organisms
4. traits that are passed from parents to their offspring
5. mating organisms to produce offspring with certain traits
6. threadlike structures that control heredity
7. traits that are inherited along with sex
8. chemical that makes up chromosomes
9. combination of genes in which a mixture of both traits shows
10. stronger gene that always shows itself
11. chart that shows possible gene combinations

Fill-in *Write the TechTerm that best completes each statement.*

1. The characteristics of an organism are called _____ .
2. The process by which DNA is duplicated is called _____ .
3. An organism is _____ if it has two like genes for a given trait.
4. A disease caused by an inherited abnormal gene is called an _____ .
5. Crossing plants with desirable traits over many generations is called _____ .
6. The process by which new forms of DNA are produced is called _____ .
7. The mating of two different kinds of organisms is called _____ .
8. The weaker gene that is hidden when a dominant gene is present is a _____ .
9. Organisms that have two unlike genes for a trait are _____ for that trait.

Content Challenges..

Multiple Choice *Write the letter of the term that best completes each statement.*

1. Mendel found that when he crossed pure tall pea plants with pure short pea plants, the offspring were all
   a. short.   **b.** tall.   **c.** medium height.   **d.** both tall and short.
2. Male organisms have
   a. two X chromosomes.   **b.** two Y chromosomes.   **c.** three Y chromosomes.   **d.** one X and one Y chromosome.

3. The nitrogen base adenine always pairs with the nitrogen base
   a. cytosine.   b. guanine.   c. thymine.   d. chromosomes.

4. Two sex-linked traits that are carried on the X chromosome are
   a. hemophilia and skin color.   b. hemophilia and color blindness.   c. color blindness and sickle-cell anemia.   d. sickle-cell anemia and PKU.

5. Humans have
   a. 46 pairs of chromosomes.   b. 23 chromosomes.   c. 21 chromosomes.   d. 23 pairs of chromosomes.

6. Triticale, produced by crossing wheat plants and rye plants, is an example of
   a. hybridization.   b. mass selection.   c. inbreeding.   d. self-pollination.

7. The field of biology that studies the inheritance of traits is
   a. heredity.   b. embryology.   c. genetics.   d. ecology.

8. When Mendel crossed pure short plants with pure tall plants, the result was tall plants. He concluded that the tall trait is
   a. dominant.   b. pure.   c. hidden.   d. recessive.

9. A special diet can be prescribed for a child with
   a. hemophilia.   b. Down's syndrome.   c. sickle-cell anemia.   d. PKU.

10. The sides of a DNA ladder are made up of
    a. nitrogen bases.   b. sugars and phosphates.   c. proteins.   d. melanin.

True/False *Write true if the statement is true. If the statement is false, change the underlined term to make the statement true.*

1. Sex-linked traits are found more often in females.
2. Organisms with two of the same genes for a particular trait are hybrids.
3. Inbreeding is the crossing of organisms that are from different species.
4. The process of transferring genes from the DNA of one organism to the DNA of another organism is called gene splicing.
5. In the labeling of traits, a capital letter represents a recessive trait.
6. The parent who determines the sex of the offspring is the female.
7. A change in a gene or chromosome is called a mutation.
8. When two hybrids are crossed, the offspring will be one-fourth pure dominant, one-fourth pure recessive, and one-half hybrid dominant.
9. During reproduction, one chromosome from each pair goes to each offspring.
10. The trait of skin color is determined by dominant genes.
11. The gene that carries sex-linked diseases is dominant.

Understanding the Features..

Reading Critically *Use the feature reading selections to answer the following. Page numbers for the features follow each question in parentheses.*

1. **Infer:** Why do you think Mendel often is referred to as the 'father of genetics?' (355)
2. What contributions has Barbara McClintock made to the field of genetics? (357)
3. **Name:** Who were James Watson and Francis Crick? (359)
4. Where are some of the places genetic counselors work? (365)

5. Why is it recommended that pregnant women over 35 get an amniocentesis taken? (367)
6. How did hemophilia spread during the 19th century? (369)
7. What is the scientific term for plant food? (373)

8. **Infer:** How do courses in biology and agriculture help an animal breeder? (375)
9. How does the production of man-made interferon help in the fight against cancer? (377)

Concept Challenges..

Critical Thinking *Answer each of the following in complete sentences.*

1. **Infer:** Why do you think some people are against genetic engineering?
2. If a genetic disorder was present in your family, would you go to a genetic counselor before having children? Why or why not?
3. **Explain:** How does the genetic code control an organism's traits?

4. How do Punnett squares help in the prediction of the inheritance of traits?
5. If you were a geneticist, what animals or plants would you study? Why?

Interpreting a Diagram *Use the diagram of DNA replication to answer the following.*

1. What does the letter A stand for?
2. What does the letter C stand for?
3. What does the letter T stand for?
4. What does the letter G stand for?
5. How do the nitrogen bases pair up during DNA replication?
6. What happens during the process of replication?

Finding Out More...

1. Some law enforcement agencies use DNA in addition to fingerprints as a method of identifying criminals. Use library references to find out more about DNA testing. Write your findings in a report.
2. Cloning is the production of organisms with identical genes. Seedless grapes and seedless oranges are two types of fruit that are produced by cloning. Find out two cloning techniques. Describe the techniques in a written report.

3. Blood type is controlled by different genes. Research the four major blood groups and which genes are dominant. Write the different gene combinations each blood type may have.
4. Use library references to research the work of Walter Sutton. Compare his findings to the findings of Mendel.

CHANGE THROUGH TIME

CONTENTS

STUDY HINT As you read each lesson in Unit 19, write the lesson title and lesson objective on a sheet of paper. After you complete each lesson, write the sentence or sentences that answer each objective.

19-1 How are fossils formed?

Objective ▶ Explain how different kinds of fossils are formed.

TechTerms

▶ **amber:** hardened tree sap
▶ **extinct** (ik-STINKT): species that is no longer found alive
▶ **fossil** (FAHS-ul): remains or traces of a once-living organism

Figure 1 Fingerprints

Fossils At the scene of a crime, detectives look for fingerprints. The criminals are gone, but they may have left behind clues. Many kinds of organisms are now **extinct** (ik-STINKT), or no longer found as a living species. However, some of these organisms left behind clues that they existed. These clues are **fossils** (FAHS-uls). Fossils are the remains or traces of once-living organisms.

Figure 2 Fossil in mud

📝 *Define:* What are fossils?

Kinds of Fossils There are many kinds of fossils. Some fossils are footprints in mud that later changed to rock. Sometimes, an entire organism is found. In March, 1990, a whole mammoth was discovered in Siberia. It had been frozen in ice. A mammoth was an elephantlike animal. Insects have been trapped and preserved in **amber.** Amber is hardened tree sap. Most fossils, however, are the remains of hard parts of organisms, such as bones, teeth, and shells.

Figure 3 A mammoth

🗨 *Hypothesize:* Would a worm, a rabbit, or a butterfly most likely leave fossil remains? Explain.

Fossils in Rocks Most fossils are found in sedimentary (sed-uh-MEN-tar-ee) rocks. Sedimentary rock is made up of layers of sediments. Sediments are bits of clay, soil, sand, and other earth materials.

Sedimentary rocks usually form in water. When an animal living in the water dies, its body may sink to the bottom. The soft parts decay quickly. The bones or shells may remain and become covered by sediments. The sediments build up, layer by layer. Great pressure and weight press the sediments together. Slowly, the sediments change to rock. The fossil bones or shells are preserved in the rock.

Figure 4 Fossils in rock layers

▶ *Name:* In what kind of rock are most fossils found?

LESSON SUMMARY

► Any remains or traces of a once-living organism is a fossil.

► Most fossils are the remains of hard parts of organisms, such as bones, teeth, and shells.

► Most fossils are formed in sedimentary rock.

► Fossils form in sedimentary rock when organisms were buried quickly beneath layers of sediment.

CHECK *Write true if the statement is true. If the statement is false, change the underlined term to make the statement true.*

1. A mammoth is an example of a <u>living</u> species.

2. Insect fossils often are found preserved in <u>tar</u>.

3. Sand and mud that are carried and deposited by a river are <u>sediments</u>.

4. Most fossils form in <u>sedimentary</u> rock.

APPLY *Complete the following.*

5. Most fossils in sedimentary rock cannot be seen unless the rock has been cut across from top to bottom. Explain why.

6. **Analyze:** Study the diagram. Explain how the bones of the animals could have been trapped in the tar.

InfoSearch

Read the passage. Ask two questions that you cannot answer from the information in the passage.

Petrified Wood The remains of an organism may decay very slowly. Minerals dissolved in water take the place of the original materials. The minerals form an exact stone copy of the remains. The fossil that forms is called petrified (PET-ruh-fyd). Many trees are petrified. You can see petrified trees in the Petrified Forest of Arizona.

SEARCH: Use library references to find answers to your questions.

ACTIVITY

MAKING FOSSILS

You will need 3 large paper cups, clay, a small object such as a shell, a key, or a coin, plaster of Paris, petroleum jelly.

1. Press some clay into each of two paper cups so that the clay is 2-3 cm high in each cup.

2. Push a small object down into the clay of one cup. Then carefully remove the object.

3. Coat the object with petroleum jelly. Very lightly press the object into the clay of the second cup.

4. Prepare the plaster of Paris according to the directions.

5. Pour some of the plaster into each cup. Let the cups stand overnight. After the plaster hardens, tear away the cups. Remove the clay from the plaster.

Questions

1. **Observe:** Describe the two fossils.

2. **Analyze:** Which fossil is the mold? Which is the cast?

19-2 What is geologic time?

Objectives ► Explain how the layering of rocks forms a record of life on the earth. ► Describe how scientists use the geologic time scale.

TechTerms

- **geologic** (jee-oh-LAHJ-ik) **time scale:** record of the earth's history based upon types of organisms that lived at different times
- **era:** largest division of geologic time
- **relative age:** age of something compared with the age of something else

Relative Age By comparing the positions of the layers of sedimentary rocks in which fossils are found, scientists can find the **relative ages** of different fossils. Relative age is the age of something compared with the age of something else. As a general rule, the lower the layer of rock in which a fossil is found, the older the fossil. Thus, a fossil found in the bottom layer of rock usually is older than a fossil found in a higher layer.

▰▰▷ **Define:** What is relative age?

The Fossil Record Scientists have studied fossils from rocks all over the world. They have arranged the fossils in order, from oldest to youngest. Scientists use this information to construct a **geologic** (jee-oh-LAHJ-ik) **time scale.** A geologic time scale is a record of the earth's history based upon the types of organisms that lived at different times.

Look at the geologic time scale. An **era** is the largest time division on the scale. Eras are divided into periods. The periods of the most recent era, the Cenozoic era, are further divided into epochs (EP-uks).

Fossil evidence shows that life first appeared on the earth about 4.5 billion years ago. This date marks the beginning of the geologic time scale. Fossils also show that life on the earth started out simple and became more and more complex.

▰▰▷ **Sequence:** Name the three time divisions in the geologic time scale, from largest to smallest.

Table 1 Geologic Time Scale				
ERA	PERIOD	EPOCH	START DATE (MILLIONS OF YEARS AGO)	ORGANISMS
Cenozoic	Quaternary	Recent	0.025	Modern Humans
		Pleistocene	1.75	Mammoths
	Tertiary	Pliocene	14	Large carnivores
		Miocene	26	Many land mammals
		Oligocene	40	Primitive apes
		Eocene	55	Early horses
		Paleocene	65	Primates
Mesozoic	Cretaceous		130	Flowering plants
	Jurassic		180	Dinosaurs, birds
	Triassic		225	Conifers
Paleozoic	Permian		275	Seed Plants
	Carboniferous		345	Reptiles
	Devonian		405	Insects, amphibians
	Silurian		435	Fishes
	Ordovician		480	Algae, fungi
	Cambrian		600	Invertebrates
Precambrian			**over 600** billion	Bacteria, blue-green algae

LESSON SUMMARY

▶ The relative age of a fossil can be found by comparing it with the age of another fossil.

▶ The geologic time scale arranges fossils in order from oldest to youngest.

▶ The geologic time scale is divided into eras, periods, and epochs.

CHECK *Answer the following.*

1. What are the largest time divisions in the geologic time scale?

2. Is a fossil older or younger than another fossil found in a higher layer?

3. How would the age of a fossil shell compare with the age of a fossil leaf found in the same layer of rock?

4. About how long ago did life first appear on the earth?

5. In what era did modern humans first appear?

APPLY *Complete the following.*

6. **Analyze:** Why are the oldest layers of a sedimentary rock found at the bottom?

Use the Geologic Time Scale to answer the following.

7. During what era did the dinosaurs live?

8. **Analyze:** During what era of geologic time are you living?

Skill Builder

Sequencing When you sequence, you place things in order. The diagram shows layers of sedimentary rocks. Use the diagram to place the fossils in the rock layers in order from oldest to youngest. How does the age of fossil B compare with the age of fossil D?

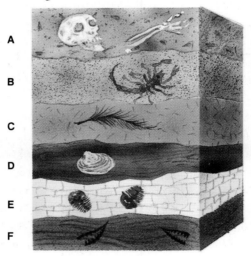

A

B

C

D

E

F

SCIENCE CONNECTION

RADIOACTIVE DATING

Scientists can measure the actual age of a fossil or a rock by a method called radioactive (ray-dee-oh-AK-tiv) dating. Radioactive dating involves the use of radioactive elements. The atoms of radioactive elements decay, or break down, into atoms of other elements. The unit used to measure the rate of decay of a radioactive element is called a half-life. A half-life is the time it takes for one-half of the atoms in a sample of a radioactive element to decay.

Carbon-14, or C-14, is the radioactive form of carbon. The half-life of C-14 is 5730 years. If you had a 10-gram sample of C-14 today, in 5730 years, only half of the C-14 atoms, or 5 grams, would remain. The other 5 grams would have broken down into nitrogen. At the end of two half-lives (11,460 years), only one fourth of the C-14 atoms, or 2.5 grams, would be present.

Carbon-14 is present in all living things. When an organism dies, the C-14 continues to decay. By measuring the amount of C-14 in a fossil, a scientist can tell how old the fossil is.

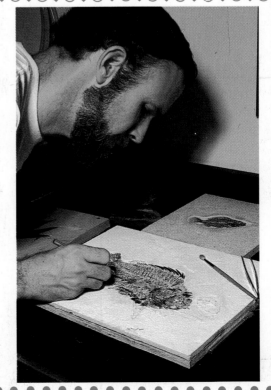

Objectives ► Define evolution. ► Explain how organisms change due to adaptations and mutations.

TechTerms

► **adaptation** (ad-up-TAY-shun): trait that helps an organism survive in its environment

► **evolution** (ev-uh-LOO-shun): process by which organisms change over time

► **mutation** (myoo-TAY-shun): change in a gene

Evolution Fossil evidence shows that the earliest living things on the earth were very simple organisms. In the billions of years that have passed, living things have become more and more complex. The study of fossils also shows that old species became extinct and new species took their place. Most scientists believe that new species develop from old species as a result of gradual change, or **evolution** (ev-uh-LOO-shun). Evolution is the process by which organisms change over time.

▐▐▶ *Define:* What is evolution?

Adaptations Living things are found everywhere on the earth, from the highest mountain tops to the deep oceans, and from hot, dry deserts to polar ice caps. How do different organisms survive such a wide variety of conditions? They often have **adaptations** (ad-up-TAY-shuns). An adaptation is a feature, or trait, that helps an organism to live in a certain environment.

The plants in the picture have needlelike leaves. They also have a thick, tough covering on their stems. These adaptations make it possible for these plants to live in the dry conditions of a desert environment.

▐ *Hypothesize:* What adaptations help birds to live in trees?

Mutations Species change, or evolve, over time. Generally, a change in a species is caused by new combinations of genes that occur with each new generation. Such changes take place over a very long period of time. Sometimes, however, something may cause a gene to change suddenly. A sudden change in a gene is called a **mutation** (myoo-TAY-shun). A mutation can produce a change in an organism. If the change is helpful to the organism, it will be kept and passed on to future generations. This may eventually result in a new species. If the variation is harmful, it will die out.

▐▐▶ *Define:* What is a mutation?

LESSON SUMMARY

► Evolution is the process by which organisms change over time.

► Adaptations are special traits that help organisms to survive in their environments.

► A mutation is a sudden change in a gene.

► Mutations can produce a change in an organism that can be passed on to future generations and may eventually result in a new species.

CHECK *Write true if the statement is true. If the statement is false, change the underlined term to make the statement true.*

1. Organisms change over time by the process of <u>evolution</u>.

2. The needlelike leaves of a cactus is an example of <u>a mutation</u>.

3. A sudden change in a gene is an <u>adaptation</u>.

APPLY *Answer the following.*

4. What do you call a trait that helps an organism to survive in its environment?

5. From what do new species develop?

6. Not all mutations are passed on to future offspring. Explain why.

Skill Builder

Designing an Experiment Desert plants, such as the cactus, have certain adaptations that help them to survive in a hot, dry environment. Design an experiment to show what would happen to a cactus if its environment suddenly changed to a cool, wet environment. Include a control in your experiment.

LOOKING BACK IN SCIENCE

THE VOYAGE OF THE *BEAGLE*

In 1831, a young biologist named Charles Darwin set sail on a British ship called the HMS *Beagle*. The ship left port on a five-year journey to the South Pacific and South America. The purpose of the trip was to prepare maps and to observe and collect specimens of various plants and animals. As the ship's naturalist (NACH-ur-ul-ist), Darwin made many observations. He kept careful records of his observations. He collected many different kinds of plants and animals. He also collected tiny fossils of different organisms.

Of all the places he visited, Darwin was most excited by the Galapagos (guh-LAH-puh-gos) Islands. The Galapagos Islands are located off the coast of South America. On these islands, Darwin discovered 13 species of finches. Darwin observed that the beak of each species of finch had a different shape. Each beak was adapted for eating a certain kind of food. Most other traits of the finches were very similar. Darwin inferred that all the finches had evolved from a common ancestor. Darwin used his observations of the finches as the basis for his ideas about natural selection.

What is natural selection?

Objective ► Explain Darwin's theory of natural selection.

TechTerms

► **natural selection:** survival of offspring that have favorable traits

► **variation** (ver-ee-AY-shun): differences in traits among individuals of a species

Explaining Evolution The fossil record shows that living things have evolved throughout the earth's history. However, knowing this fact does not explain how the changes happened. Offspring are supposed to be the same species as their parents. How then, can a new species develop from an old one? Over 100 years ago, the English biologist Charles Darwin suggested a theory of evolution as an answer to this question. Darwin's theory is accepted by most scientists today.

▟▶ *State:* What evidence shows that organisms have changed throughout the earth's history?

Natural Selection Darwin used the term **natural selection** to describe his theory of evolution. Natural selection is the survival of organisms that have favorable traits. The theory of natural selection includes the following ideas:

► **Overproduction** Each species produces more offspring than can survive. There is not enough food or living space for all of the offspring.

► **Struggle for Existence** The young of each generation compete for those things they need to live. Only a few will live long enough to reproduce. The others will die.

► **Variation** The young of each generation are not exactly alike. For example, some individuals are bigger or stronger than others. Differences in traits among individuals of a species are called **variations** (ver-ee-AY-shuns).

► **Survival of the Fittest** Some variations make individuals better suited for survival in their environments. These individuals are more likely to survive and reproduce than others.

► **Evolution of New Species** Individuals with favorable variations survive and reproduce. They pass their favorable traits to their offspring. Therefore, their offspring are more likely to survive and reproduce in the next generation. Unfavorable variations die out. In this way, favorable variations stay in the species. Over many generations, these changes can result in the appearance of a new species.

▟▶ *List:* List the main ideas of the theory of natural selection.

LESSON SUMMARY

▶ A new species can develop from an old one by the process of evolution.

▶ The modern theory of evolution is called the theory of natural selection.

▶ Differences among individuals in a species are called variations.

CHECK *Write true if the statement is true. If the statement is false, change the underlined term to make the statement true.*

1. The idea which states that a species produces more offspring than can survive is called <u>competition</u>.

2. Individuals with favorable <u>variations</u> survive and reproduce.

3. Natural selection is a theory of <u>evolution</u>.

APPLY *Use the pictures of the giraffes to complete the following.*

4. **Sequence:** Each picture of the giraffes shows one idea of natural selection. Identify the three ideas in the order in which they are shown.

Skill Builder

Researching Use library references to find out how Jean Baptiste Lamarck's ideas of evolution differ from Darwin's. Present your findings in a report.

ACTIVITY

MODELING NATURAL SELECTION

You will need 3 sheets of brown construction paper, 11″ × 17″, 1 sheet of green construction paper, 11″ × 17″, and scissors.

1. Work with a partner.

2. Copy the Data Table shown.

3. Cut out 15 brown squares and 15 green squares, 3 cm × 3 cm. **Caution: Be very careful when working with sharp objects.** These squares represent brown and green insects.

4. Have your partner place 2 sheets of brown construction paper on the floor to represent the environment.

5. While you turn your head, have your partner scatter the 30 insects on the brown paper.

6. At the signal "Ready," pick up 15 insects as fast as you can.

7. Count the number of insects of each color that you picked up. Record the results in the Data Table.

8. Repeat Steps 5 through 7 two more times.

9. Find the total number of brown insects and the total number of green insects in the three trials.

Data Table: Insect Trials		
TRIAL	NUMBER OF INSECTS	
	Brown Insects	Green Insects
1.		
2.		
3.		
Total		

Questions

1. Which color insect is most visible in a brown environment?

2. **Hypothesize:** Which insects would more likely be picked from a green leaf by a bird? Why?

What evidence supports evolution?

Objective ▶ Understand how fossils are used to support the theory of evolution.

A Record of Change The fossil record clearly shows that changes have taken place throughout the earth's history. There is evidence that the earth's climate has changed many times. There are fossils of many species of organisms that no longer exist. Perhaps the best-known group of extinct organisms is the dinosaurs. Fossils of about 400 different species of dinosaurs have been found. Some of these species roamed the earth for more than 100 million years. Yet, dinosaurs died out and have been extinct for more than 65 million years. Many possible theories have been suggested to explain why such a large and successful group of organisms died out. However, nobody knows for sure.

▶ **Describe:** What does the fossil record indicate about the earth's past?

The Changing Horse Evolution deals with changes within a species. The most complete fossil record of evolutionary change is that of the horse. Through fossils, scientists have traced animals related to the modern horse as far back as 60 million years. The earliest member of this family was a four-toed animal about the size of a small dog. Later fossils show how this animal changed through time. There were changes in the animal's teeth, its legs, its toes, and its size.

👁 **Observe:** In what ways have horses changed through time?

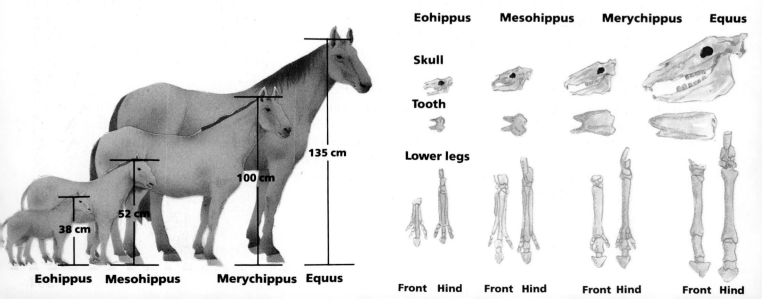

LESSON SUMMARY

▶ The fossil record clearly shows that living things have changed throughout the earth's history.

▶ The most complete fossil record of evolutionary change is that of the horse.

CHECK *Find the sentence in the lesson that answers each question. Then write the sentence.*

1. Why is the horse used to support the history of evolution?

2. In addition to changes in living things, what other changes are shown in the fossil record?

3. What caused the dinosaurs to die out?

4. How many species of dinosaurs have been identified?

5. About how big was the earliest horse?

APPLY *Complete the following.*

▶ 6. **Infer:** How can the fossil record show changes in climate?

👁 7. **Observe:** How has the size of the horse's jaw changed through time?

Skill Builder

🧪 *Measuring and Calculating* Measure the height of each horse in the scale drawings of the evolutionary stages of the horse. Using the heights given in each drawing, find out what scale was used for the drawings.

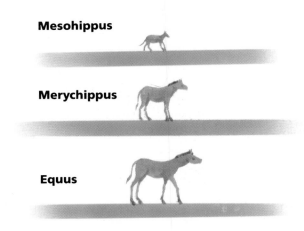

Mesohippus

Merychippus

Equus

LOOKING BACK IN SCIENCE

THE MOTHS OF MANCHESTER

A species of moth known as the British peppered moth lived during the early 1800s. This moth can be used to support the idea that natural selection can help an organism to survive in its environment. In 1800, most of the peppered moths in England were light gray in color. A few of the moths were black. The moths spent a good deal of time on the tree trunks in the area. The trunks were the same color as the light gray peppered moths. Birds could not easily see these moths. The black moths were easily seen. As a result, many black moths were eaten by the birds. Few survived to reproduce.

Many large coal-burning factories were built near where the moths lived. The soot given off by these factories soon changed the color of the tree trunks in the area from gray to black. Now the gray moths were easily seen and eaten by the birds. The black moths blended in with the color of the tree trunks. Many black moths survived and reproduced. In a short period of time, most British peppered moths were black.

19-6 How have humans changed through time?

Objective ▶ Describe some of the ways that humans have changed through time.

TechTerm

▶ **anthropology** (an-thruh-PAHL-uh-jee): science that deals with the study of human beings

Early Humanlike Species The fossil record of human evolution is not complete. It is still being pieced together by scientists who work in the field of **anthropology** (an-thruh-PAHL-uh-jee) Anthropology is the science that deals with the study of human beings.

The oldest humanlike fossil that has been found is thought to be about 3.5 million years old. The skeleton indicates that the individual was a little more than 1 m tall and walked upright. Later humanlike fossils of different species of humans also have been found. The ages of these fossils range from about 3 million years to as late as 0.5 million years. Fossils of each species show more humanlike traits and behaviors than the species that lived before them. These changes include increased body size and larger skulls. Evidence also indicates that some of the later species lived in caves, used fire, and made tools.

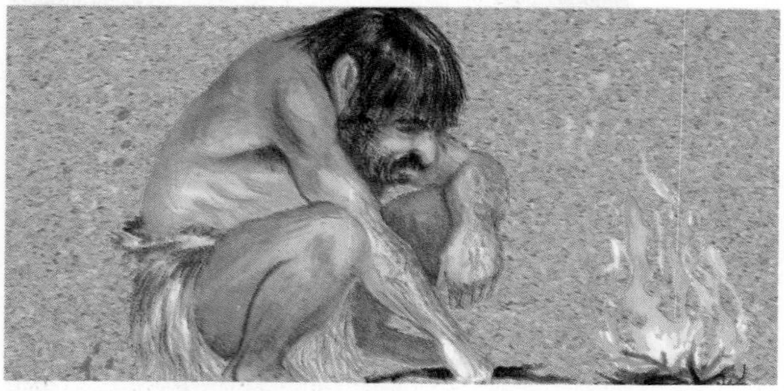

Figure 1 Early human

Interpret: What does fossil evidence show about the evolution of humanlike species?

Modern Humans All modern humans belong to the species *Homo sapiens* (say-PEE-uns), which means "wise human." Fossils of two early types of *Homo sapiens* have been found. The two types are the Neanderthals (nee-AN-dur-thals) and Cro-Magnons (kroh-MAG-nunz). Neanderthals lived from 130,000 to 35,000 years ago. They were shorter than modern humans and they had much larger skulls than did earlier humanlike species. The skulls had sloping foreheads and heavy brow ridges.

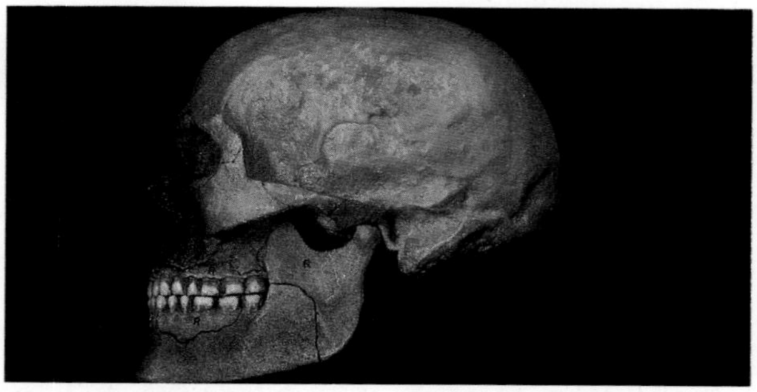

Figure 2 Cro-Magnon skull

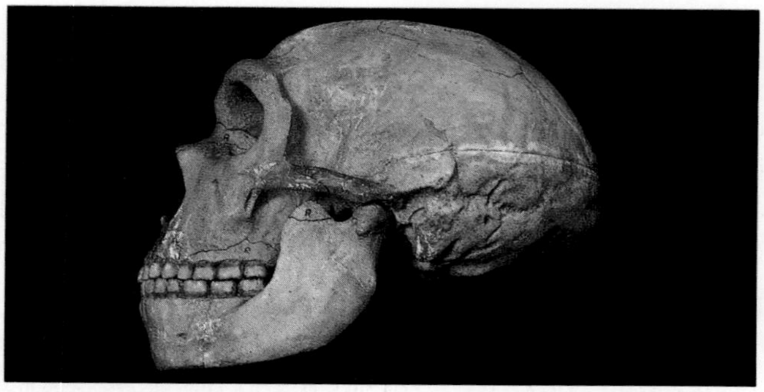

Figure 3 Neanderthal skull

Cro-Magnons replaced Neanderthals when they died out. Cro-Magnons looked like modern humans. They were tall and had large brain cases and rounded skulls. Fossil evidence shows that Cro-Magnons lived together in large groups. They were skilled hunters and toolmakers.

Contrast: How are the fossils of Neanderthals different from fossils of earlier humanlike species?

LESSON SUMMARY

▶ Anthropology is the science that deals with the study of human beings.

▶ The oldest humanlike fossil found to date is thought to be about 3.5 billion years old.

▶ Modern humans belong to the species *Homo sapiens*. Two early examples of this species are Neanderthals and Cro-Magnons.

CHECK *Write true if the statement is true. If the statement is false, change the underlined term to make the statement true.*

1. The science that deals with the study of human beings is <u>biology</u>.

2. The oldest humanlike fossil found to date is about <u>3.5 billion</u> years old.

3. <u>Neanderthals</u> looked like modern humans.

Answer the following.

4. What evidence tells anthropologists about early humanlike activities?

5. Which *Homo sapiens* skulls had sloped foreheads and heavy brow ridges?

6. **Infer:** Why do you think modern humans were named *Homo sapiens?*

APPLY *Complete the following.*

7. **Hypothesize:** A fossil skeleton is found in a layer of sedimentary rock. The skull has a sloped forehead and heavy brow ridges. Several years earlier, a Cro-Magnon skeleton was found several layers deeper in the same sedimentary rock. Why might an anthropologist be puzzled by the arrangement of the two fossil skeletons? How might this situation be explained?

Skill Builder

Building Vocabulary Anthropology is the science that deals with the study of human beings. Find out where the word "anthropology" comes from. In the dictionary, look up the prefix "anthropo-" and the suffix "-logy". Write the meanings on a piece of paper along with their origins.

Ideas in Action

IDEA: Anthropologists have discovered that Cro-Magnons were skilled tool makers. People today also use many tools.

ACTION: List three kinds of tools that you use often. Explain how you use each tool.

PEOPLE IN SCIENCE

THE LEAKEYS

In the field of anthropology, no name is more familiar than that of the Leakey family. For more than 50 years, Mary and Louis Leakey searched for and studied fossil remains of early ancestors of humans. Like most anthropologists, the goal of the Leakeys was to trace the evolutionary development of humans.

In 1959, Mary Leakey found a humanlike skull that was more than 1.75 million years old. In 1972, Richard Leakey, son of Mary and Louis, discovered a skull similar to the one found by his mother. This second skull was determined to be about 2 million years old. Crude stone tools found with fossils led the Leakeys to call their discovery "Handy Man."

In 1984, Richard Leakey and his co-workers made a very significant discovery in Kenya, Africa. They found a skull and almost complete skeleton of a 12 year old male. The skeleton, which was about 1 million years old, was of a species known as "Upright Human." This species is the closest known ancestor to modern humans. The skeleton is one of the most complete specimens discovered to date.

Challenges

STUDY HINT Before you begin the Unit Challenges, review the TechTerms and Lesson summary for each lesson in this unit.

TechTerms

adaptation (386)
amber (382)
anthropology (392)
era (384)

evolution (386)
extinct (382)
fossil (382)
geologic time scale (384)

mutation (386)
natural selection (388)
relative age (384)
variation (388)

TechTerm Challenges

Matching *Write the TechTerm that matches each description.*
1. largest division of the geologic time scale
2. change in organisms over time
3. sudden change in a gene
4. remain of a once-living organism
5. survival of offspring that have favorable traits
6. differences in traits among individuals of a species
7. no longer found as a living species
8. hardened tree sap
9. science that deals with the study of human beings
10. special trait that helps an organism survive in its environment

Fill in *Write the TechTerm that best completes each statement.*
1. A record of the earth's history based on the kinds of organisms that lived at different times is a _____ .
2. The age of a fossil compared with that of a different fossil is _____ .
3. The changes shown in the fossil record of the horse illustrate _____ .
4. Dinosaurs are an example of a group of organisms that are _____ .

Content Challenges

Multiple Choice *Write the letter of the term that best completes each statement.*
1. An insect in amber is an example of
   a. a mutation. **b.** a fossil. **c.** variation. **d.** evolution.
2. Organisms that are no longer found as living species are
   a. mammals. **b.** fossils. **c.** extinct. **d.** endangered.
3. The shortest time division on a geologic time scale is
   a. an epoch. **b.** a century. **c.** an era. **d.** a period.
4. A sudden change in a gene is called
   a. an adaptation. **b.** a variation. **c.** evolution. **d.** a mutation.
5. Most fossils are found
   a. frozen in ice. **b.** in igneous rocks. **c.** in sedimentary rocks. **d.** in tar.

6. The long neck of a giraffe is an example of
   a. an adaptation. **b.** a mutation. **c.** a variation. **d.** extinction.

7. The animal with the most complete fossil record is the
   a. elephant. **b.** finch. **c.** human. **d.** horse.

8. Fossil evidence indicates that the first living things appeared on the earth about
   a. 10 million years ago. **b.** 4.5 billion years ago. **c.** 10 thousand years ago. **d.** 1 billion years ago.

9. The science that deals with the study of humans is
   a. geology. **b.** genetics. **c.** anthropology. **d.** biology.

10. Of the organisms listed, the one least likely to be found as a fossil is a
    a. dinosaur. **b.** horse. **c.** jellyfish. **d.** finch.

True/False *Write true if the statement is true. If the statement is false, change the underlined term to make the statement true.*

1. Rocks that usually form in water are metamorphic.
2. In a geologic time scale, eras are divided into periods.
3. The change in a species over time is called variation.
4. Survival of the fittest also is known as natural selection.
5. All modern humans belong to the species *Homo sapiens*.
6. Dinosaurs are examples of a living species.
7. The smallest time division in a geologic time scale is an era.
8. The age of something compared to the age of something else is absolute age.
9. In a sedimentary rock, the bottom layer usually is the oldest layer.
10. Differences in traits among organisms of the same species is called variation.

Understanding the Features...

Reading Critically *Use the feature reading selections to answer the following. Page numbers for the features follow each question in parentheses.*

1. What was the purpose of the voyage of the HMS *Beagle?* (387)
2. **Infer:** How might the story of the British peppered moths have been different if the factories burned natural gas? (391)
3. Why is carbon-14 an important tool of anthropologists? (385)
4. What discoveries have been made by members of the Leakey family in the field of anthropology? (393)

Critical Thinking *Answer each of the following in complete sentences.*

1. Explain why carbon-14 cannot be used to find the ages of fossils of the earliest living things.
2. Explain the difference between variation and mutation.
3. At one time there were as many short-necked giraffes as there were long-necked giraffes. Using the theory of natural selection, explain why the long-necked giraffes survived , while the short-necked giraffes died out.
4. Why are few fossils found in metamorphic and igneous rocks?
5. Artifacts (AHR-tuh-fakts) are items made by humans. Explain why the study of artifacts is an important part of an anthropologist's work.

Concept Challenges

Interpreting a Table *Use the geologic time scale to answer the following questions.*

Table 1 Geologic Time Scale				
ERA	PERIOD	EPOCH	START DATE (MILLIONS OF YEARS AGO)	ORGANISMS
Cenozoic	Quaternary	Recent	0.025	Modern Humans
		Pleistocene	1.75	Mammoths
	Tertiary	Pliocene	14	Large carnivores
		Miocene	26	Many land mammals
		Oligocene	40	Primitive apes
		Eocene	55	Early horses
		Paleocene	65	Primates
Mesozoic	Cretaceous		130	Flowering plants
	Jurassic		180	Dinosaurs, birds
	Triassic		225	Conifers
Paleozoic	Permian		275	Seed Plants
	Carboniferous		345	Reptiles
	Devonian		405	Insects, amphibians
	Silurian		435	Fishes
	Ordovician		480	Algae, fungi
	Cambrian		600	Invertebrates
Precambrian			over 600 billion	Bacteria, blue-green algae

1. What are the three divisions of geologic time?
2. **Analyze:** What is the name of the most recent era in geologic time?
3. How long ago did dinosaurs roam the Earth?
4. **Explain:** What is the geologic time scale?
5. **Analyze:** During which epoch did reptiles first appear?
6. **Analyze:** During which period did flowering plants first appear?
7. **Analyze:** What organisms are found in the Oligocene epoch?
8. What are the four eras?
9. **Analyze:** When did modern humans first appear?

Finding Out More

1. Visit a zoo and observe the behavior of some primates, such as monkeys, apes, and gorillas. List several kinds of behavior, and compare that behavior in each of the primates you observe.
2. Use library references to find out more about Charles Darwin and his voyage on the *Beagle.* Include a discussion on his theory of evolution. Present your findings in a report.
3. Write a report on one of the following scientists: Charles Lyell, Thomas Huxley, or Joseph Hooker. Include in your report how the scientist you chose influenced Charles Darwin.
4. Write to your state wildlife agency and get a list of organisms that are endangered and threatened. Pick five endangered animals and five threatened animals and find out what is being done to protect these animals. Compare your findings with those of your classmates.
5. Use library references to find out about Donald Johanson and his theory of evolution. Develop a chart illustrating your findings.

Appendix A

THE METRIC SYSTEM AND SI UNITS

The metric system is an international system of measurement based on units of 10. More than 90 percent of the nations of the world use the metric system. In the United States, both the English or Imperial Measurement System and the metric system are used.

Systeme International, or SI, has been used as the international measurement system since 1960. SI is a modernized version of the metric system. Like the metric system, SI is a decimal system based on units of 10.

In both SI and the metric system, prefixes are added to base units to form larger or smaller units. Each unit is 10 times larger than the next smaller unit, and 10 times smaller than the next larger unit. For example, the meter is the basic unit of length. The next larger unit is a dekameter. A dekameter is 10 times larger than a meter. The next smaller unit is a decimeter. A decimeter is 10 times smaller than a meter. Ten decimeters is equal to one meter. How many meters equal one dekameter? 10

When you want to change from one unit in the metric system to another unit, you multiply or divide by a multiple of 10.

• When you change from a smaller unit to a larger unit, you divide.

• When you change from a larger unit to a smaller unit, you multiply.

SI UNITS	
The basic unit is printed in capital letters.	
Length	**Symbol**
kilometer	km
METER	m
centimeter	cm
millimeter	mm
Area	**Symbol**
square kilometer	km²
SQUARE METER	m²
square millimeter	mm²
Volume	**Symbol**
CUBIC METER	m³
cubic millimeter	mm³
liter	L
milliliter	mL
Mass	**Symbol**
KILOGRAM	kg
gram	g
tonne	t
Temperature	**Symbol**
KELVIN	K
degree Celsius	°C

SOME COMMON METRIC PREFIXES		
Prefix		**Meaning**
micro-	=	0.000001, or 1/1,000,000
milli-	=	0.001, or 1/1000
centi-	=	0.01, or 1/100
deci-	=	0.1, or 1/10
deka-	=	10
hecto-	=	100
kilo-	=	1000
mega-	=	1,000,000

SOME METRIC RELATIONSHIPS	
Unit	**Relationship**
kilometer	1 km = 1000 m
meter	1 m = 100 cm
centimeter	1 cm = 10 mm
millimeter	1 mm = 0.1 cm
liter	1 L = 1000 mL
milliliter	1 mL = 0.001 L
tonne	1t = 1000 kg
kilogram	1 kg = 1000 g
gram	1 g = 1000 mg
centigram	1 cg = 10 mg
milligram	1 mg = 0.001 g

SI-ENGLISH EQUIVALENTS

	SI to English	*English to SI*
Length	1 kilometer = 0.621 mile (mi)	1 mi = 1.61 km
	1 meter = 0.914 yards (yd)	1 yd = 1.09 m
	1 meter = 3.28 feet (ft)	1 ft = 0.305 m
	1 centimeter = 0.394 inch (in)	1 in = 2.54 cm
	1 millimeter = 0.039 inch	1 in = 25.4 mm
Area	1 square kilometer = 0.3861 square mile	$1 \text{ mi}^2 = 2.590 \text{ km}^2$
	1 square meter = 1.1960 square yards	$1 \text{ yd}^2 = 0.8361 \text{ m}^2$
	1 square meter = 10.763 square feet	$1 \text{ ft}^2 = 0.0929 \text{ m}^2$
	1 square centimeter = 0.155 square inch	$1 \text{ in}^2 = 6.452 \text{ cm}^2$
Volume	1 cubic meter = 1.3080 cubic yards	$1 \text{ yd}^3 = 0.7646 \text{ m}^3$
	1 cubic meter = 35.315 cubic feet	$1 \text{ ft}^3 = 0.0283 \text{ m}^3$
	1 cubic centimeter = 0.0610 cubic inches	$1 \text{ in}^3 = 16.39 \text{ cm}^3$
	1 liter = .2642 gallon (gal)	1 gal = 3.79 L
	1 liter = 1.06 quart (qt)	1 qt = 0.94 L
	1 liter = 2.11 pint (pt)	1 pt = 0.47 L
	1 milliliter = 0.034 fluid ounce (fl oz)	1 fl oz = 29.57 mL
Mass	1 tonn = .984 ton	1 ton = 1.016 t
	1 kilogram = 2.205 pound (lb)	1 lb = 0.4536 kg
	1 gram = 0.0353 ounce (oz)	1 oz = 28.35 g
Temperature	Celsius = 5/9 (°F −32)	Fahrenheit = 9/5°C + 32
	0°C = 32°F (Freezing point of water)	72°F = 22°C (Room temperature)
	100°C = 212°F (Boiling point of water)	98.6=F = 37°C
		(Human body temperature)

Appendix B

SAFETY IN THE SCIENCE CLASSROOM

Safety is very important in the science classroom. Science classrooms and laboratories have equipment and chemicals that can be dangerous if not handled properly. To avoid accidents in the science laboratory, always follow proper safety rules. Listen carefully when your teacher explains precautions and safety rules that must be followed. You should never perform an activity without your teacher's direction. By following safety rules you can help insure the safety of yourself and your classmates. Safety rules that should be followed are listed below. Read over these safety rules carefully. Always look for caution statements before you perform an activity.

Clothing Protection • Wear your laboratory apron. • Confine loose clothing.

Eye Safety • Wear safety goggles in the laboratory. • If anything gets in your eyes, flush them with plenty of water. • Be sure you know how to use the emergency wash system in the laboratory.

Heat and Fire Safety • Be careful when handling hot objects. • Use proper procedures when lighting Bunsen burners. • Turn off all heat sources when they are not in use. • Tie back long hair when working near an open flame. • Confine loose clothing. • Turn off gas valves when not in use.

Electrical Safety • Keep all work areas clean and dry. • Never handle electrical equipment with wet hands. • Do not overload an electric circuit. • Do not use wires that are frayed.

Glassware Safety • Never use chipped or cracked glassware. • Never pick up broken glass with your bare hands. • Allow heated glass to cool before touching it. • Never force glassware into a rubber stopper.

Chemical Safety • Never taste chemicals as a means of identification. • Never transfer liquids with a mouth pipette. Use a suction bulb. • Be very careful when working with acids or bases. Both can cause serious burns. • Never pour water into an acid or base. Always pour an acid or base into water. • Inform your teacher immediately if you spill chemicals or get any chemicals on your skin. • Use a waving motion of your hand to observe the odor of a chemical. • Never put your nose near a chemical. • Never eat or drink in the laboratory.

Sharp Objects • Use knives, scissors, and other sharp instruments with care. • Cut in the direction away from your body.

Cleanup • Clean up your work area before leaving the laboratory. • Follow your teacher's instructions for disposal of materials. • Wash your hands after an activity.

Appendix C   *The Microscope*

One of the most important tools in biology is the microscope. A microscope enables scientists to view and study objects or structures not visible to the unaided eye. The compound microscope is used most often in biology classes.

A compound microscope has two or more lenses. One lens, called the ocular lens, is located in the eyepiece. The ocular lens usually has a magnification of 10X. An object viewed through this lens would appear 10 times larger than it would look with the unaided eye.

The second lens is called the objective lens. Compound microscopes may have many objective lenses. Most compound microscopes, however, have two objective lenses. Each objective lens has a different magnification. The magnification is printed on each objective lens.

To find the total magnification of a microscope, multiply the magnification of the ocular lens by the magnification of the objective lens that you are using. For example, the ocular lens magnification of 10X multiplied by the objective lens magnification of 10X equals 100X. The total magnification of the microscope is 100X. Looking at an object under this total magnification, you would be seeing the object 100 times larger than the object would look with the unaided eye.

A microscope is a delicate, but relatively uncomplicated, easy-to-use tool. Before you try to use a microscope, however, you should know the parts of a microscope and what each part does. Study the parts of the microscope and their functions.

Parts of the Microscope

1. Eyepiece: holds the ocular lens
2. Body tube: keeps the ocular and objective lenses the proper distance apart
3. Arm: supports the body tube
4. Stage: platform that supports the slide
5. Stage clips: hold slide in position
6. Diaphragm: adjusts the amount of light entering the stage opening
7. Base: supports the microscope
8. Illuminator: mirror or electric light that is the light source of the microscope
9. Revolving nosepiece: holds the objectives and can be turned to change objectives
10. Low-power objective: shorter objective that contains a lens with a magnification of 10X
11. High-power objective: longer objective that contains a lens with a magnification of 40X
12. Coarse adjustment knob: moves the body tube up and down for bringing the object into view
13. Fine adjustment knob: moves the body tube up and down for focusing clearly

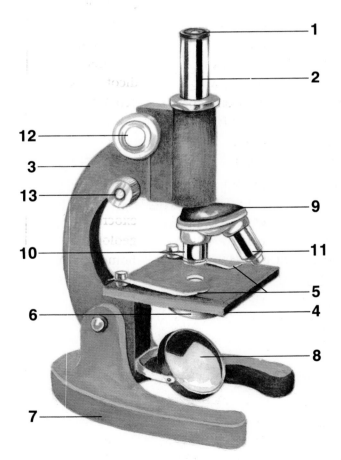

Appendix D

PREFIXES AND SUFFIXES

Prefixes and suffixes are word parts that can be helpful in determining the meaning of an unfamiliar term. Prefixes are found at the beginning of word. Suffixes are found at the end of words. Both prefixes and suffixes have meanings that mainly come from Latin and Greek words. Some meanings of prefixes and suffixes commonly used in life science words are listed below.

PREFIX	MEANING	EXAMPLE
a-	not; without	abiotic
aero-	air	aerobic
anti-	against	antibodies
bi-	two	biceps
bio-	life	biotechnology
centi-	hundred	centimeters
chemo-	of, with, or by chemicals	chemosynthesis
chloro-	green	chloroplasts
cyto-	cell	cytoplasm
deca-	ten	decameter
deci-	one tenth	deciliter
di-	twice; double	dicot
eco-	environment; habitat	ecosystem
ecto-	outer	ectoderm
endo-	inside	endospore
epi-	on; on the outside	epidermis
exo-	outside	exocrine
geo-	earth	geologic
hemo-	blood	hemoglobin
hydro-	water	hydroponics
kilo-	thousand	kilometer
leuco-	white	leucocyte
mono-	one	monocot
photo-	light	photosynthesis
syn-	together with; by means of	synthetic
thigmo-	touch	thigmotropism
tri-	three	triceps
trop-	turn; respond to	tropism
uni-	one	univalve
zoo-	animal	zoology

SUFFIX	MEANING	EXAMPLE
-derm	skin, covering	echinoderm
-itis	disease of	appendicitis
-logy	study of; science of	biology
-ose	carbohydrate	glucose
-phyte	a plant; to grow in a certain way or place	bryophyte
-vore	to eat	carnivore

Glossary

Pronunciation and syllabication have been derived from *Webster's New World Dictionary,* Second College Edition, Revised School Printing (Prentice Hall, 1985). Syllables printed in capital letters are given primary stress (Numbers in parentheses indicate the page number, or page numbers, on which the term is defined.)

PRONUNCIATION KEY

Symbol	Example	Respelling	Symbol	Example	Respelling
ah	molecule	(MAHL-uh-kyool)	k	Calorie	(KAL-uh-ree)
aw	absorption	(ab-SAWRP-shun)	ks	thorax	(THOR-aks)
ay	adaptation	(ad-up-TAY-shun)	oh	embryo	(EM-bree-oh)
ee	ecology	(ee-KAHL-uh-jee)	oo	asexual	(ay-SEK-shoo-wul
ew	nucleus	(NEW-klee-us)	sh	circulation	(sur-kyuh-LAY-shun)
f	phloem	(FLOH-em)	uh	data	(DAY-tuh)
g	gamete	(GAM-eet)	y	migration	(my-GRAY-shun)
ih	specialization	(SPESH-uh-lih-zay-shun)	yoo	cellulose	(SEL-yoo-lohs)
j	genes	(JEENS)	z	organism	(AWR-guh-niz-um)

A

abdomen (AB-duh-mun): third section of an insect's body (182)

absorption (ab-SAWRP-shun): movement of food from the digestive system to the blood (234)

adaptation (ad-up-TAY-shun): trait of a living thing that helps it live in its environment (36, 386)

addiction (uh-DIK-shun): uncontrollable dependence on a drug (326)

AIDS: viral disease that attacks a person's immune system (324)

alcoholic (al-kuh-HOWL-ik): person who is dependent on alcohol (328)

alveoli (al-VEE-uh-ly): microscopic air sacs in the lungs (270)

amber: hardened tree sap (382)

amino acid: building block of proteins (216)

amnion (AM-nee-on): fluid-filled membrane that surrounds an embryo (346)

amphibian (am-FIB-ee-un): animal that lives part of its life in water and part on land (198)

angiosperms (AN-jee-uh-spurms): flowering plants (140)

anther: part of the stamen that produces pollen (156)

anthropology (an-thruh-PAHL-uh-jee): science that deals with the study of human beings (392)

antibiotic (an-ti-by-AHT-ik): chemical made by a living organism that kills bacteria (318)

antibodies (AN-ti-bahd-eez): substances the body makes to protect itself from disease (312)

aorta (ay-OWR-tuh): largest artery in the body (258)

arteries (ART-ur-ees): blood vessels that carry blood away from the heart (258)

asexual reproduction (ay-SEK-shoo-wul ree-pruh-DUK-shun): reproduction needing only one parent (42, 126, 162)

atherosclerosis (ath-ur-oh-skluh-ROH-sis): buildup of fat deposits on artery walls (320)

atrium (AY-tree-um): upper chamber of the heart (256)

axon: fiber that carries messages away from a nerve cell (288)

B

bacilli (buh-SIL-y): rod-shaped bacteria (116)

bacteriology (bak-tir-ee-AHL-uh-jee): study of bacteria (118)

bacteriophage (bak-TIR-ee-uh-fayj): virus that infects bacteria (106)

behavior (bi-HAYV-yur): ways in which living things respond to stimuli (38)

benign (bi-NYN) **tumor:** mass of cells that is usually harmless (322)

bile: green liquid that breaks down fats and oils (232)

biome (BY-ohm): large region of the earth with particular plant and animal communities (68)

blade: wide, flat part of a leaf (150)

blending: combination of genes in which a mixture of both traits shows (364)

bronchi (BRAHN-kee): tubes leading to the lungs (270)

bryophytes (BRY-uh-fyts): plant division that includes mosses, liverworts, and hornworts (108); plant that reproduces by spores and has no transport tubes (134)

budding: kind of asexual reproduction in which a new organism forms from a bud on a parent (126)

bulb: underground stem covered with fleshy leaves (162)

C

Calorie (KAL-uh-ree): unit used to measure energy from foods (224)

cap: umbrella-shaped top of a mushroom (124)

capillaries (KAP-uh-ler-ees): tiny blood vessels that connect arteries to veins (258)

capsid (KAP-sid): protein covering of a virus (106)

carbohydrate (kar-buh-HY-drayt): nutrient that supplies energy (214)

cardiac (KAHR-dee-ak) **muscle:** type of muscle found only in the heart (248)

carrying capacity: largest amount of a population that can be supported by an area (58)

cartilage (KART-ul-idj): tough, flexible connective tissue (196, 240)

cell: basic unit of structure and function in living things (34, 78)

cell division: process by which cells reproduce (88)

cell membrane (MEM-brayn): thin structure that surrounds a cell (80)

cell wall: outer, nonliving part of a plant cell (84)

cellulose (SEL-yoo-lohs): hard, nonliving material that makes up the cell wall of a plant cell (84)

cerebellum (ser-uh-BELL-um): part of the brain that controls balance and body motion (290)

cerebrum (suh-REE-brum): large part of the brain that controls the senses and thinking (290)

cervix (SUR-viks): narrow end of the uterus (338)

chemical digestion: process by which large food molecules are broken down into smaller food molecules (228)

chitin (KYT-in): hard material that makes up the exoskeleton of arthropods (180)

chlorophyll (KLOR-uh-fil): green material in chloroplasts that is needed for plants to make food using photosynthesis (84, 122, 152)

chloroplast: organelle of green plant cells where photosynthesis takes place (84, 152)

chromosomes (KROH-muh-sohms): cell parts that determine what traits a living thing will have (88); threadlike structures in the nucleus of a cell that control heredity (356)

chyme (KYM): thick liquid form of food (230)

cilia (SIL-ee-uh): microscopic hairlike structures (120, 274)

circulation (sur-kyuh-LAY-shun): movement of blood through the body (262)

cirrhosis (suh-ROH-sis): liver disorder caused by damaged liver cells (328)

classification (klas-uh-fi-KAY-shun): grouping things according to similarities (100)

climate: overall weather in an area over a long period of time (68)

climax (KLY-maks) **community:** last community in a succession (66)

closed circulatory (SUR-kyuh-luh-towr-ee) **system:** organ system in which blood moves through blood vessels (176, 262)

cnidocytes (NY-duh-syts): stinging cells (172)

cocci (KAK-sy): spherical bacteria (116)

cochlea (KOK-lee-uh): part of the ear that changes vibrations into nerve signals (298)

cocoon (kuh-KOON): protective covering around the pupa (184)

coldblooded: having a body temperature that changes with the temperature of the surroundings (196)

colon: large intestine (234)

community: all the populations that live in a certain place (54)

compact bone: bone cells that make up the hardest part of bones (242)

conditioned (kun-DISH-und) **response:** learned behavior in which a new stimulus causes the same response that an old one did (304)

conifer (KAHN-uh-fur): tree that produces cones and has needlelike leaves (138)

connective tissue: tissue that holds parts of the body together (92)

conservation (kon-sur-VAY-shun): wise use of natural resources (70)

consumer (kun-SOO-mur): organism that obtains food by eating other organisms (62)

control: part of the experiment in which no change is made (20)

controlled breeding: mating organisms to produce offspring with certain traits (374)

controlled experiment: two experiments exactly alike except for one change in one of them (20)

cornea (KOR-nee-uh): clear covering at the front of the eye (296)

coronary (KOWR-uh-ner-ee) **arteries:** arteries that carry blood and oxygen to the tissues of the heart (320)

cotyledon (kaht-LEED-on): seed structure that contains food for the developing plant (140)

cycle (SY-kul): something that happens over and over in the same way (60)

cytoplasm (SYT-uh-plaz-um): all the living material inside a cell except the nucleus (80)

D

data (DAY-tuh): information (18)

daughter cells: new cells produced by cell division (88)

decomposer (dee-kum-POHZ-er): organism that breaks down the wastes or remains of other organisms (62)

deficiency (di-FISH-un-see) **disease:** disease caused by the lack of a certain nutrient (218)

degree Celsius: metric unit of temperature (24)

dendrite (DEN-dryt): fiber that carries messages from other neurons to the nerve cell body (288)

depressant (di-PRES-unt): drug that slows down the central nervous system (326)

dermis: living, inner layer of skin (250)

diaphragm (DY-uh-fram): sheet of muscle below the lungs (278)

dicots: seed plants with two cotyledons (140)

diffusion (dih-FYOO-shun): movement of material from an area where molecules are crowded to an area where they are less crowded (86)

digestion (di-JES-chun): process of breaking down food into forms that can be used by living things (44, 226)

DNA: chemical contained in chromosomes (358)

dominant (DOM-uh-nuhnt) **gene:** stronger gene that always shows itself (360)

E

eardrum: sheet of tissue that vibrates when sounds strike it (298)

echinoderms (ee-KY-noh-durms): spiny-skinned animals (186)

ecology (ee-KAHL-uh-jee): study of living things and their environments (52)

ecosystem (EE-koh-sis-tum): living and nonliving things in an environment, together with their interactions (54)

ectoderm (EK-tuh-durm): outer tissue layer (174)

egg: female sex cell (208)

embryo (EM-bree-oh): undeveloped plant or animal (156); hollow ball of cells formed by cell division of the zygote (346)

emulsification (i-mul-suh-fi-KAY-shun): process of breaking down large droplets of fat into small droplets of fat (232)

endangered species (in-DAYN-jurd SPEE-sheez): living things that are in danger of dying out (72)

endocrine (EN-duh-krin) **gland:** gland that does not have ducts (300)

endoderm (EN-duh-durm): inner tissue layer (174)

endoskeleton (en-doh-SKEL-uh-tun): internal skeleton (110, 186, 194, 240)

endospore: inactive cell surrounded by a thick wall (116)

energy pyramid (PIR-uh-mid): way of showing how energy moves through a food chain (64)

environment (in-VY-run-munt): everything that surrounds a living thing (36, 52)

enzyme (EN-zym): protein that controls chemical activities (216)

epidermis (ep-uh-DUR-mis): outer, protective layer of the leaf (150); dead, outer layer of skin (250)

epiglottis (ep-uh-GLAT-is): flap of tissue that prevents food from entering the windpipe (226)

epithelial (ep-uh-THEEL-ee-uhl) **tissue:** tissue that covers and protects parts of the body (92)

era: largest division of geologic time (384)

esophagus (i-SAF-uh-gus): tube that connects the mouth to the stomach (226)

evaporation (i-VAP-uh-ray-shun): changing of a liquid to a gas (276)

evolution (ev-uh-LOO-shun): process by which organisms change over time (386)

excretion (ik-SKREE-shun): process of removing waste products from the body (44, 276)

exhale: to breathe out (272)

exoskeleton (ek-so-SKEL-uh-tun): external skeleton (110, 180, 240)

experiment (ik-SPER-uh-munt): something that is done to test a hypothesis or prediction (20)

extensor (ik-STEN-sur): muscle that straightens a joint (246)

extinct (ik-STINKT): species that is no longer found alive (382)

F

fertilization (fur-tul-i-ZAY-shun): union of a male sex cell and a female sex cell (156, 208, 344)

fetus (FEET-us): developing baby (346)

fibrous (FY-brus) **root system:** root system made up of many thin, branched roots (146)

filament: stalk of the anther (156)

flagella (fluh-JEL-uh): whiplike structures on a cell (116, 170)

flexor (FLEK-sur): muscle that bends a joint (246)

food chain: way of showing the food relationships among a group of organisms (64)

food web: way of showing how food chains are related (64)

fossil (FAHS-ul): remains or traces of a once-living organism (382)

frond: featherlike leaf of a fern (136)

fungi (FUN-gy): plantlike organisms that lack chlorophyll (104)

G

gamete (GAM-eet): reproductive cell (336)

gastric juice: juice produced in the stomach that contains mucus, pepsin, and hydrochloric acid (230)

gene splicing (SPLYS-ing): moving a section of DNA from the genes of one organism to the genes of another organism (376)

genes (JEENS): parts of a chromosome that control inherited traits (356)

genetic engineering (juh-NET-ik en-juh-NEER-ing): methods used to produce new forms of DNA (375)

genetics (juh-NET-iks): study of heredity (354)

genus (JEE-nus): classification group made up of related species (102)

geologic (jee-oh-LAHJ-ik) **time scale:** record of the earth's history based upon types of organisms that lived at different times (384)

gills: structures that produce mushroom spores (124); organs that absorb dissolved oxygen from water (196)

gland: organ that makes chemical substances used or released by the body (300)

gram: basic metric unit of mass (22)

guard cell: cell in a plant that helps control the passage of materials into and out of the stomates (90)

gymnosperms (JIM-nuh-spurms): group of land plants with uncovered seeds (138)

H

habitat (HAB-i-tat): place where an organism lives (56)

hallucinogen (hul-LOO-suh-nuh-jen): drug that causes a person to see, hear, smell, and taste things in a strange way (326)

heart attack: lack of blood and oxygen to a part of the heart (320)

hemoglobin (HEE-moh-gloh-bin): iron compound in red blood cells (260)

herbaceous (hur-BAY-shus) **stem:** stem that is soft and green (148)

heredity (huh-RED-uh-tee): passing of traits from parents to offspring (354)

hibernation (HY-bur-nay-shun): inactive state of some animals during winter months (38)

hilum (HY-lum): mark on the seed coat where the seed was attached to the ovary (160)

homeostasis (hoh-mee-oh-STAY-sis): ability of a living thing to keep conditions inside its body constant (46)

hormone (HAWR-mohn): chemical substance that regulates body functions (302)

hybrid (HY-brid): having two unlike genes (360)

hybridization (hy-brid-ih-ZAY-shun): mating two different kinds of organisms (374)

hypothesis (hy-PAHTH-uh-sis): suggested answer to a problem (16)

I

immune system: body system made up of cells and tissues that help a person fight disease (324)

immunity (im-MYOON-i-tee): resistance to a specific disease (314)

inbreeding: mating closely related organisms (374)

ingestion (in-JEST-shun): process of taking in food (44)

inhale: to breathe in (272)

inherited disease: disease caused by an inherited gene (370)

inherited (in-HER-uh-ted) **traits:** traits that are passed from parents to their offspring (354)

innate behavior: behavior you are born with (304)

interact: to act upon each other (52)

invertebrates (in-VUR-tuh-brayts): animals without backbones (110)

iris (Y-ris): part of the eye that controls the amount of light entering the eye (296)

J

joint: place where two or more bones meet (244)

K

kingdom: classification group made up of related phyla (102)

L

larva (LAHR-vuh): wormlike stage of insect development (184)

learned behavior: behavior you practice and learn (304)

legumes (LEG-yooms): group of plants that includes beans and peas (132)

lens (LENZ): piece of curved glass that causes light rays to come together or spread apart as they pass through (26); part of the eye that forms an image on the retina (296)

ligaments (LIG-uh-ments): tissue that connects bone to bone (244)

limiting factors: conditions in the environment that put limits on where an organism can live (58)

lipase (LY-pays): enzyme that digests fats and oils (232)

liter: basic metric unit of volume (22)

M

magnify: to make something look larger than it is (26)

malignant (muh-LIG-nunt) **tumor:** harmful mass of cells that can spread throughout the body (322)

mammary (MAM-ur-ee) **glands:** glands where milk is produced in female mammals (206)

mantle: thin membrane that covers a mollusk's organs (178)

mass: amount of matter in an object (22)

mass selection: crossing plants with desirable traits (374)

mechanical digestion: process by which large pieces of food are cut and crushed into smaller pieces (228)

medulla (muh-DULL-uh): part of the brain that controls heartbeat and breathing rate (290)

medusa (muh-DOO-suh): umbrellalike form of a cnidarian (172)

meiosis (my-OH-sis): cell division that produces gametes (336)

menstrual (MEHN-struhl) **cycle:** monthly cycle of change that occurs in the female reproductive system (342)

menstruation (men-stroo-WAY-shun): process by which blood and tissue from the lining of the uterus break apart and leave the body (342)

mesoderm (MES-uh-durm): middle tissue layer (174)

mesophyll (MES-uh-fil): middle layer of leaf tissue in which food-making occurs (150)

metamorphosis (met-uh-MOWR-fuh-sis): changes during the stages of development of an organism (184, 200)

meter (MEE-tur): basic metric unit of length (22)

microscope (MIKE-roh-scope): tool that makes things look larger than they really are (26)

migration (my-GRAY-shun): movement of animals from one living place to another (38)

mineral: nutrient needed by the body to develop properly (220)

mitochondria (myt-uh-KAHN-dree-uh): rice-shaped structures that produce energy for a cell (82)

mitosis (my-TOH-sis): division of the nucleus (88)

molecule (MAHL-uh-kyool): smallest part of a substance that has all the properties of that substance (216)

molting: process by which an animal sheds its outer covering (180)

monerans (muh-NER-uns): single-celled organisms that do not have a nucleus (104)

monocots: seed plants with only one cotyledon (140)

mucus (MYOO-kus): sticky liquid (274)

mutation (myoo-TAY-shun): change in a gene (386)

N

natural resources (REE-sowrs-ez): materials found in nature that are used by living things (70)

natural selection: survival of offspring that have favorable traits (388)

neuron (NOOR-ahn): nerve cell (288)

niche (NICH): organism's role, or job, in its habitat (56)

nicotine (NIK-uh-teen): stimulant found in tobacco (330)

nitrogen-fixing bacteria: bacteria that can use nitrogen in soil to make nitrogen compounds (60)

nonrenewable resources: natural resources that cannot be renewed or replaced (70)

notochord (NOHT-uh-kowrd): strong, rodlike structure in chordates that can bend (194)

nucleus (NEW-klee-us): control center of a cell (80)

nutrient (NOO-tree-unt): chemical substances in food needed by the body for growth, energy, and life processes (214)

nymph (NIMF): young insect that looks like the adult (184)

O

offspring: new organisms produced by a living thing (42)

organ (OWR-gun): group of tissues that work together to do a special job (94)

organ system: group of organs that work together (94)

organelles (or-guh-NELS): small structures in the cytoplasm that do special jobs (82)

organism (AWR-guh-niz-um): any living thing (34)

osmosis (ahs-MOH-sis): movement of water through a membrane (86)

ovaries (OH-vuhr-eez): main organs of the female reproductive system (338)

ovary (OH-vuhr-ee): bottom part of the pistil (158)

oviduct (OH-vuh-dukt): long tube between the ovary and the uterus (338)

ovulation (oh-vyuh-LAY-shun): release of a mature egg from the ovary (342)

ovule (OH-vyool): part of the ovary that develops into a seed after fertilization (158)

oxidation (ahk-suh-DAY-shun): slow burning of foods in your body (224)

P

pepsin (PEP-sin): enzyme that digests proteins (230)

periosteum (per-i-AS-tee-um): thin membrane that covers a bone (242)

peristalsis (per-uh-STAWL-sis): wavelike movement that moves food through the digestive tract (226)

perspiration (pur-spuh-RAY-shun): waste water and salts that leave the body through the skin (276)

petal: kind of leaf that is often brightly colored (152)

phloem (FLOH-em): tissue that carries food from the leaves throughout the plant (148)

photosynthesis (foht-uh-SIN-thuh-sis): food-making process in leaves that uses sunlight (108, 152)

phylum (FY-luhm): classification group made up of related classes (102)

pistil: female reproductive organ in a flower (152)

placenta (pluh-SEN-tuh): organ in female placental mammals that connects the mother and embryo through which the embryo receives nourishment and gets rid of wastes (206, 346)

plankton (PLANK-tun): organisms that float at the water's surface (122)

plasma (PLAZ-muh): liquid part of blood (92, 260)

platelets (PLAYT-lits): tiny, colorless pieces of cells (260)

pollen grains: male plant reproductive cells (156)

pollination (pahl-uh-NAY-shun): movement of pollen from a stamen to a pistil (156)

pollution: release of harmful materials into the environment (72)

polyp (PAL-ip): cuplike form of a cnidarian (172)

population: group of the same kind of organism living in a certain place (54)

pore: tiny opening (170, 250, 276)

producer (pruh-DOOS-ur): organism that makes its own food (62)

protein (PRO-teen): nutrient needed to build and repair cells (214)

protists (PROHT-ists): single-celled organisms that have a nucleus (104)

protozoans (proh-tuh-ZOH-uns): one-celled, animallike protists (120)

pseudopod (SOO-duh-pod): fingerlike projections of cytoplasm used for movement and food-getting (120)

pulmonary (PUL-muh-ner-ee) **artery:** artery that brings blood from the heart to the lungs (264)

Punnett square: chart that shows possible gene combinations (362)

pupa (PYOU-puh): resting stage during complete metamorphosis (184)

pupil (PYOO-pil): opening in the center of the iris (296)

pure: having two like genes (360)

R

radula (RAJ-oo-luh): rough, tonguelike organ of a snail (178)

range (RAYNJ): area where a type of animal or plant population is found (58)

receptor (ree-SEP-tor): organ that receives stimuli from the environment (294)

recessive (ri-SES-iv) **gene:** weaker gene that is hidden when the dominant gene is present (360)

red blood cell: blood cell that carries oxygen (90)

reflex: automatic response to a stimulus (292)

reflex arc: path of a message in a reflex (292)

regeneration (ri-jen-uh-RAY-shun): ability to regrow lost parts (188)

relative age: age of something compared with the age of something else (384)

renewable resources: natural resources that can be renewed or replaced (70)

replication (rep-luh-KAY-shun): process by which DNA is duplicated (358)

reproduction: process by which living things produce new organisms like themselves (42)

respiration (res-puh-RAY-shun): process of carrying oxygen to cells, getting rid of carbon dioxide, and releasing energy (44, 272)

response (ri-SPAHNS): reaction to a change, or stimulus (34, 292)

retina (RET-in-uh): part of the eye on which images form (296)

rhizoids (RY-zoidz): rootlike structures that anchor fungi (124); fine, hairlike structures that act as roots (134)

rhizome (RY-zohm): underground stem (136)

ribosomes (RY-buh-sohms): small, round structures that make proteins (82)

root cap: cup-shaped mass of cells that covers and protects a root tip (146)

root hair: thin, hairlike structure on the outer layer of the root tip (146)

S

saliva: liquid in the mouth that helps in digestion (228)

scavenger (SKAV-in-jur): animal that eats only dead organisms (62)

scientific method: model, or guide, used to solve problems and to get information (18)

scrotum (SKROH-tum): pocket of skin that protects and holds the testes (340)

seed: structure that contains a tiny living plant and food for its growth (138)

seed coat: outside covering of a seed (160)

sepal (SEE-pul): special kind of leaf that protects the flower bud (152)

septum: thick tissue wall that separates the left and right sides of the heart (256)

setae (SEET-ee): tiny, hairlike bristles (176)

sex-linked traits: traits that are controlled by the sex chromosomes (368)

sexual (SEK-shoo-wul) **reproduction:** reproduction needing two parents (42)

smooth muscle: muscle that causes movements that you cannot control (248)

specialization (SPESH-uh-lih-zay-shun): studying or working in only one part of a subject (14)

species (SPEE-sheez): group of organisms that look alike and can reproduce among themselves (102)

sperm: male sex cell (208)

spicules (SPIK-yools): small, hard, needlelike structures of a sponge (170)

spiracles (SPIH-ruh-kuls): openings to air tubes of a grasshopper (182)

spirilla (spy-RIL-uh): spiral-shaped bacteria (116)

spongy bone: bone cells that make up the soft and spongy ends of bones (242)

spontaneous generation (spahn-TAY-nee-us jen-uh-RAY-shun): idea that living things come from nonliving things (40)

spore: reproductive cell of bryophytes (134)

spore cases: structures that contains spores (126)

sporulation (spowr-yoo-LAY-shun): kind of asexual reproduction in which a new organism forms from spores released from a parent (126)

stamen (STAY-mun): male reproductive organ in a flower (152)

stigma (STIG-muh): top part of the pistil (158)

stimulant (STIM-yuh-lent): drug that speeds up the central nervous system (326)

stimulus (STIM-yuh-lus): something that causes a reaction to take place (38, 162, 292)

stomata (STOH-muh-tuh): tiny openings in the upper and lower epidermis of the leaf (150)

striated (STRY-ayt-ed) **muscle:** muscle attached to the skeleton, making movement possible (248)

succession (suk-SESH-un): gradual change in organisms that occurs when the environment changes (66)

T

tadpole: larval stage of a frog (200)

taproot system: root system made up of one large root and many small, thin roots (146)

taxonomy (tak-SAHN-uh-mee): science of classifying living things (100)

temperature: measure of how hot or cold something is (24)

tendons (TEN-duns): tissue that connects muscle to bone (246)

tentacles: long, armlike structures (172)

testes (TES-teez): main organs of the male reproductive system (340)

testosterone (tes-TAWS-tuhr-ohn): hormone produced in the testes (340)

theory (THEE-uh-ree): idea that explains something and is supported by data (78)

thorax (THOR-aks): middle section of an insect's body (182)

tissue: group of cells that look alike and work together (92)

tracheophytes (TRAY-kee-uh-fyts): division of plants that have conducting tubes (108)

trait: characteristic (354)

transport: process of moving nutrients and wastes in a living thing (44)

tropism: growth of a plant in response to something in the environment (162)

tube feet: small, suckerlike structures of echinoderms (186)

tuber: underground stem (162)

tumor (TOO-mur): mass or lump of cells (322)

tympanum (TIM-puh-num): hearing organ in a grasshopper (182)

U

umbilical (um-BIL-ih-kul) **cord:** structure that connects the embryo to the placenta (346)

urea (yoo-REE-uh): nitrogen compound found as a waste product (278)

urethra (yoo-REETH-ruh): tube that carries sperm to the outside of the male's body (340)

urine (YOOR-in): liquid waste formed in the kidneys (278)

uterus (YOOT-ur-us): organ in which an embryo develops (338)

V

vacuoles (VAK-yoo-wohls): liquid-filled spaces in the cytoplasm (82)

vagina (vuh-JY-nuh): birth canal (338)

valve: thin flap of tissue that acts like a one-way door (256)

variable: anything that can be changed in an experiment (20)

variation (ver-ee-AY-shun): differences in traits among individuals of a species (388)

vegetative propagation (VEJ-uh-tayt-iv prahp-uh-GAY-shun): kind of asexual reproduction that uses parts of plants to grow new plants (162)

veins (VANES): bundles of tubes that contain the xylem and phloem in a leaf (150); blood vessels that carry blood to the heart (258)

ventricle (VEN-tri-kul): lower chamber of the heart (256)

vertebrate (VUR-tuh-brit): animal with a backbone (110, 194)

villi: fingerlike projections on the lining of the small intestine (234)

virus (VY-rus): piece of nucleic acid covered with a protein (106)

vitamin (VYT-uh-min): nutrient found naturally in many foods (218)

W

warmblooded: having a body temperature that remains about the same (204)

white blood cells: blood cells that protect the body against disease (90, 312)

woody stem: stem that contains wood and is thick and hard (148)

X

xylem (ZY-lum): tissue that carries water and dissolved minerals upward from the roots (148)

Z

zygote (ZY-gote): fertilized egg produced by fertilization (344)

Index

Photo Credits

Unit 1 p. 13 Ron Austing/Photo Researchers, p. 15 J. Howard/Stock, Boston, p. 21 UPI/Bettmann Newsphotos, p. 29 J. Nettis/Photo Researchers

Unit 2 p. 33 FourByFive, p. 36 Mark Antman/The Image Works, p. 37 D. Rusten/Stock, Boston, p. 36 John Elk III/Stock, Boston, p. 36 John Elk III/Stock, Boston, p. 38 Monkmeyer, p. 41 The Granger Collection, p. 42 F. Gohier/Photo Researchers, p. 43 L. Lefever/Grant Heilman, p. 45 CNRI/SPL/Photo Researchers, p. 47 B. Daemmrich/Stock, Boston

Unit 3 p. 51 Runk/Schoenberger/Grant Heilman, p. 53 D. Davidson/Tom Stack, p. 55 NASA, p. 57 Carolina Biological Supply, p. 57 Carolina Biological Supply, p. 57 L. L. Rue/Animals, Animals, p. 58 Leeson/Photo Researchers, p. 58 Leszczynski/Animals, Animals, p. 59 Hansen/NAS/Photo Researchers, p. 63 Hank Morgan, p. 65 G. Kansler/Photo Researchers, p. 67 Carolina Biological Supply, p. 67 Leszczynski/Animals, Animals, p. 68 Carolina Biological Supply, p. 69 J. McCann/Photo Researchers, p. 69 R. Planck/Photo Researchers, p. 71 Carolina Biological Supply, p. 72 Z. Gaal/Photo Researchers, p. 72 R. Comport/Animals, Animals, p. 73 M. Adams/NAS/Photo Researchers, p. 73 NYPL Picture Collection

Unit 4 p. 77 Burgess/SPL/Photo Researchers, p. 78 Omikron/Photo Researchers, p. 78 Grant Heilman/Grant Heilman, p. 78 Runk/Schoenberger/Grant Heilman, p. 78 Walker/SS/Photo Researchers, p. 79 UPI/Bettmann Newsphotos, p. 78 UPI/Bettmann Newsphotos, p. 80 Biophoto/SS/Photo Researchers, p. 85 Runk/Schoenberger/Grant Heilman, p. 85 Runk/Schoenberger/Grant Heilman, p. 86 Runk/Schoenberger/Grant Heilman, p. 89 Biology Media/Photo Researchers, p. 90 Biophoto/SS/Photo Researchers, p. 91 L. Migdale/Photo Researchers, p. 93 R. Eagle/Photo Researchers

Unit 5 p. 99 G. Dimijian/Photo Researchers, p. 100 Donald Deitz/Stock, Boston, p. 105 Cornell University Archives, p. 109 Rhoda Sidney, p. 110 Rota & Singer/AMNH

Unit 6 p. 115 Rod Planck/Photo Researchers, p. 117 Runk/Schoenberger/Grant Heilman, p. 117 Runk/Schoenberger/Grant Heilman, p. 117 ASM/SS/Photo Researchers, p. 118 T. Brain/SPL/Photo Researchers, p. 118 D. Gurvich/Photo Researchers, p. 119 Courtesy of Fresh Co., Inc., p. 120 M. Abbey/Photo Researchers, p. 120 M. I. Walker/Photo Researchers, p. 123 LeFever/Grushow/Grant Heilman, p. 123 Runk/Schoenberger/Grant Heilman, p. 125 G. Harrison/Grant Heilman, p. 125 Cavagnaro, Peter Arnold, p. 125 JPSS/Photo Researchers

Unit 7 p. 131 Diane Rawson/Photo Researchers, p. 133 Grant Heilman/Grant Heilman, p. 133 The Granger Collection, p. 135 Grant Heilman/Grant Heilman, p. 135 Barry Runk/Grant Heilman, p. 134 Runk/Schoenberger/Grant Heilman, p. 134 Runk/Schoenberger/Grant Heilman, p. 136 Grant Heilman, p. 136 Grant Heilman, p. 137 Dan Clark/Grant Heilman, p. 138 Runk/Schoenberger/Grant Heilman, p. 138 Grant Heilman, p. 138 Runk/Schoenberger/Grant Heilman, p. 139 Grant Heilman, p. 139 Grant Heilman/Grant Heilman

Unit 8 p. 145 Runk/Schoenberger/G. Heilman, p. 147 Jane Grushow/Grant Heilman, p. 146 Barry Runk/Grant Heilman, p. 146 Grant Heilman, p. 148 LeFever/Grushow/Grant Heilman, p. 149 Grant Heilman, p. 151 Del Mulkey/Photo Researchers, p. 151 Runk/Schoenberger/Grant Heilman, p. 154 Rhoda Sidney Stock Photography, p. 155 Stephen Swinburne/Stock, Boston, p. 157 Runk/Schoenberger/Grant Heilman, p. 159 Blair Seitz/Photo Researchers, p. 162 Barry Runk/Grant Heilman, p. 163 G. Whiteley/Animals, Animals, p. 165 Alfred Thomas/Animals, Animals, p. 165 Runk/Schoenberger/Grant Heilman

Unit 9 p. 169 Runk/Schoenberger/G. Heilman, p. 171 Walker/Photo Researchers, p. 171 G. Harrison/Grant Heilman, p. 172 Grant Heilman, p. 173 Runk/Schoenberger/Grant Heilman, p. 177 Runk/Schoenberger/Grant Heilman, p. 179 Runk/Schoenberger/Grant Heilman, p. 181 Comstock, p. 183 Runk/Schoenberger/Grant Heilman, p. 185 Mutrux & Associates, p. 186 A.J.Martinez/Photo Researchers, p. 186 Runk/Schoenberger/Grant Heilman, p. 187 Biophoto/Photo Researchers, p. 188 Carolina Biological Supply, p. 189 H. Morgan/SS/Photo Researchers, p. 189 Fred Winner/Photo Researchers

Unit 10 p. 193 L. Lefever/Grant Heilman, p. 195 Susan Kuklin/Photo Researchers, p. 194 Runk/Schoenberger/Grant Heilman, p. 197 Runk/Schoenberger/Grant Heilman, p. 197 Runk/Schoenberger/Grant Heilman, p. 198 Steinhart Aquarium/Photo Researchers, p. 198 Runk/Schoenberger/Grant Heilman, p. 199 Juan Renjifo/Animals, Animals, p. 199 Don Kelly/Grant Heilman, p. 202 Brian Enting/Photo Researchers, p. 202 Runk/Schoenberger/Grant Heilman, p. 205 American Egg Board, p. 207 Rhoda Sidney Stock Photography, p. 208 Mark Stouffer/Animals, Animals

Unit 11 p. 213 Zerschling/Photo Researchers, p. 221 Jerry Berndt/Stock, Boston, p. 212 Barry Runk/Grant Heilman, p. 222 Barry Runk/Grant Heilman, p. 222 Barry Runk/Grant Heilman, p. 222 Runk/Schoenberger/Grant Heilman, p. 224 Ken Levinson/Monkmeyer, p. 224 Stephen Frisch/Stock, Boston, p. 225 W. R. Normark/USGS, p. 229 Paul Conklin/PhotoEdit, p. 231 Lester Bergman & Associates, p. 231 The Granger Collection, p. 234 Dourmashkin/SPL/Photo Researchers

Unit 12 p. 239 J. Sullivan/Photo Researchers, p. 241 PMROH/SPL/Photo Researchers, p. 242 Eric Grave/Photo Researchers, p. 245 S. Camazine/Photo Researchers, p. 245 Thompson/BRI/SPL/Photo Researchers, p. 248 M. I. Walker/Photo Researchers, p. 248 Biophoto/Photo Researchers, p. 248 M. Abbey/Photo Researchers, p. 249 T. England/SS/Photo Researchers

Unit 13 p. 255 L. Lefever/Grant Heilman, p. 256 Biophoto/SS/Photo Researchers, p. 256 Biophoto/SS/Photo Researchers, p. 260 CNRI/Photo Researchers, p. 261 NYPL Picture Collection, p. 263 Archives/Photo Researchers

Unit 14 p. 269 Curtsinger/Photo Researchers, p. 270 Comstock, p. 273 Jerry Howard/Stock Boston, p. 274 Carolina Biological Supply, p. 275 Longcore/SPL/Photo Researchers, p. 276 B. Seacrest/Photo Researchers, p. 279 M. I. Walker/Photo Researchers, p. 280 M. Gadomski/Photo Researchers, p. 283 SIU/Photo Researchers

Unit 15 p. 287 E. Harris/Photo Researchers, p. 289 SPL/Photo Researchers, p. 292 Bergman Associates, p. 293 McConnaughey/Photo Researchers, p. 299 Richardson/Photo Researchers, p. 299 S. Frisch/Stock, Boston, p. 299 Bergman & Associates, p. 303 Bob Daemmrich/The Image Works, p. 302 B. Cirone/Photo Researchers, p. 301 NIH/SS/Photo Researchers, p. 304 T. Tucker/Monkmeyer, p305 Michal Heron/Monkmeyer, p. 305 Sam Falk/Monkmeyer, p. 306 Barry Runk/Grant Heilman, p. 306 J. W. Myers/Stock, Boston, p. 306 N. & M. Jansen/SuperStock

Unit 16 p. 311 P. Slimbroom/Photo Researchers, p. 313 Bob Daemmrich/The Image Works, p. 312 Liepins/SPL/Photo Researchers, p. 314 Stephen Feld, p. 315 UPI/Bettmann Newsphotos, p. 318 SMHMS/SPL/Photo Researchers, p. 319 UPI/Bettmann Newsphotos, p. 320 Ward's Natural Science, p. 320 Glauberman/Photo Researchers, p. 322 SPL/Photo Researchers, p. 323 Mulvehill/SS/Photo Researchers, p. 324 A. Gauberman/Photo Researchers, p. 327 R. Hutchings/Photo Researchers, p. 328 Robert Capece/Monkmeyer, p. 328 Glauberman/Photo Researchers, p. 330 S.B.Hosp./SS/Photo Researchers

Unit 17 p. 335 E. Harris/Photo Researchers, p. 337 Bob Daemmrich/Stock, Boston, p. 339 Brad Bower/Stock, Boston, p. 341 M. Abbey/Photo Researchers, p. 343 Bob Daemmrich/Stock, Boston, p. 344 Fawcett/Phillips/Photo Researchers, p. 345 McIntyre/SS/Photo Researchers, p. 346 Nestle/SS/Photo Researchers, p. 347 Lester Bergman & Associates, p. 348 Jacques Charla/Stock, Boston, p. 348 Thomas Hovland/Grant Heilman, p. 348 Joseph Nettis/Stock, Boston, p. 348 J. Wachter/Photo Researchers, p. 349 M. McVay/Stock, Boston

Unit 18 p. 353 R. Rowan/Photo Researchers, p. 354 Donna Jernigan/Monkmeyer, p. 354 R. Pearcy/Animals, Animals, p. 355 Bob Daemmrich/Stock, Boston, p. 355 UPI/Bettmann Newsphotos, p. 356 Runk/Schoenberger/Grant Heilman, p. 357 UPI/Bettmann Newsphotos, p. 365 Stacy Pick/Stock, Boston, p. 367 SIU/Photo Researchers, p. 369 Lipman/The Image Works, p. 370 Eric Grave/Photo Researchers, p. 370 Rhoda Sidney, p. 372 Runk/Schoenberger/Grant Heilman, p. 372 Ed Lettau/SS/Photo Researchers, p. 373 John Colwell/Grant Heilman, p. 374 John Colwell/Grant Heilman, p. 375 Martin Rogers/Stock, Boston, p. 376 Palailly/SPL/Photo Researchers

Unit 19 p. 381 Runk/Schoenberger/Heilman, p. 385 Breck Kent/Earth Scenes, p. 386 Grant Heilman, p. 386 McHugh/Photo Researchers, p. 387 G. Dimijian/Photo Researchers, p. 388 The Granger Collection, p. 388 Lindblad/Photo Researchers, p. 391 Breck Kent/Animals, Animals, p. 392 Degginger/Animals, Animals, p. 393 Latham/The Image Works